REDOX - PRINCIPLES AND ADVANCED APPLICATIONS

Edited by **Mohammed Awad Ali Khalid**

Redox - Principles and Advanced Applications
http://dx.doi.org/10.5772/66005
Edited by Mohammed Awad Ali Khalid

Contributors

Ruiyong Chen, Sangwon Kim, Zhenjun Chang, Xin Wang, Gunnar Jacks, Hao Zhu, Qiliang Li, Mauricio Díaz Muñoz, Isabel Méndez, Francisco Vázquez-Cuevas, Rolando Hernández-Muñoz, Héctor Valente-Godínez, Olivia Vázquez-Martínez, Mohammed Awad Khalid, Makoto Tominaga, Leonardo Ontiveros, Gastón Orlando Suvire, Pedro Enrique Mercado, Agustín G. Asuero, Julia Martín, Laura Ortega Estévez, Tadeusz Michalowski, Anna Maria Michałowska-Kaczmarczyk, Aneta Spórna-Kucab

Published by InTech
Janeza Trdine 9, 51000 Rijeka, Croatia

Publishing Process Manager Nina Kalinic
Technical Editor SPi Global
Cover InTech Design team

First published September, 2017
Printed in Croatia
Legal deposit, Croatia: National and University Library in Zagreb

Additional hard copies can be obtained from orders@intechopen.com

Redox - Principles and Advanced Applications, Edited by Mohammed Awad Ali Khalid
p. cm.
Print ISBN 978-953-51-3393-3
Online ISBN 978-953-51-3394-0

For EU product safety concerns:
IN TECH d.o.o., Prolaz Marije Krucifikse Kozulić 3, 51000 Rijeka, Croatia,
info@intechopen.com or visit our website at intechopen.com.

Meet the editor

Mohammed Khalid received his BS degree in Chemistry in 2000 and PhD degree in Physical Chemistry in 2007 from the University of Khartoum, Sudan. He moved to the School of Chemistry, Faculty of Science, University of Sydney, Australia, in 2009 and joined Dr. Ron Clarke as a postdoctoral fellow where he worked on the interaction of ATP with the phosphoenzyme of the Na+/K+-ATPase and dual mechanisms of allosteric acceleration of the Na+/K+-ATPase by ATP; then he joined the Department of Chemistry, University of Khartoum as an assistant professor, and, few years later, he was promoted as an associate professor. In 2011, he joined the staff of the Department of Chemistry at the Taif University, Saudi Arabia, where he is currently an assistant professor; his research interests include the following:

- P-Type ATPase enzyme kinetics and mechanisms
- Kinetics and mechanism of redox reactions
- Autocatalytic reactions
- Computational enzyme kinetics
- Allosteric acceleration of P-type ATPases by ATP
- Exploring of allosteric sites of ATPases
- Interaction of ATP with ATPases located in the cell membranes

Contents

Preface

Oxidation-reduction (redox) reactions are among the most important and interesting chemical reactions that occur in life forms including respiration and metabolism; aquatic systems including soils, sediments, aquifers, rivers, lakes, and water treatment systems; atmospheric processes including ozone-oxygen reverse reaction; many industrial processes; and electrochemical processes. Redox reactions are central to the major element cycling, many cell cycles, many chemisorption and physisorption processes, trace element mobility from rocks and sediments toward wells, aquifers, trace element toxicity toward life forms, and most remediation schemes including water treatments; over the last three decades, the field has attracted a lot of scientists, and a great deal of researches has been done in redox chemistry, but the field is only beginning to converge around a unified body of knowledge.

This book provides a very broad overview of the state of the art of understanding redox process. The book starts with giving a concise introduction that describes the origin, historical background, and the development of the redox definitions. The book is organized into two sections that include ten chapters and introduces, in Section 1, generalized electron balance theory and its applications in electrolytic redox systems, redox-active molecules and its applications in device memory, fundamentals and applications of flow batteries and its integration into antidirect current, and donor acceptor titrations of displacement and electronic transference. Section 2 introduces redox in biological processes, including roles of reactive oxygen species in respiration, metabolism, and regulations, and redox in physiological processes as redox-sensitive TRP channels TRPA1 and TRPM2. All chapters are written by different authors (with the exception of Chapter 1 [Introduction]). This clearly reflects the broad range of topics that have been covered by experts in the field.

The book is primarily aimed at an academic audience but will also be useful for a wide range of professionals at different career stages from master"s students to any scientist who wants to investigate and understand the electrochemical and coupled properties of electroactive materials.

I thank all the chapter authors for their contributions to the book and the book processing manager Nina Kalinic for her valuable comments and assistance in the preparation of the book for publication. I hope that readers with an interest in redox will find the book interesting and of value for their own research.

Mohammed A. A. Khalid
University of Khartoum, Faculty of Science
Department of Chemistry, Sudan
University of Taif, College of Applied Medical and Sciences
Department of Chemistry, Saudi Arabia

Section 1

Electrical Apllications

Chapter 1

Introductory Chapter: Historical and Newest Perspectives

Mohammed Awad Ali Khalid

Additional information is available at the end of the chapter

http://dx.doi.org/10.5772/intechopen.68917

1. Historical background

The history of oxidation reduction reactions can be traced back to the early time of the human development, since the first time that the human knew the fire and used it in their daily life, especially the Copper-Bronze age of the human development, the early time, around 4000–8000 years ago. In that era, people were benefited by the use of copper in their life; they heated copper ores and coal in an oxidation reduction reaction by which copper ores are reduced to copper metal and coal is oxidized to carbon dioxide, and besides the production of copper, the Bronze age is also known to use clay in the production of pottery. Greeks are the first to use oxidation reduction reaction as oxidizing and reducing fire conditions for pottery making; this can be summarized as firing clay in a rich or indigent atmosphere of oxygen. Clay containing iron will turn to orange-red if fired under rich atmosphere of oxygen due to the presence of red iron oxide (higher oxidation state) or will turn to black if fired under indigent atmosphere of oxygen when black iron oxide forms (lower oxidation state). Then, the human development jumped to the Iron Age, which is around 3000–5000 years ago, when the human for the first time used iron mainly in production of knives and bayonets which were used in their daily lives and wars. A great historic jump in the oxidation reduction reactions is the use of explosives, first used by Chinese as early as 950 A.D. It was the Chinese who first developed this deadly weapon, but around the thirteenth century, the Europeans would have jumped on the technological band wagon, using it to devastate the natives of the New World during the Age of Exploration.

A modern exploration of the oxidation reduction reaction starts formally with Georg Ernst Stahl [1] in 1697 when he proposed the phlogiston theory [2], which was based on the premise that the metals often produce a calx when heated (calx is defined by Stahl as the crumbly residue left after a mineral or metal is roasted), and phlogiston is given off whenever metals or something were burned; moreover, the calxes form metals when heated with charcoal and wood, and charcoal is particularly rich in phlogiston because they leave very little ash when they are burned;

however, the theory of phlogiston was not widely accepted in the scientific media. Seventy-five years later, Antoine-Laurent Lavoisier (1743–1794) came with solid explanation of combustion [3]. In 1772, Lavoisier discovered that when phosphorus or sulfur is burned in air, the products are acidic in nature, and the products also weigh more than the original phosphorus or sulfur, and he came to the conclusion that the elements combine with something in the air to produce acids [4], but he could not recognize what was in the air that combined with phosphorus or sulfur. In 1974, he met Joseph Priestley (the father of oxygen discovery, 1733–1804) during his visit to Paris; he told Lavoisier about the gas produced when he decomposed the compound which we now call mercury oxide. This gas supported combustion much more powerfully than normal air. Priestley believed the gas was a particularly pure version of air; he started calling it dephlogisticated air, believing its unusual properties were caused by the absence of phlogiston. In 1779, Lavoisier coined the name oxygen for the element released by decomposition of mercury oxide, and from here, explanation of certain reactions as oxidation reduction officially started [5].

Coming back to the explosive materials, the year 1964 was the year that explosives, nitrocellulose and nitroglycerin, were both discovered, and later on trinitrotoluene (TNT), involved in weapon production and widely used in the First World War (1914–1919). The cheap mixture of ammonium nitrate and fuel oil was recognized as a powerful explosive in 1955, and this was used to bomb the Federal Building in Oklahoma City in 1995; finally, the explosives that were used in the fireworks are believed to be used for the first time in China in the sixth century.

Now, the five main types of redox reactions are combinations, decompositions, displacements, combustions, and disproportionations. In combination redox reactions, two elements are combined whereas one element becomes oxidant and the other reluctant; in decomposition redox reactions, a compound is broken down into its constituent parts; in displacement redox reactions, one or more atoms is swapped out for another; in combustion reactions, a compound reacts with oxygen to produce carbon dioxide, water, and heat; and in disproportionation redox reactions, a molecule is both reduced and oxidized. These types of reactions are rare, and many reactions are considered in the interface between these areas.

Concluding this historical background that chemists worldwide later recognized that other elements reacted in the same general manner as oxygen, the concepts of oxidation and reduction were extended to include other elements; electrochemistry as a new field is further broadening the definition of the oxidation reduction reaction. Investigators observed that the ferric ions could be formed from the ferrous ions by the action of oxygen gas. This consumption of oxygen, oxidation, involved a loss of electrons by the ferrous ions species, and hence, an oxidation reaction could refer also to a transfer of electrons.

2. Redox reactions in biological processes

An understanding of the redox reactions of inorganic and organic compounds is central to understanding the metabolism of living things. One of the most important processes that occurs in living organisms is photosynthesis, which consists of a series of oxidation reduction reactions; the series begins when the chlorophyll in barks or leaves of plant cells absorbs sunlight with certain wavelengths and converts carbon dioxide into carbon and oxygen in a reduction process and ends the series with the production of glucose molecules. In other organisms, glucose is being consumed to generate energy in a long series of enzyme-catalyzed reactions; in simple words, electrons can be transferred from glucose to molecular oxygen, oxidizing the carbon molecules to carbon dioxide and reducing O_2 to water.

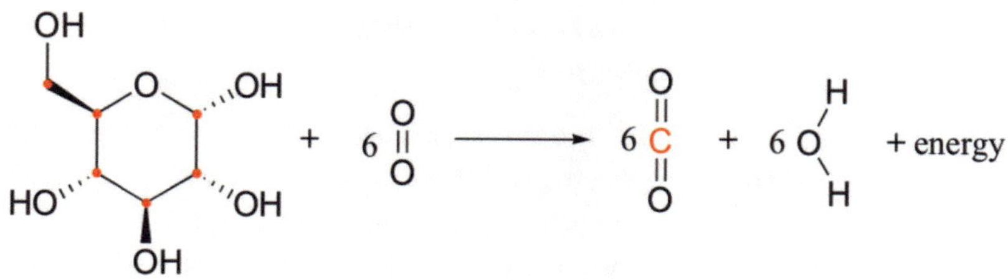

This aspect of redox reactions in living organisms is called cellular respiration by which cells break down molecules of food (glucose) in a series of chemical reactions to produce energy, carbon dioxide, and water; the process depends heavily on the reduction of NAD^+ to NADH and the reverse oxidation reaction of NADH to NAD^+ as intermediate steps [6]. The oxidation of glucose is a thermodynamically favored process, meaning the transfer of electrons from

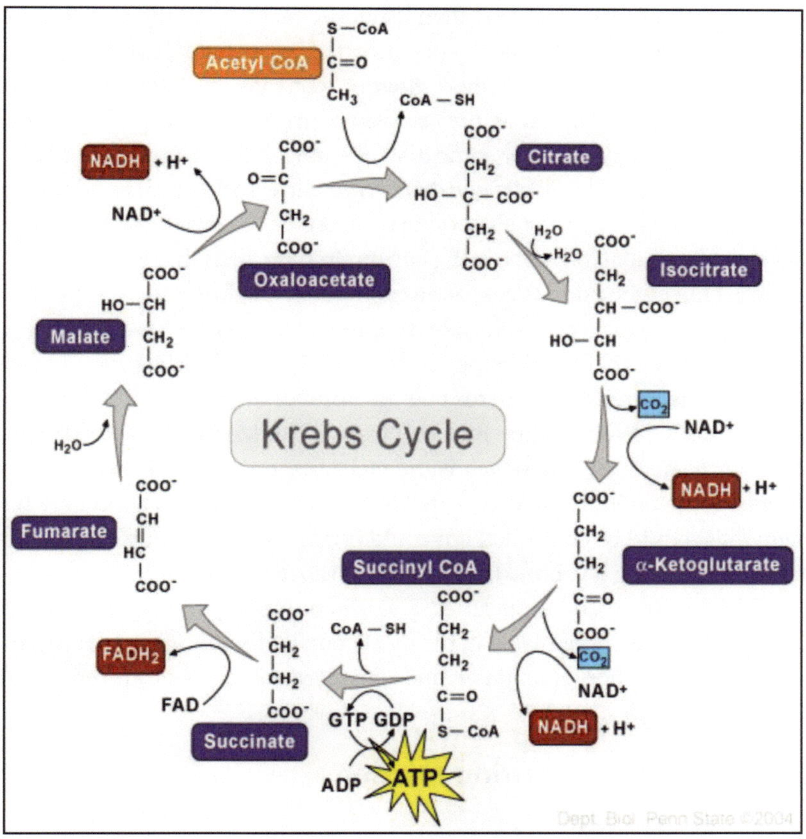

Figure 1. An illustrated diagram for Krebs cycle, copied from the website, https://wikispaces.psu.edu.

glucose to O_2 is thermodynamically downhill, and cells use this released energy to carry out a wide variety of energy-requiring activities. **Figure 1** illustrates how glucose is burned in a series of redox reactions and ends up in the formation of carbon dioxide and energy that is stored as adenosine triphosphate (ATP); in the diagram called Krebs Cycle which describes cell burning of glucose, enzymes are used in each step to lower the activation energy for each step and aid in breaking and formation of bonds; the overall reaction is a redox reaction, that is, electrons are lost or gained in each step.

Other biological processes that involve the redox reaction is the production of free radicals, which can be produced by detaching electrons from certain type of molecules and reattaching to another type of molecule instantaneously; free radicals play an important role for the programmed cell death (apoptosis), and any uncontrolled production of free radicals may lead to cause cancer [7].

3. Redox reactions in corrosion reactions

Corrosion is another type of redox reaction; it occurs when a metal comes in contact to a solution or at least moisture; the metal corrodes with evolving of electrons that move to cathodic part of the so-called localized galvanic cell, and then, cathodic reaction starts with the production of hydrogen gas if the electrolyte is acidic or conversion of water to hydroxide if the electrolyte is neutral or basic. In this case, the intensity of flow of electrons from the anodic part (metal) to cathodic part (electrolyte) is defined as the corrosion current; there may be some microscopic galvanic cells with adjacent distance or some distance apart if the electric current in the galvanic cells is huge and is more than the electrolyte capacity to allow the current to pass, then the operation is governed by the movement of electrolyte ions; on the other hand, if the electric current is less than the electrolyte capacity to allow the current to pass, then the operation is governed by activation energy. One of the famous corrosion examples is the iron rust, and in this case, iron is oxidized at the beginning to ferrous ions releasing two electrons, and the reaction will proceed as long as the metal is capable of releasing electrons and electrolytic solution to carry the ions; the corrosion current is increased by increasing the number of oxidized iron atoms, and if there is excess of oxygen in electrolyte, then ferrous ions are oxidized further to ferric ions that can give ferric oxide or ferric carbonate which is the main constituent of the iron rust. Besides iron, most of elements in the periodic table are capable to corrode, and corrosion now has become a global problem that should be controlled if it could not be stopped (according to the second law of thermodynamics); the biggest breakthrough that has been achieved in the corrosion research is the invention of the electrochemical series, a series in which ordering the periodic table elements depends on the redox potential, and the most benefit of this is trying not to gather two elements of far different reduction potential in one alloy because that is produced in fast corrosion.

4. Redox reactions in combustion reactions

To start a discussion on this, let us first ask this question, is the combustion reactions is a redox reaction? Answering this, as the oxidation state is changed from 0 in the molecular oxygen to −2 in the species that produce in the reaction, the reaction is a redox in nature; combustion in the form of fire produces flame and a considerable amount of heat, which can make combustion self-sustaining. In case of burning metals such as mercury, copper, zinc, and so on, the product is the metal oxide; in case of burning organic molecules, the products are carbon dioxide, water, and heat, and the combustion reaction is not as easy as it looks; probably the reaction takes place in a series of more than 10 steps, and hence finding the overall rate of reaction becomes extremely complicated, and the computer softwares are the only logical solution for this. MatLab, Avogadro, Copasi, and Kintecus are some of the most powerful softwares used in this regard.

5. Redox reactions in explosion reactions

Potassium nitrate, when mixed with carbon and sulfur in correct ratios which are the constituents of gunpowder, nitrate is reduced to form nitrogen, mono and dioxides, while carbon is oxidized to form carbon mono and dioxides, and sulfur is oxidized to form di and trioxides. The reaction will not start unless it is initiated; it has been found that such reactions can be initiated by electric shocks, spark, or electric current, and the reaction is maintained in a series of complicated steps; production of all these gases increase suddenly the pressure, the contents of the reaction come to explode to relieve the pressure, and besides increase in pressure, temperature is also increased tremendously; the main mechanism of explosive reaction is the chain reaction by which one product of the reaction, called free radical, is initiated and activates other molecules in the reaction mixture, and the reaction is proceeded till all free radicals are used up.

Although nuclear explosion is one of the massive explosions on earth, but is not itself a redox reaction, and is something more complicated, as in the case of uranium, the nuclei split and form two different elements and release energy more than any ordinary explosive.

Nowadays, redox reactions fuel the most advanced form of the space transportation and the space shuttle; powdered aluminum and ammonium perchlorate are used to undergo redox reactions that produce the gases hydrogen and oxygen and give the shuttle enormous amount of extra thrust; the redox reaction is represented as follows:

$$3NH_4ClO_{4(s)} + 3Al_{(s)} \rightarrow Al_2O_{3(s)} + AlCl_{3(s)} + 3NO_{(g)} + 6H_2O_{(g)} + \text{energy} \tag{1}$$

It produces temperatures of about 5700 F and 3.3 million pounds of thrust in each rocket; thus, the redox reactions furnish the energy to launch the space shuttle.

6. Conclusions

Besides the above examples, so many examples can be drawn to prove the importance of applications of redox reactions in general life; in electrochemical cells, electrons formed from the oxidation of one element and pass through a conductor to the reduction element; bleaching solutions that are used to brightening clothes are made of oxidizing agent (clorox), and this oxidizes any constituent that is capable of being oxidized and then make clothes clean and bright, and so on.

In this summarized introduction, we aimed to draw the reader's attention to the wide range of applications as well as the importance of redox reactions; luckily, chapters of this book can be categorized into two main parts: Part (1) batteries and computer applications and part (2) drugs and biological applications, and the diverse of chapters exhibit clearly the wide range of researches in the field of redox reactions.

Author details

Mohammed Awad Ali Khalid

Address all correspondence to: yasinawad@gmail.com

1 Department of Chemistry, College of Applied Medical Sciences, University of Taif, Turabah, Saudi Arabia

2 Department of Chemistry, Faculty of Science, University of Khartoum, Khartoum, Sudan

References

[1] Stahl GE. Encyclopedia Britannica Inc.; 2013. Web

[2] Conant JB. The Overthrow of the Phlogiston Theory: The Chemical Revolution of 1775-1789. Cambridge, MA: Harvard University Press; 1964

[3] Lavoisier LA. Elements of Chemistry: In a New Systematic Order, Containing all the Modern Discoveries, Illustrated with Thirteen Copperplates. 4th ed. 1799

[4] Lavoisier LA. 1775 memoir on the nature of the principle which combines with metals during calcination and increases their weight. A Source Book in Chemistry 1400-1900. New York: McGraw Hill; 1952

[5] Jaffe B. Crucibles: The Story of Chemistry from Ancient Alchemy to Nuclear Fission. 4th ed. Courier Dover Publications; 1976. ISBN 9780486233420

[6] Rich PR. The molecular machinery of Keilin's respiratory chain. Biochemical Society Transactions. 2003;**31**(Pt 6):1095-1105

[7] Hopkins WG. Photosynthesis and Respiration. New York: Chelsea House, imprint of Infobase Publishing; 2006. ISBN 9780791085615

Chapter 2

Generalized Electron Balance (GEB) as the Law of Nature in Electrolytic Redox Systems

Anna Maria Michałowska-Kaczmarczyk,
Aneta Spórna-Kucab and Tadeusz Michałowski

Additional information is available at the end of the chapter

http://dx.doi.org/10.5772/intechopen.69341

Abstract

This chapter refers to fundamental/general/obligatory regularities of electrolytic systems. The linear combination $2 \cdot f(O) - f(H)$ of elemental balances, $f(H)$ for H and $f(O)$ for O, provides a rigorous criterion distinguishing between redox and non-redox systems is presented as the general relation distinguishing between electrolytic redox and non-redox systems in aqueous media. As the linearly independent equation for a redox system, $2 \cdot f(O) - f(H)$ is considered as the primary form of the generalized electron balance (GEB), perceived as a law of nature, as the hidden connection of physicochemical laws and the breakthrough in thermodynamic theory of electrolytic redox systems. GEB completes the set of 2+K equations necessary for thermodynamic resolution of redox systems according to generalized approach to electrolytic systems (GATES) applying all relevant, physicochemical knowledge available. GATES/GEB, perceived as an example of excellent paradigm, provides the best thermodynamic approach to electrolytic redox systems of any degree of complexity, in aqueous, non-aqueous, and mixed-solvent media. The formulation of GEB does not need prior knowledge of oxidation numbers for all elements in components forming any electrolytic system, within GATES/GEB, the stoichiometry, oxidation number, oxidant, reductant and equivalent mass are as derivative concepts.

Keywords: electrolytic redox systems, approach I to GEB, approach II to GEB, GATES/GEB

1. Introduction

According to the principles assumed in the generalized approach to electrolytic systems (GATES) introduced/formulated by Michałowski in 1992 [1], a balancing of any electrolytic system is based on the rules of conservation of particular elements/cores Y_g (g = 1,..., G), and on a charge balance (ChB) expressing the rule of electroneutrality of this system. The closed systems, separated from its environment by diathermal walls, are considered for modelling purposes. The elemental $f(E_g)$ or/and core $f(core_g)$ balances are denoted briefly as $f(Y_g)$ ($f(E_g)$ or $f(core_g)$). The balance for the gth element (E_g) or core ($core_g$) is expressed by the equation interrelating the numbers of gth atom or core in components as units composing the system with the numbers of atoms or cores of gth kind in the species of the system thus formed. For ordering purposes, we assume E_1 = H (hydrogen) and E_2 = O (oxygen); then, we have f(H) for $Y_1 = E_1$ = H, f(O) for $Y_2 = E_2$ = O, etc. Free water particles and water bound in hydrates are included in balances for f(H) and f(O).

The ChB interrelates the charged species (ions) in this system. A core is a cluster of elements with defined composition, expressed by its chemical formula, structure and external charge, which remains unchanged in a system considered, e.g. SO_4^{-2} in Eq. (54).

The rules of conservation, formulated according to GATES principles [1–36], have the form of algebraic equations related to closed systems, composed of condensed (e.g. liquid, liquid + solid, etc.) phases separated from its environment by diathermal (freely permeable by heat) walls; it enables the heat exchange between the system and its environment. Any chemical process, such as titration, is carried out under isothermal conditions, in a *quasistatic* manner; constant temperature (T = constant) is one of the conditions securing constancy of equilibrium constant values. Any exchange of the matter (H_2O, CO_2, O_2,...) between the system and its environment is thus forbidden for modelling purposes.

We refer first to aqueous media, where the species $X_i^{z_i}$ exist as hydrates $X_i^{z_i} \cdot n_{iW}$; $z_i = 0, \pm1, \pm2$, ... is a charge, expressed in terms of elementary charge unit, $e = F/N_A$ (F = 96485.333 C/mol, Faraday's constant; $N_A = 6.022141 \cdot 10^{23}$ mol^{-1}, Avogadro's number; n_i ($\equiv n_{iW} \equiv n_i H_2O$) ≥ 0 is a mean number of water (W = H_2O) molecules attached to $X_i^{z_i}$; the case n_{iW} = 0 is then also admitted. For ordering purposes, we assume $X_2^{z_2} \cdot n_{2W} = H^{+1} \cdot n_{2W}$, $X_3^{z_3} \cdot n_{3W} = OH^{-1} \cdot n_{3W}$, ..., i.e. $z_2 = 1$, $z_3 = -1$,... . Molar concentration of the species $X_i^{z_i} \cdot n_{iW}$ is denoted as $[X_i^{z_i}]$. The $n_i = n_{iW}$ values are virtually unknown, even for $X_2^{z_2} = H^{+1}$ [37] in aqueous media, and depend on ionic strength (I) of the solution. The $X_i^{z_i}$s with different numbers of H_2O molecules involved, e.g. H^{+1}, H_3O^{+1}, $H_9O_4^{+1}$; $H_4IO_6^{-1}$, IO_4^{-1}; $H_2BO_3^{-1}$, $B(OH)_4^{-1}$; AlO_2^{-1} and $Al(OH)_4^{-1}$ are considered equivalently [27], i.e. as the same species in this medium. The ChB interrelates charged species (ions, $z_j \neq 0$) in the system.

From f(H) and f(O), the linear combination 2·f(O) – f(H) is formulated and termed as the *primary* form of generalized electron balance (GEB), *pr*-GEB = 2·f(O) – f(H), obtained according to approach II to GEB; this leitmotiv will be extended in further parts of this chapter. The GEB is the immanent part of GATES; the computer software applied to redox systems is denoted as GATES/GEB [1]. When related to redox systems, GATES is based on the generalized electron balance (GEB) [1–36] concept, perceived as a law of nature [1, 2, 14, 15, 22], as the hidden connection of physicochemical laws, as a breakthrough in the theory of electrolytic redox systems. The GATES refers to mono- and poly-phase, redox and non-redox, equilibrium and metastable [38–42] static and dynamic systems, in aqueous, non-aqueous and mixed-solvent media [26, 29] and in liquid-liquid extraction systems [16, 43].

The generalized electron balance (GEB) concept, discovered by Michałowski as the approach I in 1992 and approach II in 2006 to GEB, plays the key role in redox systems; both approaches are equivalent:

$$\therefore \text{approach II to GEB} \Leftrightarrow \text{aapproach I to GEB}$$

The GEB is fully compatible with charge balance (ChB) and concentration balances $f(Y_g)$ ($g = 3, \ldots, G$) formulated for different elements and cores Y_g. The elemental $f(E_g)$ or/and core $f(\text{core}_g)$ balances are transformed into concentration balances: CB(Y_g) for $g = 3, \ldots, G$.

To avoid redundant terms/naming, the acronyms ChB and GEB are applied both to equations, expressed in terms of particular units (N_i, N_{0j}), or in terms of molar concentrations. On the basis of Eqs. (1a) and (1b) exemplified in Eqs. (2a) and (2b), this should not cause any misinterpretations. Then, $i \in <1, I>$ enumerate species, $j \in <1, J>$ enumerate components, $g \in <0, G>$ enumerate equations for ChB ($g = 0$) and elements/cores: $g = 1$ for H, $g = 2$ for O, $g \in <3, G>$ for other elements/cores, i.e. $Y_1 = H$, $Y_2 = O, \ldots$, for ordering purposes.

The terms components and species are distinguished. In the notation applied here, N_{0j} ($j = 1, 2, \ldots, J$) is the number of molecules of components of jth kind composing the static or dynamic D + T system, whereby the D and T are composed separately, from defined components, including water. The mono- or two-phase electrolytic system thus obtained involve N_1 molecules of H_2O and N_i species of ith kind, $X_i^{z_i} \cdot n_{iW}$ ($i = 2, 3, \ldots, I$), specified briefly as $X_i^{z_i}$ (N_i, n_i), where $n_i \equiv n_{iW} \equiv n_i H_2O$; then, we have H^{+1} (N_2, n_2), OH^{-1} (N_3, n_3),…. Thus, the components form a (sub)system, and the species $X_i^{z_i} \cdot n_{iW}$ enter the system thus formed. A solid $\mathbf{X_i}{\cdot}n_{iW}$ (precipitate, $z_i = 0$), as a species in a two-phase system, is marked by bold letters, e.g. $\mathbf{I_{2(s)}}$ and **AgCl**.

In *Example* 1 (Section 4.1), $CuSO_4{\cdot}5H_2O$ is one of components, and $Cu(OH)_3^{-1}{\cdot}n_9H_2O$ is one of the species in the system. N_{01} molecules of $CuSO_4{\cdot}5H_2O$ involve $10N_{01}$ atoms of H, $9N_{01}$ atoms of O and N_{01} atoms of Cu, and N_{01} atoms of S. The notation $Cu(OH)_3^{-1}$ (N_9, n_9) refers to N_9 ions of Cu $(OH)_3^{-1}{\cdot}n_9H_2O$ involving $N_9(3 + 2n_9)$ atoms of H, $N_9(3 + n_9)$ atoms of O and N_9 atoms of Cu.

Molar concentration $[X_i^{z_i}]$ of $X_i^{z_i} \cdot n_{iW}$ is as follow:

$$\begin{aligned} &(a)\ [X_i^{z_i}] = 10^3 \cdot (N_i/N_A)/V_0 \text{ for a static system, or} \\ &(b)\ [X_i^{z_i}] = 10^3 \cdot (N_i/N_A)/(V_0 + V) \text{ for a dynamic D + T system} \end{aligned} \quad (1)$$

In a static or dynamic system, the balances are ultimately expressed in terms of molar concentrations of compounds and species, like the expressions for equilibrium constants. In particular, the charge balance (ChB) formulated as follows:

$$\sum_{i=2}^{I} z_i \cdot N_i = 0 \tag{2}$$

interrelates charged ($z_i \neq 0$) species of this system. In terms of molar concentrations [mol/L] (Eq. (1a) or (1b)), the charge balance has the form

$$\sum_{i=2}^{I} z_i \cdot [X_i^{z_i}] = 0 \tag{2a}$$

where $z_1 = 0$ for $X_1^{z_1} = H_2O$, $z_2 = +1$ for $X_2^{z_2} = H^{+1}$, $z_3 = -1$ for $X_3^{z_3} = OH^{-1}$,...

In non-aqueous and mixed-solvent media, with amphiprotic (co)solvent(s) involved, we assume/allow the formation of mixed solvates $X_i^{z_i} \cdot n_{i1}n_{i2}...n_{iS}$, where $n_{is} = n_{iA_s}(\geq 0)$ are the mean numbers of A_s ($s = 1,..., S$) molecules attached to $X_i^{z_i}$. We apply the notation $X_i^{z_i}(N_i, n_{iA_1}, n_{iA_2},..., n_{iA_S})$, where N_i is a number of entities of these species in the system [25, 27, 28, 44–46].

2. Preliminary information

The balances $f(Y_g)$ and ChB will be related to some dynamic redox and non-redox systems for comparative purposes. The balances for a given system are combined according to linear combination principles, and some general properties of the resulting balances are indicated.

The components and species in redox systems are involved in GEB, charge (ChB) and elemental balances. All the balances are founded on the well-established physical, physicochemical and chemical rules involved with charge and all elements conservation, and on the so-named mass action law (MAL), with its 'old-fashioned' principles. However, to be fully compatible with the GATES idea, introduced to avoid the stoichiometric reasoning, the equilibrium law (EL), suggested by Michałowski in 2016, was put instead of MAL; the principle of EL formulation is based on the idea of Lagrange multipliers for searching the local extrema of a function that subjects some constraints, expressed by GEB, ChB and concentration balances. The Gibbs free energy G is applied here as a function of the measurable, intensive properties: p and T and the numbers of constituents are convenient in the study of chemical reaction equilibrium. The problem in question is then consistent with the GATES formulation.

The generalized equivalent mass (GEM) concept, suggested by Michałowski in 1979 [11, 19, 47] is also based on the GATES principles, contrary to IUPAC recommendations [48] based on the stoichiometry of chemical reactions. Within GATES, the stoichiometric reaction notations are used only to formulate the expressions for the set of equilibrium constants related to the system in question, not to one 'representative' reaction, selected arbitrarily.

Within GATES, the mass conservation law is not limited only to components/species of one, specified/ 'responsible' chemical reaction equation but relates to all components/species of an electrolytic system. The mass change involved with an (exo- or endothermal) effect is negligible when compared with the total mass of the system. For example, the mass change, Δm, involved with enthalpy ΔH° of the reaction $H_{2(g)} + 0.5O_{2(g)} = H_2O_{(l)}$ ($\Delta H^\circ = -286$ kJ/mol H_2O), equal $\Delta m = \Delta H^\circ/c^2 = -3.18 \cdot 10^{-9}$ g, is negligible (not measurable) when compared with 18 g of H_2O; c = 299,792,458 m/s is the speed of light in vacuum. The neutralization or dilution gives much smaller heat effects. The resulting law of mass preservation is then fulfilled, irrespectively on whether stoichiometric or non-stoichiometric chemical reactions occur (or do not occur) in the system.

All considerations made in this chapter refer, in principle, to the systems where the elements are formed by stable/non-radioactive isotopes [2], i.e. where no nuclear (α, β^+, β, or electron capture) transformations occur [7], with emission of γ and/or X-ray radiation. However, it is possible to extend the description of redox systems on the systems with radioactive elements, in which the concentrations of the respective components are expressed by dependencies, identified as Bateman's system of linear differential equations [49–52], binding quantitatively the radioactive elements with products of their decay. Radioactive elements and their decay products are also included in the balances for other elements and dependencies for the equilibrium constants.

3. Linear dependence of algebraic equations and transformation into identity

Linear combination of algebraic equations plays a fundamental/decisive role in thermodynamics of electrolytic systems, considered according to GATES [1]. An elementary information related to linear combination, perceived (mainly) from mathematical/algebraic viewpoint, can be found in Ref. [23]. It should be noted that the results of simple addition, i.e. $f_1 + f_2$ and simple subtraction, i.e. $f_1 - f_2$ or $f_2 - f_1$, are also linear combinations of any equations f_1 and f_2.

We refer here to the problem of linear dependency of balances, analogous to the problem of dependency of linear equations, considered in elementary algebra, see the picture below. In this context, the general property of linear independency, inherently involved with redox systems, will be emphasized.

(1) For the beginning, let us take the set of linear equations:

$$a_{11}x_1 + a_{12}x_2 + a_{13}x_3 = b_1 \tag{3}$$

$$a_{21}x_1 + a_{22}x_2 + a_{23}x_3 = b_2 \tag{4}$$

completed by the linear combination of these equations, i.e.

$$\begin{aligned} &c_1(a_{11}x_1 + a_{12}x_2 + a_{13}x_3) + c_2(a_{21}x_1 + a_{22}x_2 + a_{23}x_3) \\ &\quad \equiv (c_1a_{11} + c_2a_{21})x_1 + (c_1a_{12} + c_2a_{22})x_2 + (c_1a_{13} + c_2a_{23})x_3 = c_1b_1 + c_2b_2. \end{aligned} \tag{5}$$

Applying the matrix algebra, we see that the determinant

$$D = \begin{vmatrix} a_{11} & a_{12} & a_{13} \\ a_{21} & a_{22} & a_{23} \\ c_1a_{11} + c_2a_{21} & c_1a_{12} + c_2a_{22} & c_1a_{13} + c_2a_{23} \end{vmatrix} \tag{6}$$

has zero value

$$D = c_1 \cdot \begin{vmatrix} a_{11} & a_{12} & a_{13} \\ a_{21} & a_{22} & a_{23} \\ a_{11} & a_{12} & a_{13} \end{vmatrix} + c_2 \cdot \begin{vmatrix} a_{11} & a_{12} & a_{13} \\ a_{21} & a_{22} & a_{23} \\ a_{21} & a_{22} & a_{23} \end{vmatrix} = c_1 \cdot 0 + c_2 \cdot 0 = 0 \tag{7}$$

irrespectively on the c_1 and c_2 values; at $D = 0$, calculation of x_1, x_2 and x_3 is then impossible.

(2) Let us consider the set of $G + 1$ equations: $f_g(\mathbf{x}) = \varphi_g(\mathbf{x}) - b_g = 0$, where $g = 0,1,\ldots, G$, $\mathbf{x}^T = (x_1,\ldots, x_I)$ – transposed (^{T}) vector $\mathbf{x}$, composed of independent (scalar) variables x_i ($i \in <1, I>$); a_{gi}, $b_g \in \mathcal{R}$ are independent (explicitly) on $\mathbf{x}$. After multiplying the equations by the numbers $\omega_g \in \mathcal{R}$ and addition of the resulting equations, we get the linear combination $\sum_{g=0}^{G} \omega_g \cdot f_g(\mathbf{x}) = 0 \Leftrightarrow \sum_{g=0}^{G} \omega_g \cdot \varphi_g(\mathbf{x}) = \sum_{g=0}^{G} \omega_g \cdot b_g$ of the basic equations.

Formation of linear combinations is applicable to check the linear dependency or independency of the balances. A very useful/effective manner for checking/stating the linear dependence of the balances is the transformation of an appropriate system of equations to the identity, 0 = 0 [2, 23]. For this purpose, we will try, in all instances, to obtain the simplest form of the linear combination. To facilitate these operations, carried out by cancellation of the terms on the left and right sides of equations after changing sides of these equations, we apply the equivalent forms of the starting equations $f_g(\mathbf{x}) = 0$:

$$f_g(\mathbf{x}): \quad \varphi_g(\mathbf{x}) - b_g = 0 \Leftrightarrow \varphi_g(\mathbf{x}) = b_g \Leftrightarrow -f_g(\mathbf{x}): \quad -\varphi_g(\mathbf{x}) = -b_g \Leftrightarrow b_g = \varphi_g(\mathbf{x}) \tag{8}$$

In this notation, $f_g(\mathbf{x})$ will be essentially treated not as the algebraic expression on the left side of the equation $f_g(\mathbf{x}) = 0$, but as an equation that can be expressed in alternative forms presented above.

We refer now to the set of linear, algebraic equations

$$\sum_{i=1}^{I} a_{gi} \cdot x_i = b_g \Leftrightarrow \sum_{i=1}^{I} a_{gi} \cdot x_i - b_g = 0, \qquad (g = 0, 1, \ldots, G) \tag{9}$$

where a_{gi} are the coefficients and b_g is the free terms. Multiplying Eq. (9) by ω_g, after subsequent summation and rearrangement, we have

$$\sum_{g=0}^{G}\omega_g \cdot \sum_{i=1}^{I} a_{gi} \cdot x_i = \sum_{g=0}^{G}\omega_g \cdot b_g \Leftrightarrow \sum_{i=1}^{I} x_i \cdot \sum_{g=0}^{G}\omega_g \cdot a_{gi} = \sum_{g=0}^{G}\omega_g \cdot b_g \tag{10}$$

Assuming

$$b_g = \sum_{j=1}^{J} b_{gj} \cdot x_{0j} \tag{11}$$

from Eqs. (10) and (11) we have

$$\sum_{i=1}^{I} x_i \cdot \sum_{g=0}^{G}\omega_g \cdot a_{gi} = \sum_{j=1}^{J} x_{0j} \cdot \sum_{g=0}^{G}\omega_g \cdot b_{gj} \tag{12}$$

Referring to the problem in question, and placing $x_i = N_i$, $x_{0j} = N_{0j}$ in Eq. (11), we write

$$b_g = \sum_{j=1}^{J} b_{gj} \cdot N_{0j} \tag{11a}$$

Then for ChB (Eq. (2a)), where the right side equals zero, we have

$$b_0 = \sum_{j=1}^{J} b_{0j} \cdot N_{0j} = 0 \quad \text{for} \quad b_{0j} = 0 \quad (j = 1, \dots, J) \tag{13}$$

$$\sum_{i=1}^{I} N_i \cdot \sum_{g=0}^{G}\omega_g \cdot a_{gi} = \sum_{j=1}^{J} N_{0j} \cdot \sum_{g=0}^{G}\omega_g \cdot b_{gj} \tag{14}$$

The charge balance (ChB) is expressed by Eqs. (2a) and (2b), where $X_i^{z_i}$ is defined by Eq. (1a) (for a static system) or Eq. (1b) (for a dynamic system). The elemental/core balances: f(H), f(O) and $f(Y_g)$ ($Y_g \neq$ H, O, $g = 3,\dots, G$) are written as follows:

$$\begin{aligned} f_1 = f(H) &= \sum_{i=1}^{I}(a_{1i} + 2n_i) \cdot N_i - \sum_{j=1}^{J} b_{1j} \cdot N_{0j} = 0 \ \text{ for } Y_1 = H, \\ f_2 = f(\mathrm{O}) &= \sum_{i=1}^{I}(a_{2i} + n_i) \cdot N_i - \sum_{j=1}^{J} b_{2j} \cdot N_{0j} = 0 \quad \text{for } Y_2 = \ O, \dots, \\ f_g = \sum_{i=1}^{I} a_{gi} \cdot N_i - \sum_{j=1}^{J} b_{gj} \cdot N_{0j} &= 0, \dots, f_G = \sum_{i=1}^{I} a_{Gi} \cdot N_i - \sum_{j=1}^{J} b_{Gj} \cdot N_{0j} = 0 \end{aligned} \tag{15}$$

where a_{gi} and b_{gj} are the numbers of atoms/cores of gth kind in the ith species as a constituent of the system and in the component of jth kind, respecively. Denoting, for a moment, $\omega_1 = -1$, $\omega_2 = 2$, we transform the balance

$$f_{12} = 2 \cdot f_2 - f_1 = 2 \cdot f(\mathrm{O}) - f(\mathrm{H}) = \sum_{i=2}^{I} (2a_{2i} - a_{1i}) \cdot N_i - \sum_{j=1}^{J} (2b_{2j} - b_{1j}) \cdot N_{0j} = 0 \Rightarrow \tag{16}$$

$$\begin{aligned} f_{12} &= \omega_1 \cdot \left(\sum_{i=2}^{I} (a_{1i} \cdot N_i - \sum_{j=1}^{J} b_{2j} \cdot N_{0j}) \right) + \omega_2 \cdot \left(\sum_{i=2}^{I} (a_{2i} \cdot N_i - \sum_{j=1}^{J} (b_{2j} \cdot N_{0j}) \right) \\ &= \omega_1 \cdot f_1^* + \omega_2 \cdot f_2^* = 0 \end{aligned} \tag{17}$$

In Eq. (17), f_1^* and f_2^* have the shape similar to the general expression for f_g ($g = 3,\ldots, G$) in Eq. (15).

In the balances related to aqueous media, the terms involved with water, i.e. N_{0j} (for j related to H_2O, as the component), N_1, and all $n_i = n_{iW}$ are not involved (in $f_0, f_3,\ldots, f_G$) or are cancelled within f_{12} (Eq. (17)). Other species, such as CH_3COOH transformable (mentally, purposefully) into $C_2H_4O_2 \equiv C_2(H_2O)_2$, and the species in which H and O are not involved in $X_i^{z_i}$ are also cancelled within f_{12}.

On the basis of relations (13) and (17), the linear combination of G + 1 balances $f_0, f_{12}, f_3,\ldots, f_G$ expressed by Eq. (14) can be presented in equivalent forms:

$$\sum_{i=2}^{I} N_i \cdot \left(z_i + \sum_{g=1}^{G} \omega_g \cdot a_{gi} \right) = \sum_{j=1}^{J} N_{0j} \cdot \sum_{g=1}^{G} \omega_g \cdot b_{gj} \tag{18}$$

$$\begin{aligned} &\sum_{i=2}^{I} N_i \cdot z_i + \sum_{g=1}^{G} \omega_g \cdot \left(\sum_{i=2}^{I} N_i \cdot a_{gi} - \sum_{j=1}^{J} N_{0j} \cdot b_{gj} \right) = 0 \\ &f_0 + f_{12} + \sum_{g=3}^{G} \omega_g \cdot f_g = 0 \end{aligned} \tag{18a}$$

$$\begin{aligned} &\mathrm{ChB} + (2 \cdot f(\mathrm{O}) - f(\mathrm{H})) + \sum_{g=3}^{G} \omega_g \cdot f(Y_g) = 0 \\ &(+1) \cdot f(\mathrm{H}) + (-2) \cdot f(\mathrm{O}) + \sum_{g=3}^{G} (-\omega_g) \cdot f(Y_g) - \mathrm{ChB} = 0 \end{aligned} \tag{18b}$$

All multipliers at N_i and N_{0j} in Eq. (18a) are cancelled simultaneously, if we have

$$z_i + \sum_{g=1}^{G} \omega_g \cdot a_{gi} = 0 \quad \text{and} \quad \sum_{g=1}^{G} \omega_g \cdot b_{gj} = 0 \tag{19}$$

for all $i = 1,\ldots, I$ and $j = 1,\ldots, J$; then Eq. (18) is transformed into identity

$$\sum_{i=1}^{I} N_i \cdot 0 = \sum_{j=1}^{J} N_{0j} \cdot 0 \Leftrightarrow 0 = 0 \tag{20}$$

Transformation of a set of the equations into the identity, 0 = 0, proves the linear dependence between the equations considered. Then from Eq. (18a), we have

$$f_{12} = \sum_{g=3}^{G} (-\omega_g) \cdot f_g - f_0 \tag{18c}$$

i.e. f_{12} is the dependent balance.

Briefly, from G + 1 starting balances: $f_0, f_1, f_2, f_3, \ldots, f_G$ we obtain G balances: $f_0, f_{12}, f_3, \ldots, f_G$. If f_{12} is the dependent balance, we have G − 1 independent balances: $f_0, f_3, \ldots, f_G$; it is the case related to non-redox systems. If f_{12} is the independent balance, we have G independent balances: $f_0, f_{12}, f_3, \ldots, f_G$, that will be rearranged, optionally, as the set ($f_{12}, f_0, f_3, \ldots, f_G$) related to GEB, ChB and $f(Y_g)$ (g = 3,..., G), respectively. The number of elemental/core balances $f(Y_g)$ ($Y_g \neq$ H, O, g = 3,..., G) and then the number of concentration balances CB(Y_g) in redox systems equals $K = G - 2$.

As stated above, the linear combination of 2·f(O) − f(H) with ChB and elemental/core balances $f(Y_g)$ (g = 3,..., G) provides the criterion distinguishing between non-redox and redox systems of different complexity [3, 23]. To obtain the simplest form of the linear combination, some useful/general rules allowing to select multipliers ω_g of the corresponding balances are suggested. Namely, the ω_g values in Eq. (10) are equal to the oxidation numbers of the electron-non-active elements in the related system. The proper linear combination of the balances is reducible (a) to the identity, 0 = 0, for non-redox systems or (b) to the simplest equation different from (not reducible to) the identity, but involving some terms related to the species and components of the system. This way we state that 2·f(O) − f(H) is linearly dependent on the other balances in non-redox systems, i.e. it is not a new/independent balance in such systems. In redox systems, the balance 2·f(O) − f(H) and its linear combinations with ChB and $f(Y_g)$ (g = 3,..., G) are the new equation, completing, as the generalized electron balance (GEB), the set of equations needed for resolution of electrolytic redox systems.

Static and dynamic non-redox and redox systems, with water as the main component, are considered. A static system of volume V_0 mL is obtained by disposable mixing different components. The dynamic system is realized according to the titrimetric mode. At defined point of the titration B (C, V) ⇒ A (C_0, V_0), denoted briefly as T ⇒ D, V mL of titrant T containing the component B (C mol/L) is added into V_0 mL of titrand D containing the component A (C_0 mol/L), and $V_0 + V$ mL of D + T system/mixture is thus obtained, if the assumption of volume additivity is valid/tolerable; then D and T are sub-systems of the D + T system. The progress of the titration is represented by the fraction titrated Φ [11, 19, 47, 53, 54] value

$$\Phi = \frac{C \cdot V}{C_0 \cdot V_0} \tag{21}$$

For comparative purposes, we first refer to non-redox systems, where the derivation of relevant formulas will be carried out in detail also for training purposes. The regularities resulting from these examples, namely linear dependency of (linear) algebraic equations stated for non-redox electrolytic systems, will provide the reference point for further parts of this chapter, where redox systems are considered. Further examples in this chapter, related only to redox systems, will be presented in a synthetic manner, indicating the similarities and differences in the respective equations. The conclusions drawn here will provide the basis for further, important generalizations.

4. Linear combination of balances for non-redox systems

4.1. Examples

Example 1 (static): V_0 mL of $CuSO_4$ solution is prepared from N_{01} molecules of $CuSO_4{\cdot}5H_2O$ and N_{02} molecules of H_2O. The resulting solution consists of the following species: H_2O (N_1), H^{+1} (N_2, n_2), OH^{-1} (N_3, n_3), HSO_4^{-1} (N_4, n_4), SO_4^{-2} (N_5, n_5), Cu^{+2} (N_6, n_6), $CuOH^{+1}$ (N_7, n_7), $Cu(OH)_2$ (N_8, n_8), $Cu(OH)_3^{-1}$ (N_9, n_9), $Cu(OH)_4^{-2}$ (N_{10}, n_{10}) and $CuSO_4$ (N_{11}, n_{11}). The components and species are involved in the balances for particular elements, Y_g: H, O, Cu, S, i.e.

$$f_1 = f(\mathrm{H}):$$

$$2N_1 + N_2(1+2n_2) + N_3(1+2n_3) + N_4(1+2n_4) + N_5 2n_5 + N_6 2n_6 + N_7(1+2n_7) \tag{22}$$

$$+ N_8(2+2n_8) + N_9(3+2n_9) + N_{10}(4+2n_{10}) + N_{11}2n_{11} = 10N_{01} + 2N_{02}$$

$$f_2 = f(\mathrm{O}):$$

$$N_1 + N_2n_2 + N_3(1+n_3) + N_4(4+n_4) + N_5(4+n_5) + N_6n_6 + N_7(1+n_7) + N_8(2+n_8) \tag{23}$$

$$+ N_9(3+n_9) + N_{10}(4+n_{10}) + N_{11}(4+n_{11}) = 9N_{01} + N_{02}$$

$$f_{12} = 2\cdot f(\mathrm{O}) - f(\mathrm{H}): \quad -N_2 + N_3 + 7N_4 + 8N_5 + N_7 + 2N_8 + 3N_9 + 4N_{10} + 8N_{11} = 8N_{01} \tag{24}$$

$$f_0 = \mathrm{ChB}: \quad N_2 - N_3 - N_4 - 2N_5 + 2N_6 + N_7 - N_9 - 2N_{10} = 0 \tag{25}$$

$$-2f_3 = -2f(\mathrm{Cu}): \quad 2N_{01} = 2N_6 + 2N_7 + 2N_8 + 2N_9 + 2N_{10} + 2N_{11} \tag{26}$$

$$-6f_4 = -6f(\mathrm{S}) = -6f(\mathrm{SO_4}): \quad 6N_{01} = 6N_4 + 6N_5 + 6N_{11} \tag{27}$$

Simple addition of the elemental/core balances (Eqs. (24)–(27)) gives the identity

$$2\cdot f(\mathrm{O}) - f(\mathrm{H}) + \mathrm{ChB} - 2f(\mathrm{Cu}) - 6f(\mathrm{S}) = 0 \Rightarrow$$
$$(+1)\cdot f(\mathrm{H}) + (-2)\cdot f(\mathrm{O}) + (+2)\cdot f(\mathrm{Cu}) + (+6)\cdot f(\mathrm{S}) - \mathrm{ChB} = 0 \Rightarrow 0 = 0 \tag{28}$$

As we see, the multipliers ω_g in the transformed identity are equal to oxidation numbers of the indicated elements.

Example 2 (dynamic): V_0 mL of $CuSO_4$ solution, as a titrand D composed of N_{01} molecules of $CuSO_4 \cdot 5H_2O$ and N_{02} molecules of H_2O, is titrated with V mL of titrant T, composed of N_{03} molecules of NaOH and N_{04} molecules of H_2O, added up to a defined point of the titration. In the resulting D + T mixture (two-phase system of volume ca. $V_0 + V$ mL), we have the species as in Example 1 and, additionally: $\mathbf{Cu(OH)_2}$ (N_{12}, n_{12}), Na^{+1} (N_{13}, n_{13}). From the balances:

$$f_1 = f(\mathrm{H}): \quad (\ldots) + N_{12}(2 + 2n_{12}) + 2N_{13}n_{13} = (\ldots) + N_{03} + 2N_{04} \tag{29}$$

$$f_2 = f(\mathrm{O}): \quad (\ldots) + N_{12}(2 + n_{12}) + N_{13}n_{13} = (\ldots) + N_{03} + N_{04} \tag{30}$$

$$\begin{aligned} f_{12} = 2 \cdot f(\mathrm{O}) - f(\mathrm{H}): \quad & -N_2 + N_3 + 7N_4 + 8N_5 + N_7 + 2N_8 + 3N_9 + 4N_{10} + 8N_{11} \\ & + 2N_{12} = 8N_{01} + N_{03} \end{aligned} \tag{31}$$

$$f_0 = \mathrm{ChB}: \quad N_2 - N_3 - N_4 - 2N_5 + 2N_6 + N_7 - N_9 - 2N_{10} + N_{13} = 0 \tag{32}$$

$$-2f_3 = -2f(\mathrm{Cu}): \quad 2N_{01} = 2N_6 + 2N_7 + 2N_8 + 2N_9 + 2N_{10} + 2N_{11} + 2N_{12} \tag{33}$$

$$-6f_4 = -6f(\mathrm{S}) = -6f(\mathrm{SO_4}): \quad 6N_{01} = 6N_4 + 6N_5 + 6N_{11} \tag{34}$$

$$-f_5 = -f(\mathrm{Na}): \quad N_{03} = N_{13} \tag{35}$$

That means we obtain the identity, 0 = 0, for

$$\begin{aligned} & 2 \cdot f(\mathrm{O}) - f(\mathrm{H}) + \mathrm{ChB} - f(\mathrm{Na}) - 2 \cdot f(\mathrm{Cu}) - 6 \cdot f(\mathrm{S}) = 0 \;\Rightarrow \\ & (+1) \cdot f(\mathrm{H}) + (-2) \cdot f(\mathrm{O}) + (+2) \cdot f(\mathrm{Cu}) + (+6) \cdot f(\mathrm{S}) - \mathrm{ChB} = 0 \end{aligned} \tag{36}$$

The terms within (…)'s in Eqs. (29) and (30) are the same as in Eqs. (22) and (23).

4.2. A comment

The equations for f(H) and f(O) in involve the numbers $n_i = n_{iW}$ of hydrating water molecules, attached to particular species. In the linear combinations $2 \cdot f(\mathrm{O}) - f(\mathrm{H})$, all the n_{iW} are cancelled, together with N_1, i.e. the n_{iW} and N_1 do not enter the equation for $2 \cdot f(\mathrm{O}) - f(\mathrm{H})$. The n_{iW} is not involved in ChB and in $f(Y_j)$ for $Y_j \neq$ H, O and then they are not introduced in the linear combinations of $2 \cdot f(\mathrm{O}) - f(\mathrm{H})$ with these equations. The multipliers ω_g in Eq. (10) are equal to oxidation numbers of particular elements involved in the related elemental/core balances. Then, the linear combination of the balances related to a non-redox system, when put in context with Eqs. (19) and (20) gives the identity 0 = 0. This regularity is obligatory for all non-redox systems considered above (*Examples* 1 and 2), and elsewhere, e.g. [22, 37, 55]. Note that $2 \cdot f(\mathrm{O}) - f(\mathrm{H})$ involves $-N_2 + N_3$, whereas ChB involves $N_2 - N_3 = -(-N_2 + N_3)$, and the related terms are cancelled within the combination $2 \cdot f(\mathrm{O}) - f(\mathrm{H})$ + ChB. Moreover, all components related to the species $X_i^{z_i} \cdot n_{iW}$, not involving H and/or O in $X_i^{z_i}$, are cancelled too within $2 \cdot f(\mathrm{O}) - f(\mathrm{H})$.

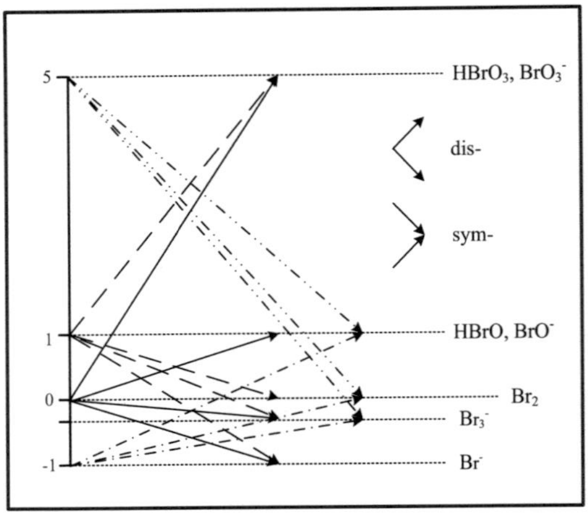

Figure 1. Schemes of disproportionation and symproportionation within bromine species [5].

In further sections, we formulate the balances for redox systems containing one, two or three electron-active elements in redox systems (aqueous media), where the complete set of expressions for independent equilibrium constants interrelating concentrations of different species is involved. We start our considerations from the systems with one electron-active element involved with disproportionation and symproportionation [5, 13, 15, 16, 19]. These properties are appropriate for the elements that form compounds and species at three or more oxidation degrees. In particular, bromine (Br) forms the species on five oxidation degrees (–1, –1/3, 0, 1, 5), see **Figure 1**, e.g. 0 (for Br_2) and 1 (for HBrO or NaBrO) ϵ (–1, 5). There are possible transitions between different bromine species, associated with changes of the oxidation states of this element, see **Figure 1**.

5. Disproportionating systems

5.1. Formulation of balances for NaOH (C, V) ⇒ Br_2 (C_0, V_0) system

5.1.1. Approach II to GEB

Note that V_0 mL of the titrand D composed of Br_2 (N_{01}) + H_2O (N_{02}) is titrated with V mL of titrant T composed of NaOH (N_{03}) + H_2O (N_{04}). For the $V_0 + V$ mL of the D + T mixture, with the set of the species completed by Na^{+1} (N_{11}, n_{11}).

The balances can be formulated as follows:

$$f_1 = f(\mathrm{H}):$$
$$2N_1 + N_2(1+2n_2) + N_3(1+2n_3) + N_4(1+2n_4) + 2N_5n_5 + N_6(1+2n_6) + 2N_7n_7 + 8N_8n_8$$
$$+ 2N_9n_9 + 2N_{10}n_{10} + N_{11}(1+2n_{11}) = 2N_{02} + N_{03} + 2N_{04} \quad (37)$$

$f_2 = f(\mathrm{O})$:

$$N_1 + N_2n_2 + N_3(1+n_3) + N_4(3+n_4) + N_5(3+n_5) + N_6(1+n_6) + N_7(1+n_7) + N_8n_8 + N_9n_9 + N_{10}n_{10} + N_{11}(1+n_{11}) = N_{02} + N_{03} + N_{04} \quad (38)$$

$$f_0 = \mathrm{ChB}: \qquad N_2 – N_3 – N_5 – N_7 – N_9 – N_{10} + N_{11} = 0 \quad (39)$$

$$–f_3 = –f(Na): \qquad N_{03} = N_{11} \quad (40)$$

$$f_4 = f(\mathrm{Br}): \qquad N_4 + N_5 + N_6 + N_7 + 2N_8 + 3N_9 + N_{10} – 2N_{01} = 0 \quad (41)$$

Then we formulate the linear combinations:

$$f_{12} = 2\cdot f(\mathrm{O}) - f(\mathrm{H}): \qquad -N_2 + N_3 + 5N_4 + 6N_5 + N_6 + 2N_7 = N_{03} \quad (42)$$

$$f_{012} = f_{12} + f_0: \qquad 5(N_4 + N_5) + N_6 + N_7 – N_9 – N_{10} + N_{11} = 0 \quad (43)$$

$$f_{0123} = f_{12} + f_0 – f_3: \qquad 5(N_4 + N_5) + N_6 + N_7 – N_9 – N_{10} = 0 \quad (44)$$

$$f_{01234} = (f_{0123} + f_4)/2: \qquad 3(N_4 + N_5) + N_6 + N_7 + N_8 + N_9 = N_{01} \quad (45)$$

For comparative purposes, we formulate also the linear combination $Z_{\mathrm{Br}}{\cdot}f_3 - f_{0123}$

$$(Z_{\mathrm{Br}} - 5)(N_4 + N_5) + (Z_{\mathrm{Br}} - 5)(N_6 + N_7) + 2Z_{\mathrm{Br}}N_8 + (3Z_{\mathrm{Br}} + 1)N_9 + (Z_{\mathrm{Br}} + 1)N_{10} = 2Z_{\mathrm{Br}}N_{01} \quad (46)$$

where $Z_{\mathrm{Br}} = 35$ is the atomic number for bromine (Br).

All the linear combinations (42)–(46) are not reducible to identity, 0 = 0; in the simplest case, we have six constituents (species, components) involved in f_{0123} (Eq. (44)). Eqs. (42)–(46) are equivalent forms of GEB, expressed in terms of particular constituents (components, species). To express them in terms of molar concentrations, we apply the relations: (1a) and $C_0{\cdot}V_0 = 10^3{\cdot}N_{01}/N_A$, $CV = 10^3{\cdot}N_{03}/N_A$. In this way, we consider the titration of V_0 mL of Br_2 (C_0 mol/L) solution with V mL of C mol/L NaOH solution added up to a defined point of the titration. In particular, from Eq. (46), we have the balance for GEB

$$(Z_{\mathrm{Br}} - 5)([\mathrm{HBrO_3}] + [\mathrm{BrO_3}^{-1}]) + (Z_{\mathrm{Br}} - 1)([\mathrm{HBrO}] + [\mathrm{BrO}^{-1}]) + 2Z_{\mathrm{Br}}[\mathrm{Br_2}] + (3Z_{\mathrm{Br}} + 1)[\mathrm{Br_3}^{-1}] + (Z_{\mathrm{Br}} + 1)[\mathrm{Br}^{-1}] = 2Z_{\mathrm{Br}}C_0V_0/(V_0 + V) \quad (46a)$$

completed by the balances for ChB, CB(Br) and CB(Na):

$$[\mathrm{H}^{+1}] - [\mathrm{OH}^{-1}] + [\mathrm{Na}^{+1}] - [\mathrm{BrO_3}^{-1}] - [\mathrm{BrO}^{-1}] - [\mathrm{Br_3}^{-1}] - [\mathrm{Br}^{-1}] = 0 \quad (46b)$$

$$[\mathrm{HBrO_3}] + [\mathrm{BrO_3}^{-1}] + [\mathrm{HBrO}] + [\mathrm{BrO}^{-1}] + 2[\mathrm{Br_2}] + 3[\mathrm{Br_3}^{-1}] + [\mathrm{Br}^{-1}] - 2\mathrm{C_0V_0}/(\mathrm{V_0} + \mathrm{V}) = 0 \quad (46c)$$

$$[Na^{+1}] = CV/(V_0 + V) \qquad (46d)$$

5.1.2. Approach I to GEB

The (optional) formulas for GEB, expressed by Eqs. (42)–(46), were obtained according to approach II to GEB, based on *pr*-GEB = 2·*f*(O) – *f*(H). In this section, we apply the approach I to GEB, known also as the 'short' version of GEB, where prior knowledge of oxidation numbers of all elements in the system is assumed/required.

In the NaOH (C, V) ⇒ Br_2 (C_0, V_0) system considered in Section 4.2, bromine is the only electron-active element, considered as the carrier of its own, bromine electrons. One atom of Br has Z_{Br} bromine electrons, and then one molecule of Br_2 has $2Z_{Br}$ bromine electrons, i.e. N_{01} molecules of Br_2 involve $2Z_{Br} \cdot N_{01}$ bromine electrons. The oxidation degree x of an atom in simple species, such as ones formed here by bromine, is calculated on the basis of known oxidation degrees: +1 for H and –2 for O and external charge of this species. Then for $HBrO_3$, we have, by turns: $1 \cdot 1 + 1 \cdot x + 3 \cdot (-2) = 0 \Rightarrow x = 5$; for BrO_3^{-1}: $1 \cdot x + 3 \cdot (-2) = -1 \Rightarrow x = 5$; for HBrO: $1 \cdot 1 + 1 \cdot x + 1 \cdot (-2) = 0 \Rightarrow x = 1$, etc.

The oxidation number is the net charge resulting from the presence of charge carriers, inherently involved in an atom: protons in nuclei and orbital electrons, expressed in elementary charge units—as +1 for protons and –1 for electrons. The number y of bromine electrons in one molecule of $HBrO_3$ is calculated from the formula: $Z_{Br} \cdot (+1) + y \cdot (-1) = 5$, i.e., bromine in $HBrO_3$ involves $y = Z_{Br} - 5$ bromine electrons, etc. On this basis, we state that [15]

N_4 species $HBrO_3 \cdot n_4H_2O$ involve	$(Z_{Br} - 5) \cdot N_4$	bromine electrons;
N_5 species $BrO_3^{-1} \cdot n_5H_2O$ involve	$(Z_{Br} - 5) \cdot N_5$	bromine electrons;
N_6 species $HBrO \cdot n_6H_2O$ involve	$(Z_{Br} - 1) \cdot N_6$	bromine electrons;
N_7 species $BrO^{-1} \cdot n_7H_2O$ involve	$(Z_{Br} - 1) \cdot N_7$	bromine electrons;
N_8 species $Br_2 \cdot n_8H_2O$ involve	$2Z_{Br} \cdot N_8$	bromine electrons;
N_9 species $Br_3^{-1} \cdot n_9H_2O$ involve	$(3Z_{Br} + 1) \cdot N_9$	bromine electrons;
N_{10} species $Br^{-1} \cdot n_{10}H_2O$ involve	$(Z_{Br} + 1) \cdot N_{10}$	bromine electrons;
N_{01} molecules of Br_2 involved	$2Z_{Br} \cdot N_{01}$	bromine electrons.

Balancing the bromine electrons, we get Eq. (46), and then Eq. (46a), considered as the GEB obtained immediately from the approach I to GEB; this proves the equivalency of approaches I and II to GEB.

5.1.3. Preparation of an algorithm and computer program

Let us refer again to balances (46a)–(46d) interrelating concentrations $[X_i^{zi}]$ of the species X_i^{zi} with the total concentrations C_0, C of indicated components of the system. In this place, one should distinguish between the terms equation and equality. The term equation is related here

No.	Reaction	Equilibrium equation	Equilibrium data
1	$BrO_3^{-1}+6H^{+1}+6e^{-1} = Br^{-1}+3H_2O$	$[Br^{-1}] = K_{e1}\cdot[BrO_3^{-1}][H^{+1}]^6[e^{-1}]^6$	$E_{04} = 1.45$ V
2	$BrO^{-1}+H_2O+2e^{-1} = Br^{-1}+2OH^{-1}$	$[Br^{-1}] = K_{e2}\cdot[BrO^{-1}][H^{+1}]^2[e^{-1}]^2/K_W^2$	$E_{03} = 0.76$ V
3	$Br_2 + 2e^{-1} = 2Br^{-1}$	$[Br^{-1}]^2 = K_{e3}\cdot[Br_2][e^{-1}]^2$	$E_{03} = 1.087$ V
4	$Br_3^{-1} + 2e^{-1} = 3Br^{-1}$	$[[Br^{-1}]^3 = K_{e4}\cdot[Br_3^{-1}][e^{-1}]^2$	$E_{04} = 1.05$ V
5	$HBrO_3 = H^{+1} + BrO_3^{-1}$	$[H^{+1}][BrO_3^{-1}] = K_{51}\cdot[HBrO_3]$	$pK_{51} = 0.7$
6	$HBrO = H^{+1} + BrO^{-1}$	$[H^{+1}][BrO^{-1}] = K_{11}\cdot[HBrO]$	$pK_{11} = 8.6$
7	$H_2O = H^{+1} + OH^{-1}$	$[H^{+1}][OH^{-1}] = K_W$	$pK_W = 14.0$

where $\log K_{e1} = 6AE_{01}$, $\log K_{ei} = 2AE_{0i}$ ($i = 2, 3, 4$); $A = 16.92$; $pK_{51} = -\log K_{51}$, $pK_{11} = -\log K_{11}$, $pK_W = -\log K_W$.

Table 1. Equilibrium data related to different bromine species.

to the balance where at least two species are involved; these species are interrelated by expressions for the corresponding equilibrium constants values (**Table 1**). Such a requirement is fulfilled by Eqs. (46a)–(46c), whereas in Eq. (46d) we have concentration of one species; at any V value, $[Na^{+1}]$ is a number (not variable) and, as such, it enters immediately ChB (Eq. (46b)).

From the interrelations obtained on the basis of equilibrium data [56] collected in **Table 1** we have

$$[H^{+1}] = 10^{-pH};\ [OH^{-1}] = 10^{pH-14};\ [BrO_3^{-1}] = 10^{6A(E-1.45)-pBr+6pH};$$
$$[BrO^{-1}] = 10^{2A(E-0.76)-pBr+2pH-28};\ [Br_2] = 10^{2A(E-1.087)-2pBr};\ [Br_3^{-1}] = 102^{A(E-1.05)-2pBr};\quad (47)$$
$$[HBrO_3] = 10^{0.7-pH}\ [BrO_3^{-1}];\ [HBrO] = 10^{8.6-pH}\cdot[BrO^{-1}]$$

where the uniformly defined (scalar) variables: E, pH and pBr, forming a vector $\mathbf{x} = (E, pH, pBr)^T$, are involved; $A\cdot E = -\log[e^{-1}]$, $pH = -\log[H^{+1}]$, $pBr = -\log[Br^{-1}]$. All the variables are in the exponents of the power for 10 in $[e^{-1}] = 10^{-AE}$, $[H^{+1}] = 10^{-pH}$, $[Br^{-1}] = 10^{-pBr}$. The number of the (independent) variables equals to the number of equations, $K = 2 + 1 = 3$; this ensures a unique solution of the equations, at a pre-set C_0, C and V_0 values, and the V-value at which the calculations are realized, at a defined step of the calculation procedure, according to iterative computer program presented below.

Computer program for the NaOH_Br2 system

```
function F = NaOH_Br2(x)

global V C0 V0 C yy

    E = x(1);

    pH = x(2);

    pBr = x(3);


    H = 10^(-pH);

    Kw = 10^-14;
```

```
    pKw = 14;
    OH = Kw/H;
    A = 16.92;
    Br = 10^-pBr;
    ZBr = 35;

    Br2=Br^2*10^(2*A*(E-1.087));
    Br3=Br^3*10^(2*A*(E-1.05));
    BrO=Br*10^(2*A*(E-0.76)+2*pH-2*pKw);
    BrO3=Br*10^(6*A*(E-1.45)+6*pH);
    HBrO=10^8.6*H*BrO;
    HBrO3=10^0.7*H*BrO3;
    Na=C*V/(V0+V);

    F = [%Charge balance
      (H-OH+Na-Br-Br3-BrO-BrO3);
      %Concentration balance for Br
      (Br+3*Br3+2*Br2+HBrO+BrO+HBrO3+BrO3-2*C0*V0/(V0+V));
      %Electron balance
      ((ZBr+1)*Br+(3*ZBr+1)*Br3+2*ZBr*Br2+(ZBr-1)*(HBrO+BrO)...
      +(ZBr-5)*(HBrO3+BrO3)-2*ZBr*C0*V0/(V0+V))];

    yy(1)=log10(Br);
    yy(2)=log10(Br3);
    yy(3)=log10(Br2);
    yy(4)=log10(HBrO);
    yy(5)=log10(BrO);
    yy(6)=log10(HBrO3);
    yy(7)=log10(BrO3);
    yy(8)=log10(Na);
end
```

5.1.4. *Preparation of the computer program for NaOH (C, V) ⇒ HBrO (C_0, V_0) system*

Modification of the computer program makes it possible to carry out calculations for other systems associated with bromine, e.g. for the system NaOH (*C*, *V*) ⇒ HBrO (C_0, V_0). In this case, the changes are made in the following lines:

Computer program for the NaOH_HBrO system

Function F = NaOH_HBrO(x)

—

```
(Br+3*Br3+2*Br2+HBrO+BrO+HBrO3+BrO3-C0*V0/(V0+V));
```

—

```
+(ZBr-5)*(HBrO3+BrO3)-(ZBr-1)*C0*V0/(V0+V))];
```

—

5.1.5. *Graphical presentation of the data and discussion*

The systems considered here are characterized by distinctly marked jumps in *E* and pH values (**Figures 2A** and **B**) at the vicinity of the equivalence points, occurring at $\Phi_{eq1} = 1$ for NaOH (*C*, *V*) → HBrO (C_0, V_0) (system I), and at $\Phi_{eq2} = 2$ for NaOH (*C*, *V*) → Br_2 (C_0, V_0) (system II), see **Table 2** [15, 16].

Reactions occurred in the systems I and II can be formulated from the analysis of the respective speciation diagrams. As results from **Figure 2C**, the disproportionation reaction in the system I occurs according to the scheme

$$3Br_2 + 6OH^{-1} = BrO_3^{-1} + 5Br^{-1} + 3H_2O \tag{48}$$

i.e. $\Phi_{eq} = 2 \Rightarrow CV_{eq1} = 2C_0V_0$. After crossing the equivalent point, the main products of this disproportionation reaction are Br^{-1} and BrO_3^{-1}, not Br^{-1} and BrO^{-1} corresponding to the reaction

$$Br_2 + 2OH^{-1} = BrO^{-1} + Br^{-1} + H_2O \tag{49}$$

with the same stoichiometry, 3 : 6 = 1 : 2. From detailed calculations at $\Phi = 2.5$, i.e. at an excess of NaOH added, we find $[BrO_3^{-1}]/[BrO^{-1}] = 10^{-2.574}/10^{-6.7584} = 1.53 \cdot 10^4$, i.e. efficiency of the reaction (48) is more than 4 orders of magnitude greater than the efficiency of reaction (49).

Scheme of Br_2 disproportionation affected by NaOH (C) can also be calculated [16]. At $\Phi = 2.5$, we have $[BrO_3^{-1}]/[Br^{-1}] = 10^{-2.574}/10^{-1.875} = 10^{-0.699} = 0.2 = 1{:}5$, i.e. at an excess of NaOH added (see **Figure 2c**), the disproportionation occurs mainly according to the scheme indicated by reaction (48) from the products side.

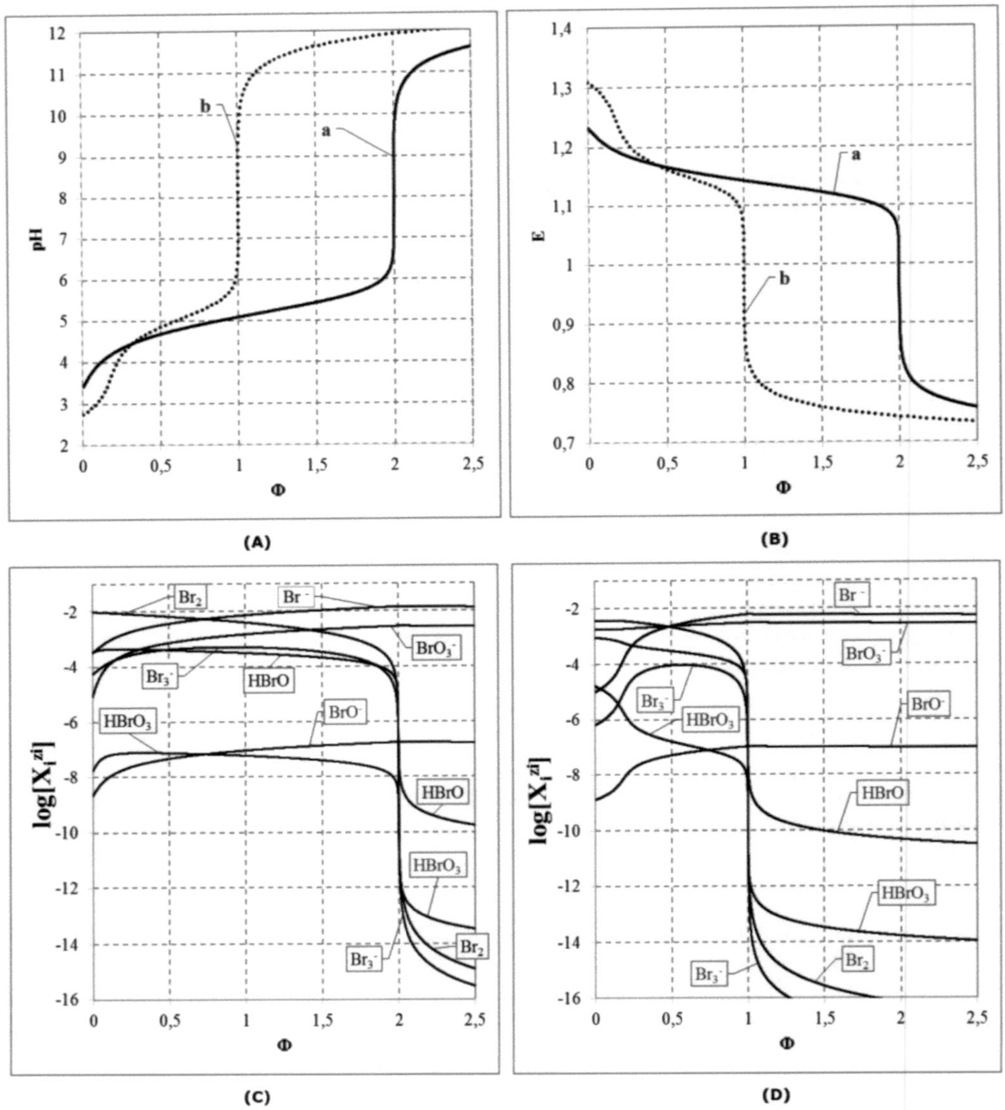

Figure 2. (A) pH = pH(Φ) and (B) E = E(Φ) relationships plotted for the systems I and II, and the related speciation diagrams for the systems: I (C) and II (D).

In the system II, disproportionation of HBrO affected by NaOH (C) added according to the titrimetric mode is presented in **Figure 2**D [16]. The $[Br^{-1}]/[BrO_3^{-1}]$ ratio equals: $10^{-2.2553}/10^{-2.5563}$ at $\Phi = 2.0$; $10^{-2.2730}/10^{-2.5740}$ at $\Phi = 2.5$, i.e. $10^{0.3010} = 2 = 2:1$, corresponding to the stoichiometric ratio of products of this reaction. As results from **Figure 2C**, the disproportionation, at an excess of NaOH added, occurs mainly according to reaction $3HBrO + 3OH^{-1} = 2Br^{-1} + BrO_3^{-1} + 3H_2O$ (stoichiometry 3:3 = 1:1), resulting from half reactions: $HBrO + 2e^{-1} + H^{+1} = Br^{-1} + H_2O$, $HBrO - 4e^{-1} + 2H_2O = BrO_3^{-1} + 5H^{+1}$, and $3H^{+1} + 3OH^{-1} = 3H_3O$. The

System II			**System I**		
NaOH (*C, V*) → HBrO (*C*$_0$, *V*$_0$)			**NaOH (*C, V*) → Br$_2$ (*C*$_0$, *V*$_0$)**		
Φ	**pH**	**E**	**Φ**	**pH**	**E**
0.995	6.347	1.0720	1.995	6.666	1.0491
0.996	6.411	1.0681	1.996	6.728	1.0455
0.997	6.498	1.0630	1.997	6.811	1.0406
0.998	6.625	1.0555	1.998	6.933	1.0334
0.999	6.866	1.0412	1.999	7.161	1.0199
1.000	8.102	0.9682	2.000	8.143	0.9619
1.001	9.002	0.9150	2.001	8.966	0.9132
1.002	9.281	0.8985	2.002	9.244	0.8968
1.003	9.450	0.8885	2.003	9.413	0.8868
1.004	9.571	0.8814	2.004	9.534	0.8797
1.005	9.666	0.8758	2.005	9.628	0.8741

Table 2. The sets of rounded (Φ, pH, *E*) values taken from the vicinity of the equivalence points; $V_0 = 100$, $C_0 = 0.01$, $C = 0.1$.

(Φ, pH, *E*) values from the close vicinity of the corresponding equivalence points on the curves in **Figures 2A** and **B** are collected in **Table 2**. The Br_2 solution is acidic, as results, e.g. from the ChB (Eq. (29a)): at $[Na^{+1}] = 0$ ($\Phi = 0$) we have $[H^{+1}] - [OH^{-1}] = [BrO_3^{-1}]+[BrO^{-1}]+[Br_3^{-1}]+[Br^{-1}] > 0$, i.e. $[H^{+1}] > [OH^{-1}]$; this inequality is also obtained from Eq. (42), at $N_{03} = 0$: $[H^{+1}] - [OH^{-1}] = 5[HBrO_3]+6[BrO_3^{-1}] + [HBrO] + 2[BrO^{-1}] > 0$. Br_2 is an acid with a strength comparable to that of acetic acid; at $C_0 = 0.01$, pH equals 3.40 for Br_2 and 3.325 for CH_3COOH ($pK_1 = 4.65$). Disproportionation of Br_2 occurs initially to a small extent (several %), according to the scheme $Br_2 + OH^{-1} = HBrO + Br^{-1}$, compare with [57].

In $C_0 = 0.01$ mol/L HBrO solution, more than 90% HBrO disproportionates occur according to the reaction $5HBrO = BrO_3^{-1} + 2Br_2 + 2H_2O + H^{+1}$; at $V = 0$, we have $[Br_2] = 10^{-2.4406}$, $[BrO_3^{-1}] = 10^{-2.7442}$, i.e. $[Br_2]/[BrO_3^{-1}] = 10^{0.3036} \approx 2$, which confirms this stoichiometry of the reaction. The H^{+1} ions formed in this reaction acidify the solution significantly: at $C_0 = 0.01$ and $V = 0$, we have pH = 2.74, although HBrO itself is a relatively weak acid.

The numerical values of the concentrations given here are taken from the corresponding files with results of iterative calculations.

5.2. Titration in NaOH (*C, V*) ⇒ HIO (C_0, V_0) system

The curves plotted in **Figure 3a–c** are related to titration of $V_0 = 10$ mL of HIO ($C_0 = 0.1$ mol/L) with *V* mL of $C = 0.1$ mol/L NaOH. At the initial part of the titration, we have the reactions:

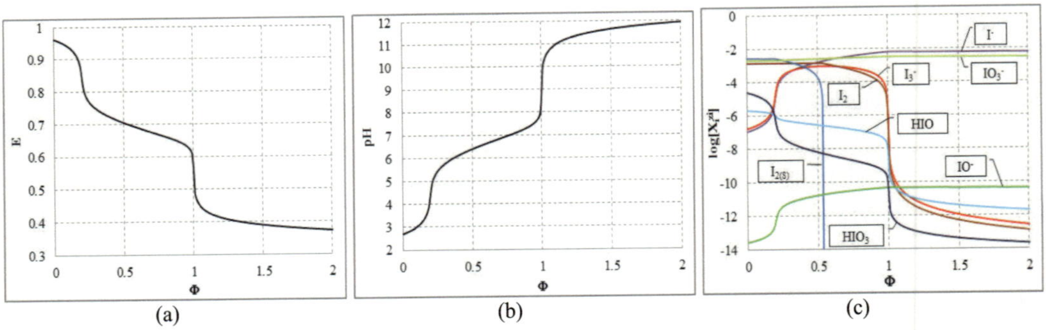

Figure 3. Plots of (a) $E = E(\Phi)$, (b) $pH = pH(\Phi)$, and (c) speciation curves for NaOH (C, V) $\Rightarrow$ HIO (C_0, V_0) system at $V_0 = 10$, $C_0 = 0.01$, $C = 0.1$.

$$5HIO + OH^{-1} = 2(I_{2(s)}, I_2) + IO_3^{-1} + 3H_2O \tag{50}$$

In the following, at Φ ca. 0.20–0.22, a pronounced increase in $[I^{-1}]$ occurs, as a result of reaction

$$3HIO + 3OH^{-1} = IO_3^{-1} + 2I^{-1} + 3H_2O \tag{51}$$

The increase in $[I^{-1}]$ is accompanied by an increase in $[I_3^{-1}]$

$$\left(\boldsymbol{I}_{2(s)}, I_2\right) + I^{-1} = I_3^{-1} \tag{52}$$

This leads to the gradual disappearance of $\boldsymbol{I}_{2(s)}$ (which is ultimately ended at Φ = 0.5347) and lowering of $[I_2]$ and $[I_3^{-1}]$; all them disproportionate

$$3(\boldsymbol{I}_{2(s)}, I_2, I_3^{-1}) + 6OH^{-1} = IO_3^{-1} + (5,\ 5,\ 8)I^{-1} + 3H_2O \tag{53}$$

At $[\mathbf{I}_{2(s)}] > 0$, we have $[I_2] = s = \text{const}$; $s = 1.33 \cdot 10^{-3}$ mol/L is the solubility of $\boldsymbol{I}_{2(s)}$ in water, at 20°C. Finally, the disproportionation of HIO, affected by NaOH, can be expressed by the equation

$$3HIO + 3OH^{-1} = IO_3^{-1} + 2I^{-1} + 3H_2O \tag{54}$$

Note that the stoichiometry of the reaction (54) is 3 : 3 = 1 : 1, which corresponds to the jump on the curves presented in **Figures 3a,b**, occurring at Φ = 1. For Φ > 1, we have $[I^{-1}]/[IO_3^{-1}] = 2$, i.e. the stoichiometry of the products of reaction (54) equals to 1 : 2. The jumps on the curves $E = E(\Phi)$ and $pH = pH(\Phi)$ (**Figures 3a,b**) occur at Φ ca. 0.2 (which corresponds to the stoichiometry 1:5 of the reaction (50)) and at Φ ca. 1 (which corresponds to the stoichiometry 3:3 = 1:1 of the reaction (54)); the maxima on the corresponding derivative curves in **Figures 4a,b** fit the stoichiometric ratios.

5.3. Titration in HCl (C, V) ⇒ NaIO (C_0, V_0) system

$V_0 = 10$ mL of $C_0 = 0.01$ mol/L NaIO is titrated with $C = 0.1$ mol/L HCl. The related curves are presented in **Figures 5a-d**. Initially, the reaction

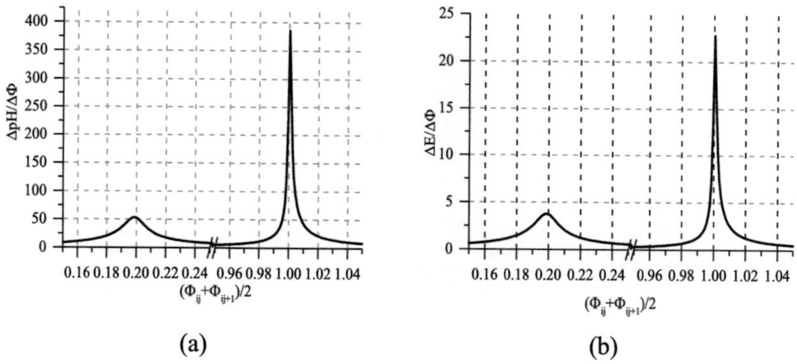

Figure 4. The (a) $\Delta pH/\Delta\Phi = (pH_{j+1} - pH_j)/(\Phi_{j+1} - \Phi_j)$, (b) $\Delta E/\Delta\Phi = (E_{j+1} - E_j)/(\Phi_{j+1} - \Phi_j)$ vs. $\Phi = (\Phi_j + \Phi_{j+1})/2$ relationships for the NaOH (C, V) ⇒ HIO (C_0, V_0) system.

$$3IO^{-1} = IO_3^{-1} + 2I^{-1} \tag{55}$$

and then, the reaction

$$5IO^{-1} + 4H^{+1} = 2I_2 + IO_3^{-1} + 2H_2O \tag{56}$$

occurs. Then I^{-1} from Eq. (55) and I_2 from Eq. (56) form I_3^{-1} in the reaction $I_2 + I^{-1} = I_3^{-1}$ and $[I_3^{-1}]$ increases. At $\Phi = 0.4654$, $\boldsymbol{I}_{2(s)}$ appears as the solid phase

$$5IO^{-1} + 4H^{+1} = 2\boldsymbol{I}_{2(s)} + IO_3^{-1} + 2H_2O \tag{57}$$

At $[\boldsymbol{I}_{2(s)}] > 0$, we have $[I_2]$ = const. The increase in $[Cl^{-1}]$, resulted from addition of HCl, causes an increase in $[I_2Cl^{-1}]$, and—to a lesser extent—the increase in $[ICl_2^{-1}]$ and [ICl]. The addition of HCl lowers pH of the solution, and then [HIO] becomes larger than $[IO^{-1}]$; $[HIO_3]$ also increases. In effect, the summary concentration [HIO] + $[IO^{-1}]$ after addition of an excess of HCl is higher than in the starting NaIO solution.

In the algorithm, we have allowed the participation of Cl^{-1} ions from HCl solution in the redox reaction. However, the concentration of Cl_2 and HClO as the main products of Cl^{-1} oxidation (**Figure 5b**) is quite negligible. This way, one can state that the Cl^{-1} ions practically do not participate in the redox reaction as a reducing agent. From linear combination of reactions (55) and

$$5I^{-1} + IO_3^{-1} + 6H^{+1} = 3\boldsymbol{I}_{2(s)} + 3H_2O \tag{58}$$

(multiplication by 5 and 2 respectively), cancellations and division by 3, we get the reaction

$$5IO^{-1} + 4H^{+1} = IO_3^{-1} + 2\boldsymbol{I}_{2(s)} + 2H_2O \tag{59}$$

with stoichiometry 4:5 = 0.8, which corresponds to Φ = 4:5 = 0.8, where the inflection point on the curves in **Figures 5c** and **d** are observed. The I^{-1} and I_3^{-1} ions are consumed in reactions (58) and

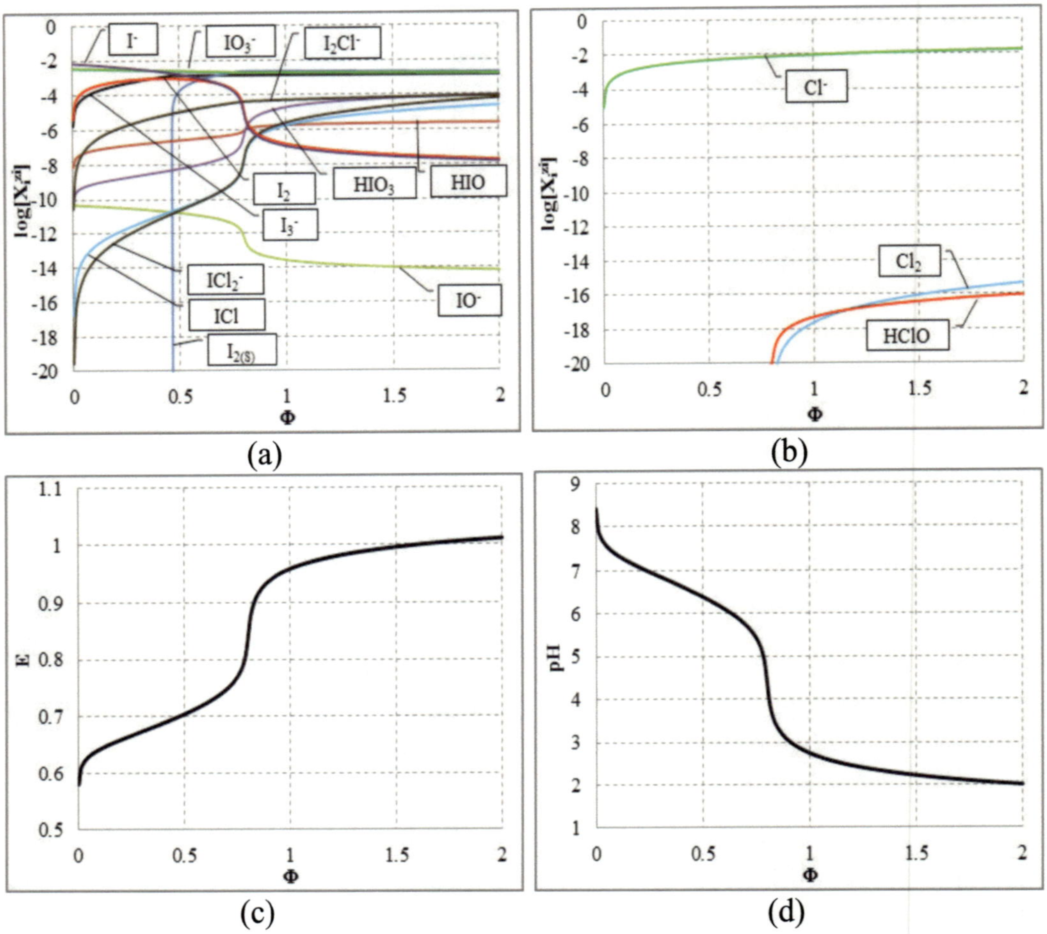

Figure 5. The speciation curves for indicated (a) iodine and (b) chlorine species $X_i^{z_i}$ and (c) $E = E(\Phi)$, (d) pH = pH(Φ) functions for HCl (C, V) ⇒ NaIO (C_0, V_0) system; V_0 = 10, C_0 = 0.01, C = 0.1.

$$5I_3^{-1} + IO_3^{-1} + 6H^{+1} = 8\boldsymbol{I}_{2(s)} + 3H_2O \quad (60)$$

See **Figure 5a**.

Computer program for HCl (C, V) ⇒ NaIO (C_0, V_0) system

```
function F = System_NaIO_HCl(x);
%NaIO<-HCl
% Titration of V0 mL of NaIO (C0) with V mL HCl (C).

global V Vmin Vstep Vmax V0 C C0 fi H OH pH E Kw pKw A aa
global I I3 I2 I2s HIO IO HI5O3 I5O3 H5I7O6 H4I7O6 H3I7O6 Na
global logI logI3 logI2 logI2s logHIO logIO logHI5O3 logI5O3 logH5I7O6
```

```
global logH4I7O6 logH3I7O6 logNa
global Cl Cl2 HClO ClO HCl3O2 Cl3O2 Cl4O2 Cl5O3 Cl7O4 I2Cl ICl ICl2
global logCl logCl2 logHClO logClO logHCl3O2 logCl3O2 logCl4O2 logCl5O3
global logCl7O4 logI2Cl logICl logICl2 pI pCl

E=x(1);
pH=x(2);
pI=x(3);
pCl=x(4);

H=10.^-pH;
pKw=14;
Kw=10.^-14;
OH=Kw./H;
I=10.^-pI;
Cl=10.^-pCl;
A=16.92;
ZCl=17;
ZI=53;

I2=I.^2.*10.^(2.*A.*(E-0.621));
I3=I.^3.*10.^(2.*A.*(E-0.545));
IO=I.*10.^(2.*A.*(E-0.49)+2.*pH-2.*pKw);
HIO=IO.*10.^(10.6-pH);
I5O3=I.*10.^(6.*A.*(E-1.08)+6.*pH);
HI5O3=I5O3.*10.^(0.79-pH);
H5I7O6=I.*10.^(8.*A.*(E-1.24)+7.*pH);
H4I7O6=H5I7O6.*10.^(-3.3+pH);
H3I7O6=I.*10.^(8.*A.*(E-0.37)+9.*pH-9.*pKw);

Cl2=Cl.^2.*10.^(2.*A.*(E-1.359));
ClO=Cl.*10.^(2.*A.*(E-0.88)+2.*pH-2.*pKw);
HClO=ClO.*10.^(7.3-pH);
```

```
Cl3O2=Cl.*10.^(4.*A.*(E-0.77)+4.*pH-4.*pKw);
HCl3O2=Cl.*10.^(4.*A.*(E-1.56)+3.*pH);
Cl4O2=Cl.*10.^(5.*A.*(E-1.5)+4.*pH);
Cl5O3=Cl.*10.^(6.*A.*(E-1.45)+6.*pH);
Cl7O4=Cl.*10.^(8.*A.*(E-1.38)+8.*pH);
I2Cl=I2.*10.^(0.2-pCl);
ICl=I2.^0.5.*10.^(A.*(E-1.105)-pCl);
ICl2=ICl.*10.^(2.2-pCl);
Na=C0.*V0./(V0+V);
if I2>1.33e-3
   I2s=I2-1.33e-3;
   I2=1.33e-3;
   aa=1;
else
   aa=0;
   I2s=0;
end;
   %Charge balance
F=[(H-OH+Na-I-I3-IO-I5O3-H4I7O6-2.*H3I7O6-Cl-ClO-Cl3O2-Cl5O3-Cl7O4...
   -I2Cl-ICl2);
   %Concentration balance for I
   (I+3.*I3+2.*(I2+aa.*I2s)+HIO+IO+HI5O3+I5O3+H5I7O6+H4I7O6+H3I7O6+...
   2.*I2Cl+ICl+ICl2-C0.*V0./(V0+V));
   %Concentration balance for Cl
   (Cl+2.*Cl2+HClO+ClO+HCl3O2+Cl3O2+Cl4O2+Cl5O3+Cl7O4+I2Cl+ICl...
   +2.*ICl2-C.*V./(V0+V));
   %Electron balance
   ((ZI+1).*I+(3.*ZI+1).*I3+2.*ZI.*(I2+aa.*I2s)+(ZI-1).*(HIO+IO)...
   +(ZI-5).*(HI5O3+I5O3)+(ZI-7).*(H5I7O6+H4I7O6+H3I7O6)+(ZCl+1).*Cl+...
```

```
    2.*ZCl.*Cl2+(ZCl-1).*(HClO+ClO)+(ZCl-3).*(HCl3O2+Cl3O2)...
    +(ZCl-4).*Cl4O2+(ZCl-5).*Cl5O3+(ZCl-7).*Cl7O4+(2.*ZI+ZCl+1).*I2Cl+...
    (ZI–ZCl).*ICl+(ZI+2.*ZCl+1).*ICl2...
    -((ZI-1).*C0.*V0+(ZCl+1).*C.*V)./(V0+V))];
logI=log10(I);
logI3=log10(I3);
logI2=log10(I2);
logI2s=log10(I2s);
logHIO=log10(HIO);
logIO=log10(IO);
logHI5O3=log10(HI5O3);
logI5O3=log10(I5O3);
logH5I7O6=log10(H5I7O6);
logH4I7O6=log10(H4I7O6);
logH3I7O6=log10(H3I7O6);
logCl=log10(Cl);
logCl2=log10(Cl2);
logHClO=log10(HClO);
logClO=log10(ClO);
logHCl3O2=log10(HCl3O2);
logCl3O2=log10(Cl3O2);
logCl4O2=log10(Cl4O2);
logCl5O3=log10(Cl5O3);
logCl7O4=log10(Cl7O4);
logI2Cl=log10(I2Cl);
logICl=log10(ICl);
logICl2=log10(ICl2);
logNa=log10(Na);
% The end of program
```

A remark: Some notations applied in this program are as follows: $[HIO_3] \rightarrow$ HI5O3, $[H_3IO_6^{-2}] \rightarrow$ H3I7O6, $[ClO_2^{-1}] \rightarrow$ Cl3O2, $[ClO_2] \rightarrow$ Cl4O2, etc.

5.4. Disproportionation in static redox systems

In disproportionating redox systems, it is advisable to check an effect of dilution of the corresponding solutes on values of the corresponding variables. It is advisable to plot the desired relationships in the figures with the values $pC = -\log C$ on the abscissa, where C is the concentration [mol/L] of the solute considered. Formally, the related plots correspond to presentation, in extended logarithmic scale, the results of titration of the initial solution of this solute with pure water as a titrant T.

The static systems with C mol/L solutions of (1) HIO and (2) NaIO are shown graphically in **Figures 6a-c** and **7a-c**, where pH, E and log $[X_i^{z_i}]$ values related to different concentrations C of the corresponding solutes with the values $pC = -\log C$ on the abscissa.

5.4.1. C mol/L HIO

As results from the speciation diagram in **Figure 6c**, in more concentrated HIO solutions, i.e. at lower pC values, the predominating reactions are as follows: $5HIO = 2(\boldsymbol{I_{2(s)}}, I_2) + IO_3^{-1} + 2H_2O + H^{+1}$ and $5HIO = 2(\boldsymbol{I_{2(s)}}, I_2) + HIO_3 + 2H_2O$; solubility of $\mathbf{I_{2(s)}}$ $s = 1.33 \cdot 10^{-3}$ mol/L (25°C). At further dilution of HIO, the reaction $3HIO = 2I^{-1} + IO_3^{-1} + 3H^{+1}$ occurs in an increasing degree. This change in disproportionation scheme, more significant at pC 4–5, resulted in a change of the shapes of the plots: $E = E(\text{pC})$ (**Figure 6a**) and pH = pH(pC) (**Figure 6b**).

5.4.2. C mol/L NaIO

The disproportionation of IO^{-1} introduced by NaIO proceeds mainly according to the scheme $3HIO = IO_3^{-1} + 2I^{-1} + 3H^{+1}$ (see **Figure 7c**), where $[I^{-1}]/[IO_3^{-1}] \cong 2$. Concentration of I_2 is lower than 10^{-6} mol/L and then solid $\boldsymbol{I_{2(s)}}$ is not formed.

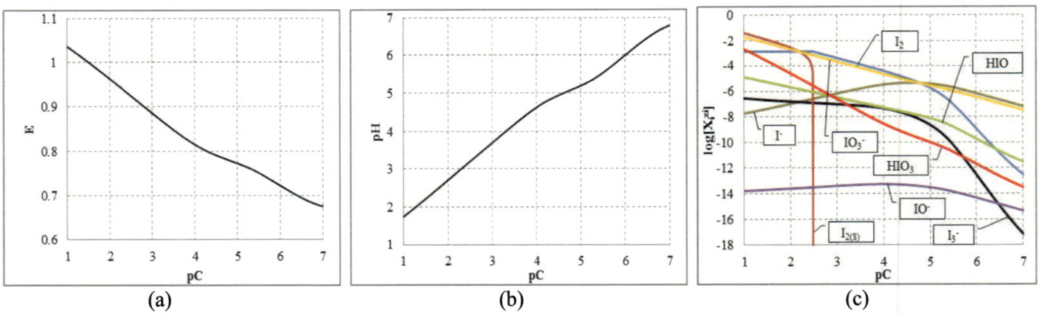

Figure 6. The plots of (a) E vs. pC, (b) pH vs. pC relationships and (c) speciation curves for indicated iodine species $X_i^{z_i}$ in C mol/L HIO.

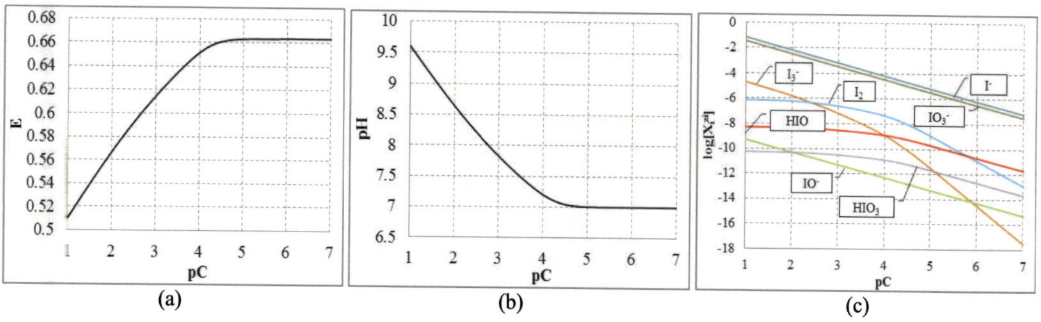

Figure 7. The functions: (a) $E = E(\text{pC})$, (b) pH = pH(pC) and (c) speciation curves for indicated iodine species $X_i^{z_i}$ plotted for *C* mol/L NaIO.

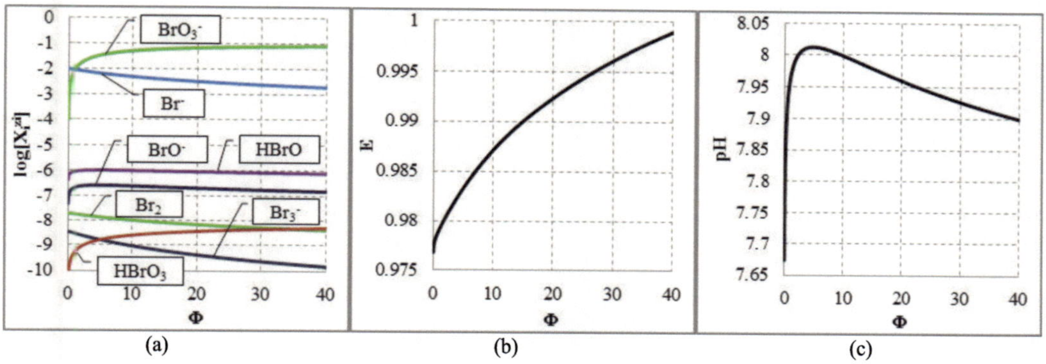

Figure 8. Plots for the $KBrO_3$ (*C, V*) ⇒ NaBr (C_0, V_0) system: (a) speciation diagram, (b) $E = E(\Phi)$, and (c) pH = pH(Φ) curves; $V_0 = 10$, $C_0 = 0.01$, $C = 0.1$.

6. Symproportionating systems

6.1. Titration in $KBrO_3$ (*C, V*) ⇒ NaBr (C_0, V_0) system

In this case, symproportionation practically does not occur (**Figure 8a**); concentration of HBrO, as the major product formed in the symproportionation reaction

$$BrO_3^{-1} + 2Br^{-1} + 3H^{+1} = 3HBrO \tag{61}$$

is ca. 10^{-6} mol/L. The potential *E* increases monotonically (**Figure 8b**), whereas pH first increases, passes through maximum and then decreases (**Figure 8c**). The relevant pH and *E* changes are small. Binding the H^{+1} ions in reaction (61) causes a weakly alkaline reaction (**Figure 8b**).

6.2. Titration in $KBrO_3$ (C, V) ⇒ NaBr (C_0) + H_2SO_4 (C_{03}) V_0 system

The stoichiometry 1:5, i.e. $\Phi_{eq} = 0.2$, stated for C_{03} values indicated at the curves plotted in **Figure 9** (column a), results from reaction

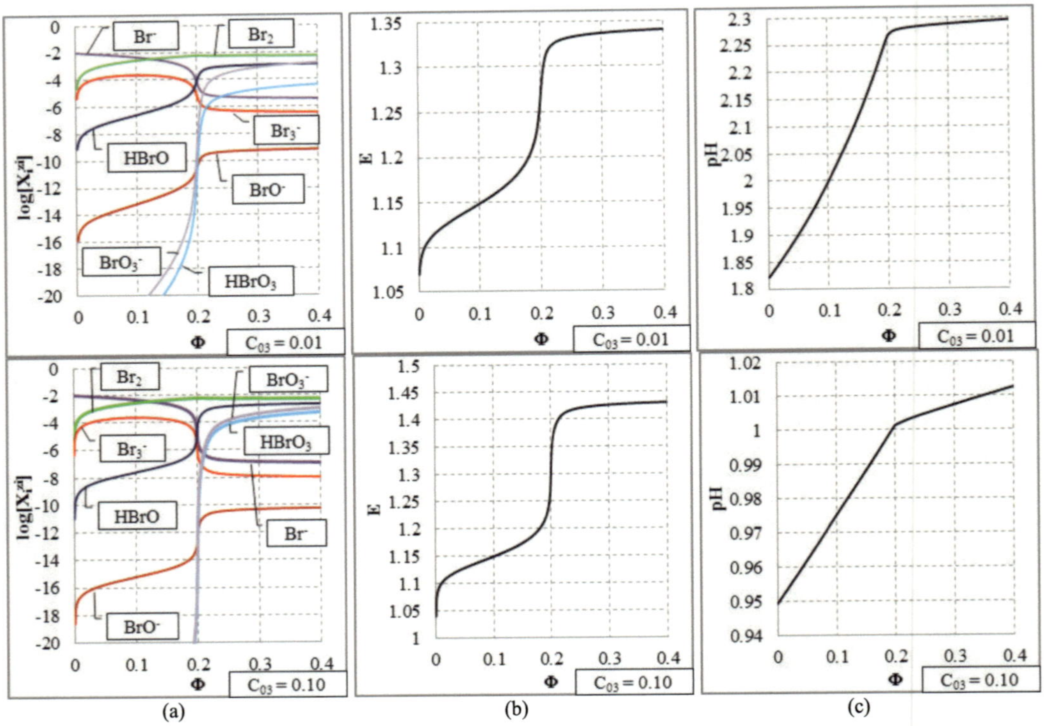

Figure 9. Plots for the $KBrO_3$ (C, V) ⇒ NaBr (C_0) + H_2SO_4 (C_{03}) V_0 system: speciation diagrams (column a); $E = E(\Phi)$ (column b) and pH = pH(Φ) (column c), at $V_0 = 100$, $C_0 = 0.01$, $C = 0.1$ and indicated C_{03} [mol/L] values for H_2SO_4.

$$BrO_3^{-1} + 5Br^{-1} + 6H^{+1} = 3Br_2 + 3H_2O \quad (62)$$

For $\Phi > 0.2$, an increase of efficiency of the competitive reaction

$$2Br^{-1} + BrO_3^{-1} + 3H^{+1} = 3HBrO \quad (63)$$

is noted. A growth of C_{03} value causes a small extension of the potential range in the jump region, on the side of higher E values (**Figure 9**, column b). With an increase in the C_{03} value, the graphs of pH vs. Φ resemble two almost straight-line segments intersecting at $\Phi_{eq} = 0.2$ (**Figure 9**, column c). However, the pH ranges covered by the titration curves are gradually narrowed (**Figure 9**, column c).

7. Redox systems with two electron-active elements

In the redox systems considered above, one electron-active element was involved in disproportionation or symproportionation reactions affected by NaOH (in dynamic systems) or water (in static systems). In the HCl (C, V) ⇒ NaIO (C_0, V_0) system, where HCl was used as disproportionating reagent, the possibility of oxidation of Cl^{-1} ions was allowed *a priori*, in which HCl was used as a disproportionating reagent, i.e. chlorine in HCl was treated as an electron-active element. It turned out, however, that the oxidation of Cl^{-1} occurred only in an extremely small extent, ca. $10^{-16}/10^{-2} \approx 10^{-14}$ part of Cl^{-1} ions was oxidized. So, we can consider that it is virtually the disproportionation of IO^{-1} ions originating from the NaIO.

In this section, we consider the systems where two electron-active elements are factually present.

7.1. Titration in $K_2Cr_2O_7$ (C, V) ⇒ KI (C_0) + H_2SO_4 (C_{01}) V_0 system

Note that V_0 mL of titrand D composed of KI (N_{03}) + H_2SO_4 (N_{04}) + H_2O (N_{05}) is titrated with V mL of $K_2Cr_2O_7$ (N_{01}) + H_2O (N_{02}) as titrant T added up to a defined point of the titration. In V_0 + V mL of D + T mixture, we have the following species:

$H_2O(N_1)$, $H^{+1}(N_2, n_2)$ $OH^{-1}(N_3, n_3)$, $K^{+1}(N_4, n_4)$, $HSO_4^{-1}(N_5, n_5)$, $SO_4^{-2}(N_6, n_6)$, $I^{-1}(N_7, n_7)$, $I_3^{-1}(N_8, n_8)$, $I_2(N_9, n_9)$, $\boldsymbol{I}_{2(s)}(N_{10}, n_{10})$, HIO (N_{11}, n_{11}), $IO^{-1}(N_{12}, n_{12})$, $HIO_3(N_{13}, n_{13})$, $IO_3^{-1}(N_{14}, n_{14})$, $H_5IO_6(N_{15}, n_{15})$, $H_4IO_6^{-1}(N_{16}, n_{16})$, $H_3IO_6^{-2}(N_{17}, n_{17})$, $HCr_2O_7^{-1}(N_{18}, n_{18})$, $Cr_2O_7^{-2}(N_{19}, n_{19})$, $H_2CrO_4(N_{20}, n_{20})$, $HCrO_4^{-1}(N_{21}, n_{21})$, $CrO_4^{-2}(N_{22}, n_{22})$, $Cr^{+3}(N_{23}, n_{23})$, $CrOH^{+2}(N_{24}, n_{24})$, $Cr(OH)_2^{+1}(N_{25}, n_{25})$, $Cr(OH)_4^{-1}(N_{26}, n_{26})$, $CrSO_4^{+1}(N_{27}, n_{27})$.

On this basis, we formulate $f_1 = f(H)$, $f_2 = f(O)$ and then:

$$f_{12} = 2 \cdot f(O) – f(H) – N_2 + N_3 + 7N_5 + 8N_6 + N_{11} + 2N_{12} + 5N_{13} + 6N_{14} + 7N_{15} + 8N_{16} + 9N_{17} + 13N_{18} + 14N_{19} + 6N_{20} + 7N_{21} + 8N_{22} + N_{24} + 2N_{25} + 4N_{26} + 8N_{27} = 14N_{01} + 6N_{04} \quad (64)$$

$$f_0 = \text{ChB} : N_2 - N_3 + N_4 - N_5 - 2N_6 - N_7 – N_8 - N_{12} - N_{14} - N_{16} - 2N_{17} - N_{18} - 2N_{19} - N_{21} - 2N_{22} + 3N_{23} + 2N_{24} + N_{25} - N_{26} + N_{27} = 0 \quad (65)$$

$$f_3 = f(K) : \quad N_4 = 2N_{01} + N_{03}) \quad (66)$$

$$f_4 = f(S) = f(SO_4) : \quad N_5 + N_6 + N_{27} = N_{04} \quad (67)$$

$$2 \cdot f(O) – f(H) + \text{ChB} – f(K) – 6 \cdot f(S) = 0 \Rightarrow \quad (68)$$

$$–N_7 – N_8 + N_{11} + N_{12} + 5N_{13} + 5N_{14} + 7N_{15} + 7N_{16} + 7N_{17} + 12N_{18} + 12N_{19} + 6N_{20} + 6N_{21} + 6N_{22} + 3N_{23} + 3N_{24} + 3N_{25} + 3N_{26} + 3N_{27} = 12N_{01} – N_{03} \Rightarrow \quad (68a)$$

$$(–1) \cdot N_7 + (–1/3) \cdot 3N_8 + 0 \cdot (N_9 + N_{10}) + (+1) \cdot (N_{11} + N_{12}) + (+5) \cdot (N_{13} + N_{14}) + (+7) \cdot (N_{15} + N_{16} + N_{17}) + (+6) \cdot (2N_{18} + 2N_{19} + N_{20} + N_{21} + N_{22}) + (+3) \cdot (N_{23} + N_{24} + N_{25} + N_{26} + N_{27}) = (+6) \cdot 2N_{01} + (–1) \cdot N_{03} \quad \Rightarrow \quad (68b)$$

$$(–1) \cdot [I^{-1}] + (–1/3) \cdot 3[I_3^{-1}] + 0\left(2[I_2] + 2[\boldsymbol{I}_{2(s)}]\right) + (+1)([HIO][IO^{-1}]) + (+5) \cdot ([HIO_3] + [IO_3^{-1}]) +(+7) \cdot ([H_5IO_6] + [H_4IO_6^{-1}] + [H_3IO_6^{-2}]) + (+6) \cdot (2[HCr_2O_7^{-1}]) + 2[Cr_2O_7^{-2}] + [H_2CrO_4] +[HCrO_4^{-1}] + [CrO_4^{-2}]) + (+3) \cdot ([Cr^{+3}] + [CrOH^{+2}] + [Cr(OH)_2^{+1}] + [Cr(OH)_4^{-1}] +[CrSO_4^{+1}]) = ((+6) \cdot 2CV + (–1) \cdot C_0V_0)/(V_0 + V) \quad (69)$$

where $C_0V_0 = 10^3 \cdot N_{03}/N_A$, $CV = 10^3 \cdot N_{01}/N_A$. Note that Eq. (69) was obtained only from linear combination of electron-non-active elements (fans) in this system. The balances for electron-active elements are as follows:

$$f_5 = f(\mathrm{I}):\ N_7 + 3N_8 + 2N_9 + 2N_{10} + N_{11} + N_{12} + N_{13} + N_{14} + N_{15} + N_{16} + N_{17} = N_{03} \quad (70)$$

$$[I^{-1}] + 3[I_3^{-1}] + 2\cdot([I_2] + [\boldsymbol{I}_{2(s)}]) + ([\mathrm{HIO}] + [\mathrm{IO}^{-1}]) + 5\cdot([\mathrm{HIO_3}] + [\mathrm{IO_3^{-1}}]) + 7\cdot([\mathrm{H_5IO_6}] + [H_4\mathrm{IO}_6^{-1}] + [\mathrm{H_3IO_6^{-2}}]) = C_0V_0/(V_0 + V) \quad (71)$$

$$f_6 = f(\mathrm{Cr}):\ 2N_{18} + 2N_{19} + N_{20} + N_{21} + N_{22} + N_{23} + N_{24} + N_{25} + N_{26} + N_{27} = 2N_{01} \Rightarrow \quad (72)$$

$$2[\mathrm{Cr_2O_7^{-2}}] + [\mathrm{H_2CrO_4}] + [\mathrm{HCrO_4^{-1}}] + [\mathrm{CrO_4^{-2}}]) + [\mathrm{Cr^{+3}}] + [\mathrm{CrOH^{+2}}] + [\mathrm{Cr(OH)_2^{+1}}] + [\mathrm{Cr(OH)_4^{-1}}] + [\mathrm{CrSO_4^{+1}}] = \mathrm{CV}/(\mathrm{V_0} + \mathrm{V}) \quad (73)$$

Subtraction of 3·*f*(Cr) (Eq. (72)) from Eq. (68a) and further operations give

$$[\mathrm{I^{-1}}] + [\mathrm{I_3^{-1}}] - [\mathrm{HIO}] + [\mathrm{IO^{-1}}]) - 5\cdot([\mathrm{HIO_3}] + [\mathrm{IO_3^{-1}}]) - 7([\mathrm{H_5IO_6}] + [\mathrm{H_4IO_6^{-1}}] + [\mathrm{H_3IO_6^{-2}}]) -3\cdot(2[\mathrm{HCr_2O_7^{-1}}] + 2[\mathrm{Cr_2O_7^{-2}}] + [\mathrm{H_2CrO_4}] + [\mathrm{HCrO_4^{-1}}] + [\mathrm{CrO_4^{-2}}]) = (C_0V_0\text{–}6\mathrm{CV})/(V_0 + V) \quad (74)$$

The simplest/shortest form of GEB, obtained from Eqs. (68), (70) and (72) is the relation

$$2\cdot f(\mathrm{O}) - f(\mathrm{H}) + \mathrm{ChB} - f(\mathrm{K}) - 6\cdot f(\mathrm{S}) + f(\mathrm{I}) - 6\cdot f(\mathrm{Cr}) = 0 \Rightarrow \quad (75)$$

$$N_8 + N_9 + N_{10} + N_{11} + N_{12} + 3\cdot(N_{13} + N_{14}) + 4\cdot(N_{15} + N_{16} + N_{17}) = 1.5\cdot(N_{23} + N_{24} + N_{25} + N_{26}) \Rightarrow \quad (75a)$$

$$[I_3^{-1}] + [I_2] + [\boldsymbol{I}_{2(s)}] + [\mathrm{HIO}] + [\mathrm{IO^{-1}}] + 3\cdot([\mathrm{HIO_3}] + [\mathrm{IO_3^{-1}}]) + 4\cdot([\mathrm{H_5IO_6}] + [\mathrm{H_4IO_6^{-1}}] + [\mathrm{H_3IO_6^{-2}}]) = 1.5\cdot([\mathrm{Cr^{+3}}] + [\mathrm{CrOH^{+2}}] + [\mathrm{Cr(OH)_2^{+1}}][\mathrm{Cr(OH)_4^{-1}}] + [\mathrm{CrSO_4^{+1}}]) \quad (75b)$$

Note that the numbers (N_{01}, N_{03}) of components forming the system are not involved in Eqs. (75a) and (75b).

Applying the atomic numbers: $Z_I = 53$ for I and $Z_{Cr} = 24$ for Cr, we obtain the linear combination

$$Z_\mathrm{I}\cdot f(I) + Z_\mathrm{Cr}\cdot f(\mathrm{Cr})\text{–}\left(2\cdot f(\mathrm{O})\text{–}f(\mathrm{H}) + \mathrm{ChB}\text{–}f(\mathrm{K})\text{–}\ 6\cdot f(\mathrm{S})\right) = 0 \Rightarrow (Z_\mathrm{I} + 1)\mathrm{N}_7 + (3Z_\mathrm{I} + 1)N_8 + 2Z_\mathrm{I}(N_9 + N_{10}) + (Z_\mathrm{I}\text{–}1)(N_{11} + N_{12}) + (Z_\mathrm{I}\text{–}5)(N_{13} + N_{14}) + (Z_\mathrm{I}\text{–}7)(N_{15} + N_{16} + N_{17}) + (Z_\mathrm{Cr}\text{–}6)(2N_{18} + 2N_{19} + N_{20} + N_{21} + N_{22}) + (Z_\mathrm{Cr}\text{–}3)(N_{23} + N_{24} + N_{25} + N_{26}) = (Z_\mathrm{I} + 1)N_{03} + 2(Z_\mathrm{Cr}\text{–}6)N_{01} \Rightarrow \quad (76)$$

$$(Z_\mathrm{I} + 1)[\mathrm{I^{-1}}] + (3Z_\mathrm{I} + 1)[\mathrm{I_3^{-1}}] + 2Z_\mathrm{I}([I_2] + [\boldsymbol{I}_2]) + (Z_\mathrm{I}\text{–}1)([\mathrm{HIO}] + [\mathrm{IO^{-1}}]) + (Z_\mathrm{I}\text{–}5)([\mathrm{HIO_3}] + [\mathrm{IO_3^{-1}}]) + (Z_\mathrm{I}\text{–}7)([\mathrm{H_5IO_6}] + [\mathrm{H_4IO_6^{-1}}] + [\mathrm{H_3IO_6^{-2}}]) + (Z_\mathrm{Cr}\text{–}6)(2[\mathrm{HCr_2O_7^{-1}}] + 2[\mathrm{Cr_2O_7^{-2}}] + [\mathrm{H_2CrO_4}] + [\mathrm{HCrO_4^{-1}}] + [\mathrm{CrO_4^{-2}}]) + (Z_\mathrm{Cr} - 3)([\mathrm{Cr^{+3}}] + [\mathrm{CrOH^{+2}}] + [\mathrm{Cr(OH)_2^{+1}}] + [\mathrm{Cr(OH_4^{-1})}] + [\mathrm{CrSO_4^{+1}}]) = ((Z_\mathrm{I} + 1)\cdot C_0V_0 + 2(Z_\mathrm{Cr}\text{–}6)\cdot CV)/(V_0 + V) \quad (77)$$

Eqs. (69), (74), (75b) and (77) (and other linear combinations, as well) are equivalent forms of GEB for this system. Note that Eq. (77) is identical with the one obtained immediately on the basis of the approach I to GEB [4]. The $E = E(\Phi)$ and $pH = pH(\Phi)$ and some speciation curves for iodine and chromium species are plotted in **Figures 10** and **11**, where Φ is the fraction titrated.

For $C_{01} = 0.01$, I_2, $\boldsymbol{I}_{2(s)}$ and I_3^{-1} are formed in reactions:

$$Cr_2O_7^{-2} + (6,6,9)I^{-1} + 14H^{+1} = 2Cr^{+3} + 3(\boldsymbol{I}_{2(s)}, I_2, I_3^{-1}) + 2Cr^{+3} + 7H_2O \quad (78)$$

$$Cr_2O_7^{-2} + (6,6,9)I^{-1} + 12H^{+1} + 2HSO_4^{-1} = 2Cr^{+3} + 3(\boldsymbol{I}_{2(s)}, I_2, I_3^{-1}) + 2CrSO_4^{+1} + 7H_2O \quad (79)$$

$$Cr_2O_7^{-2} + (6,6,9)I^{-1} + 14H^{+1} + 2SO_4^{-2} = 2Cr^{+3} + 3(\boldsymbol{I}_{2(s)}, I_2, I_3^{-1}) + 2CrSO_4^{+1} + 7H_2O \quad (80)$$

where predominating products are involved. Binding the H^{+1} ions corresponds to the pH increase, which is largest for low C_{01} value (relatively low buffer capacity of the solution, compare with [53, 54, 58–60]). In reactions (78)–(80), protons are consumed and then $dpH/d\Phi > 0$; moreover, $dE/d\Phi > 0$ for $\Phi > 0$.

At $C_{01} = 0.02$, $E = E(\Phi)$ and $[\boldsymbol{I}_{2(s)}]$ pass through maximum at Φ ca. 0.2. The plot of $pH = pH(\Phi)$ shows a slight distortion of the course at $\Phi = 1/6$, and $DpH/D\Phi > 0$ for $\Phi > 0$. The $[IO_3^{-1}]$ is comparable with $[\boldsymbol{I}_{2(s)}]$ and $[I_2]$; $[I_3^{-1}]$ is small here because $[I^-] < 10^{-6}$. The reactions

$$3(I_2, \boldsymbol{I}_{2(s)}) + 5Cr_2O_7^{-2} + 34H^{+1} = 6IO_3^{-1} + 10Cr^{+3} + 17H_2O \quad (81)$$

$$3(I_2, \boldsymbol{I}_{2(s)}) + 5Cr_2O_7^{-2} + 24H^{+1} + 10HSO_4^{-1} = 6IO_3^{-1} + 10CrSO_4^{+1} + 17H_2O \quad (82)$$

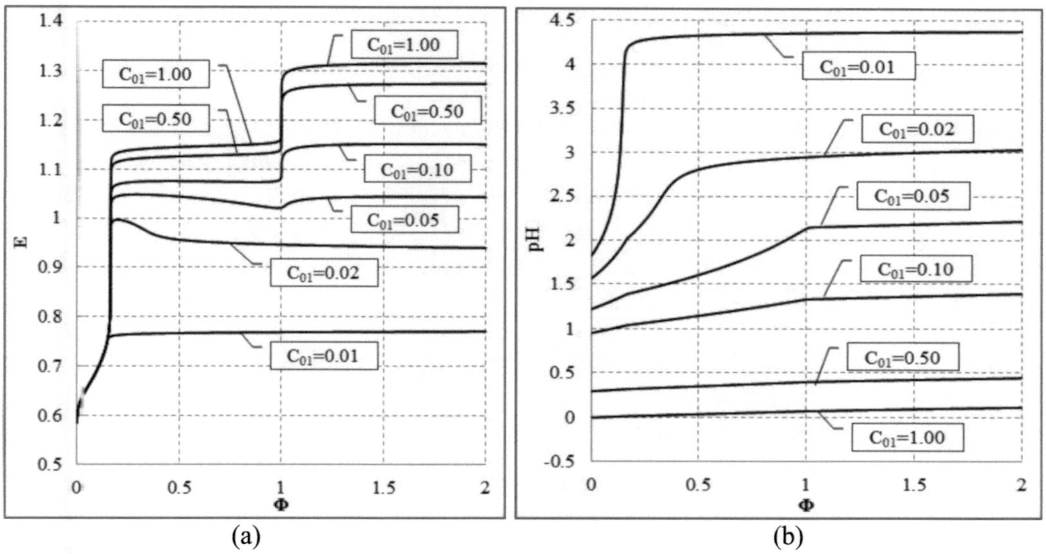

Figure 10. The plots $K_2Cr_2O_7$ $(C, V) \Rightarrow KI$ $(C_0) + H_2SO_4$ (C_{01}) V_0 system for (a) $E = E(\Phi)$ and (b) $pH = pH(\Phi)$ functions, at $V_0 = 100$, $C_0 = 0.01$ and C_{01} values indicated at the corresponding curves.

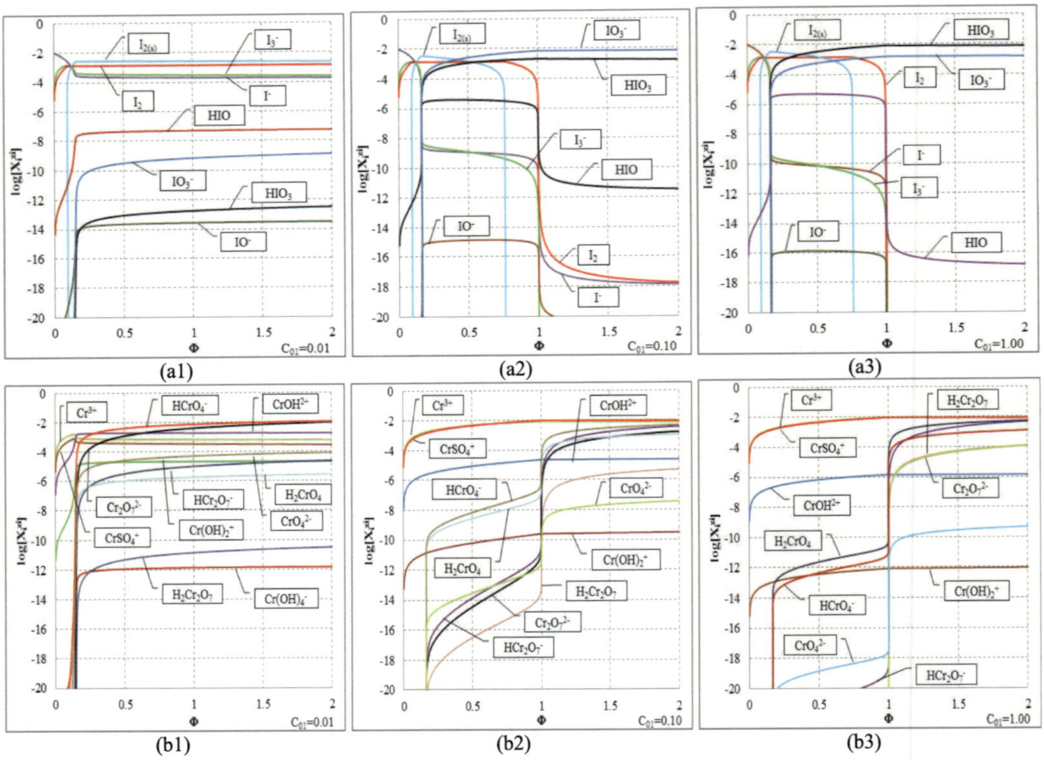

Figure 11. Speciation diagrams plotted for iodine (*a*1, *a*2, *a*3) and chromium (*b*1, *b*2, *b*3) species, at indicated C_{01} values.

$$3(I_2, \boldsymbol{I}_{2(s)}) + 5Cr_2O_7^{-2} + 34H^{+1} + 10SO_4^{-2} = 6IO_3^{-1} + 10CrSO_4^{+1} + 17H_2O \tag{83}$$

are clearly indicated.

At $C_{01} = 0.05$, the function $E = E(\Phi)$ has a complex course: it first increases, reaches a maximum at $\Phi = 0.295$, then decreases, reaches a minimum at $\Phi = 1$, increases again, passes through a flat maximum at Φ ca. 1.6 and decreases. The curve pH = pH(Φ) breaks at $\Phi = 1/6$ and $\Phi = 1$. The $\boldsymbol{I}_{2(s)}$ exists as the equilibrium solid phase at $0.0934 < \Phi < 0.762$. A decrease in $[I_2]$ value is more expressed at $\Phi > 1$. The ratio $[HIO_3]/[IO_3^{-1}]$ grows with growth of C_{01} value. The stoichiometry at $\Phi = \Phi_{eq2} = 1$ is described by the reaction

$$Cr_2O_7^{-2} + I^{-1} + 8H^{+1} = 2Cr^{+3} + IO_3^{-1} + 4H_2O \tag{84}$$

and by reactions where HSO_4^{-1}, SO_4^{-2}, $CrSO_4^{+1}$ and HIO_3 are involved.

At $C_{01} \geq 0.1$, a jump on the $E = E(\Phi)$ curve at $\Phi = 1/6$ is clearly marked. The growth of jump at $\Phi = 1$ results from a more significant decrease in the $[I_2]$ value at $\Phi > 1$. The E-range covered by the jump at $\Phi = 1$ extends with an increase in the C_{01} value (**Figure 10a**). For more details, see Ref. [4].

7.2. A comment

Redox systems with two electron-active elements were widely discussed from the GATES/GEB viewpoint. From the earlier literature, one can recall the simulated systems: $KMnO_4$ (C, V) ⇒ VSO_4 (C_0) + H_2SO_4 (C_{01}) V_0 [18]; $K_2Cr_2O_7$ (C, V) ⇒ VSO_4 (C_0) + H_2SO_4 (C_{01}) V_0 [18]; $KMnO_4$ (C, V) ⇒ $FeSO_4$ (C_0) + H_2SO_4 (C_{01}) [18]; $Ce(SO_4)_2$ (C) + H_2SO_4 (C_1) V ⇒ $FeSO_4$ (C_0) + H_2SO_4 (C_{01}) V_0 [18]; Cl_2 (C, V) ⇒ KI (C_0, V_0) [19]; $KMnO_4$ (C, V) ⇒ KBr (C_0), H_2SO_4 (C_{01}) V_0 [19]. Except the NaOH ⇒ Br_2 and NaOH ⇒ HBrO systems, in [16] were simulated also the systems: NaOH (C) + CO_2 (C_1) V ⇒ I_2 (C_0) + KI (C_{01}) + CO_2 (C_{02}); NaOH (C) + CO_2 (C_1) V ⇒ I_2 (C_0) + KI (C_{01}) + CO_2 (C_{02}) V_0 + CCl_4 (V^*) (dynamic liquid-liquid extraction system). In [16], an interesting qualitative reaction where $MgSO_4$ solution was added into I_2 + KI solution previously alkalinized with an excess of NaOH. This addition causes precipitation of **$Mg(OH)_2$** and a decrease of pH value, a growth of solubility of this precipitate and a shift of equilibrium (symproportionation) involved with formation of iodide (I_2) that adsorbes on the **$Mg(OH)_2$** precipitate turning it red-brown (similarity with starch action in visual iodometric titrations). The dynamic solubility curve was plotted, i.e. the $MgSO_4$ addition is formally treated as titration. The speciation diagrams in all the systems considered were plotted. Among others, it is stated that the disproportionation of iodine ($\boldsymbol{I}_{2(s)}$, I_2) occurs according to the scheme (A) $3(\boldsymbol{I}_{2(s)}, I_2) + 6OH^{-1} = IO_3^{-1} + 5I^{-1} + 3H_2O$, not the scheme (B) $(\boldsymbol{I}_{2(s)}, I_2) + 2OH^{-1} = IO^{-1} + I^{-1} + H_2O$, as stated in almost all contemporary textbooks. One can calculate [16] that the yield of the reaction (A) is $2.5{\cdot}10^9$ times higher than the yield of reaction (B) of the same stoichiometry, 3:6 = 1:2. It is interesting to add that 80 years ≈ $2.5{\cdot}10^9$ s. Thus, the yield of the reaction B relative to A is in such a ratio as 1 s in relation to the average length of human's life.

It should also be added that GATES/GEB enables to verify experimental results. Such a case was mentioned in Ref. [18] to the system, where KIO_3 (C_0) + HCl (C_{01}) + H_2SeO_3 (C_{02}) + $HgCl_2$ (C_{03}) was titrated with ascorbic acid ($C_6H_8O_6$). Authors of the work [61] cited in there committed, among other things, a simple mistake resulting from improper recalculation of numerical values of potentials; for more details, see Refs. [1, 17, 18].

In all instances, full attainable (quantitative+qualitative) physicochemical knowledge was involved in the related algorithms. The quantitative knowledge was related to the complete set of equilibrium constants values, whereas qualitative knowledge was helpful in aspect of metastable and kinetic systems.

7.3. An illustrative presentation of the approach I to GEB

A redox reaction can also participate two or more electron-active elements. In this convention, the approach I to GEB can be perceived as the card game, with electron-active elements as 'players', electron-non-active elements—as 'fans', and electrons—as 'money' transferred between 'players', see the picture from Ref. [62]. The 'players' provide a common pool of their own electrons into the system in question. The 'money' is transferred between 'players'. As a result of the game, the players' accounts are changed, while the fans' accounts remain intact/unchanged, in this convention.

In C_0 mol/L aqueous solution of Br_2, bromine as one electron-acitve element in Br_2 can be perceived as a 'distributor' of its own electrons, and H, O as 'fans'. For example, Eq. (42) involves H and O as 'fans', Eq. (43) involves Na as 'fan', whereas Eq. (44) does not involve 'fans'. Generally, 'fans' are eliminated after the proper (i.e. indicated above) combination of $2 \cdot f(O) - f(H)$ with the balances for elements considered as 'fans'. Further simplification can also result from further combination with the balance for 'player(s)'. But, in any case, a linear combination of balances related to a redox system does not lead to the identity 0 = 0.

As usually happens in the 'card game' practice, the players devote to the game only a part of their cash resources. Similarly, in redox reactions, electrons may participate from the valence shells of atoms of electron-active elements; the electrons from the valence shell of the reductant atoms are transferred onto the valence shell of the oxidant atoms. However, this restriction to the valence electrons is not required here. For example, replacing Z_{Br} in Eq. (46a) by ζ_{Br} ($\zeta_{Br} < Z_{Br}$), we have

$$(\zeta_{Br} - 5)([HBrO_3] + [BrO_3^{-1}]) + (\zeta_{Br} - 1)([HBrO] + [BrO^{-1}]) + 2\zeta_{Br}[Br_2] + (3\zeta_{Br} + 1)[Br_3^{-1}] + (\zeta_{Br} + 1)[Br^{-1}] = 2\zeta_{Br}C_0V_0/(V_0 + V) \quad (85)$$

Replacing Z_I by ζ_I and Z_{Cr} by ζ_{Cr} ($\zeta_I < Z_I$, $\zeta_{Cr} < Z_{Cr}$) in Eq. (77), we have

$$(\zeta_I+1)[I^{-1}]+(3\zeta_I+1)[I_3^{-1}]+2\zeta_I([I_2]+[I_2])+(\zeta_I-1)([HIO]+[IO^{-1}])+(\zeta_I-5)([HIO_3] +[IO_3^{-1}])+(\zeta_I-7)([H_5IO_6]+[H_4IO_6^{-1}]+[H_3IO_6^{-2}])+(\zeta_{Cr}-6)(2[HCr_2O_7^{-1}]+2[Cr_2O_7^{-2}] +[H_2CrO_4]+[HCrO_4^{-1}]+[CrO_4^{-2}])+(\zeta_{Cr}-3)([Cr^{+3}]+[CrOH^{+2}]+[Cr(OH)_2^{+1}] +[Cr(OH)_4^{-1}]+[CrSO_4^{+1}])=((\zeta_I+1)C_0V_0+2(\zeta_{Cr}-6)CV)/(V_0+V) \quad (86)$$

In particular, we can put $\zeta_{Cr} = \zeta_{Cr} = 0$ in Eq. (86). Obviously, we get the relation

$$[I^{-1}]+[I_3^{-1}]-(HIO)+[IO^{-1}]-5([HIO_3])+[IO_3^{-1}]-7([H_5IO_6]+[H_4IO_6^{-1}]+[H_3IO_6^{-2}]) -6(2[HCr_2O_7^{-1}])+2[Cr_2O_7^{-2}]+[H_2CrO_4]+[HCrO_4^{-1}]+[CrO_4^{-2}]-3([Cr^{+3}])+[CrOH^{+2}] +[Cr(OH)_2^{+1}]+[Cr(OH)_4^{-1}]+[CrSO_4^{+1}]=C_0V_0-12CV)/(V_0+V)-[I^{-1}]-[I_3^{-1}]+([HIO] +[IO^{-1}])+5([HIO_3]+[IO_3^{-1}])+7([H_5IO_6]+[H_4IO_6^{-1}]+[H_3IO_6^{-2}])+12([HCr_2O_7^{-1}] +[Cr_2O_7^{-2}])+6([H_2CrO_4]+[HCrO_4^{-1}]+[CrO_4^{-2}])-3([Cr^{+3}]+[CrOH^{+2}]+[Cr(OH)_2^{+1}]) +[Cr(OH)_4^{-1}]+[CrSO_4^{+1}])=(12CV-C_0V_0)/(V_0+V) \quad (87)$$

identical to the one obtained from Eq. (68a). If we put $\zeta_{Br} = 0$ in Eq. (85), we get the relation

$$-5([HBrO_3]+[BrO_3^{-1}])-([HBrO]+[BrO^{-1}])+[Br_3^{-1}]+[Br^{-1}]=0$$
$$5([HBrO_3]+[BrO_3{}^{-1}])+([HBrO]+[BrO^{-1}])-[Br_3{}^{-1}]-[Br^{-1}]=0 \quad (88)$$

identical to the ones obtained from Eq. (44). This way, we recall the card game without 'live cash' but with 'debt of honor'—in not accidental reference to the title of the thriller novel by T. Clancy; btw, *nota bene*, the "Debt of honor" was published in 1994, like the papers [15, 16, 18].

8. Conclusions

The GEB formulated according to approach I was named first as electron prebalance and presented, in totally mature form, in three papers issued in 1994 [15, 16, 18] and then followed by further articles and other communications. Currently, in context with the approach II to GEB, it is named as the approach I to GEB, fully equivalent to the approach II to GEB, explicitly related to the law of conservation of H and O.

The linear combination 2·f(O) – f(H) is a keystone for the overall thermodynamic knowledge on electrolytic systems. It can be formulated both for non-redox and redox systems, in aqueous, non-aqueous and mixed-solvent systems, with amphiprotic (co)solvent(s) involved. The 2·f(O) – f(H) is linearly independent on ChB and other balances, for elements/cores $f(Y_m) \neq$ H, O, in any redox system. For any non-redox system, 2·f(O) – f(H) is linearly dependent on those balances. Then, the linear independency/dependency of $2\cdot f(O) - f(H)$ on the other balances is the general criterion distinguishing between redox and non-redox systems. The equation for $2\cdot f(O) - f(H)$, considered as the *primary* form of GEB, $pr - \text{GEB} = 2\cdot f(\text{O}) - f(\text{H})$, is the basis of GEB formulation for redox systems according to approach II to GEB.

Ultimately, GEB, ChB and elemental/core balances are expressed in terms of molar concentrations—to be fully compatible/congruent with expressions for equilibrium constants, interrelating molar concentrations of defined species, on the basis of the mass action law applied to correctly written reaction equation. The mass action law is the one and only chemical law applied in GATES. A complete set of independent (non-contradictory [22]) relations for the equilibrium constants is needed for this purpose.

When compared with the approach I, the approach II to GEB offers several advantages. Although derivation of GEB according to the approach II is more laborious, it enables to formulate this balance without prior knowledge of oxidation numbers for the elements involved frequently in complex components and species of the system. Only the composition (expressed by chemical formula) of components forming the system and composition of the species formed in the system, together with their external charges, are required, i.e. it provides an information sufficient to formulate the GEB; it is the paramount advantage of the approach II to GEB. Anyway, the oxidation number, representing the degree of oxidation of an element in a compound or a species, is commonly perceived as a contractual concept. In this regard, formulation of GEB according to approach II is more useful than the approach I when applied to complex organic species in redox systems of biological origin [63–67]. The approach II to GEB is advantageous/desired, *inter alia,* for redox systems where radical and ion-radical species are formed. What is more, the 'players' and 'fans', as ones perceived from the approach I to GEB viewpoint, are not indicated *a priori* within the approach II to GEB. The approach I, considered as a 'short' version of GEB, is more convenient when oxidation numbers for all elements of the system are known beforehand. Within the approaches I and II to GEB, the roles of oxidants and reductants are not ascribed *a priori* to particular components forming the redox system and to the species formed in this system. In other words, full 'democracy' is established *a priori* within GATES/GEB, where the oxidation number, oxidant, reductant, equivalent mass and stoichiometric reaction notation are the redundant concepts only.

The 1 + K balances composed of ChB and K concentration balances related to equations (not equalities, see e.g. Eq. (40)) for elemental balances form a complete set of equations related to a non-redox system. The set of 1 + K variables in $\mathbf{x} = (\text{pH}, \text{pX}_1,\ldots, \text{pX}_K)^T$ is involved in the algorithm applied for calculation purposes. For example, 1 + K = 1 + 2 = 3 balances are related to non-redox systems presented in Examples 1 and 2, with independent variables represented by components of the vector $\mathbf{x} = (\text{pH}, \text{pCu}, \text{pSO}_4)^T$. In this context, Eq. (35) is considered as equality, not equation.

The 2 + K balances composed of GEB, ChB and K concentration balances related to equations (not equalities) for elemental balances form a complete set of equations related to the redox system. The set of 2 + K variables in $\mathbf{x} = (E, \text{pH}, \text{pX}_1,\ldots, \text{pX}_K)^T$ is involved in the algorithm applied for calculation purposes. In this chapter, some results of the simulations of electrolytic redox systems, made according to the GATES/ GEB principles, with use of iterative computer programs, are graphically presented and discussed. The computer simulation realized within GATES with use of iterative computer programs, e.g. MATLAB, provides quite a new quality in knowledge gaining. It enables to follow the details of the process, registered with use of measurable quantities, such as pH and/or potential E. All these calculations are made under assumption that the relevant reactions take place in *quasi*-static manner under isothermal conditions. The reactions proceeding in the respective systems were formulated under assumption that all equilibrium constants found in the relevant tables and then used in the calculations are correct.

The number of electron-active elements (considered as 'players', in terms of the approach I to GEB) in a redox system, is practically unlimited and adapted according to current needs; among others, the systems with one, two, three [13] or four [1, 18] 'players' were considered.

The GATES and GATES/GEB (in particular) can be applied for thermodynamic resolution of systems of any degree of complexity. An example is the four-step process involved with iodometric method of copper(+2) determination, considered in detail in Refs. [14, 68], where 47 species are involved in the system with three electron-active elements, eight equations and two equalities, interrelated in 35 independent expressions for equilibrium constants. The systems with similar complexity were resolved. The complexity is limited, however, by factors of the qualitative and quantitative nature, that is, the knowledge of species the knowledge of equilibrium constants interrelating concentration of complex species with concentrations of their constituting parts. The physicochemical data are incomplete or not reliable, in many cases. This problem was raised lately in Ref. [12].

Each of the components of $\mathbf{x} = (\mathrm{pH}, \mathrm{pX}_1,\ldots, \mathrm{pX}_K)^T$ is informally ascribed to the corresponding balance: ChB, and the corresponding concentration balances for elements/cores $\neq$ H, O. Each of the components of $\mathbf{x} = (E, \mathrm{pH}, \mathrm{pX}_1,\ldots,\mathrm{pX}_k)^T$ is informally ascribed to the corresponding balance: GEB, ChB, and the corresponding concentration balances for elements/cores $\neq$ H, O. A unequivocal solution of equations is obtained when the number of independent equations equals to the number of independent variables; it is the 'iron rule' obligatory in mathematics; $pe = -\log[e^{-1}] = A \cdot E$, where $1/A = RT/F \cdot \ln 10$, $\mathrm{pH} = -\log[H^{+1}]$, $pX_i = -\log[X_i^{z_i}]$.

GATES avoids the necessity of quantitative inferences based on fragile/rachitic chemical reaction notation, involving only some of the species existing in the system; it is only a faint imitation of a true, algebraic notation. From the GATES viewpoint, the 'stoichiometry' can be perceived as a mnemonic term only. In calculations, the metastable state is realized by omission of potential products in the related balances, whereas 'opening' a reaction pathway in metastable state is based on insertion of possible (from equilibrium viewpoint) products in the related balances [17, 18]. One can also test the interfering effects of different kinds.

All the inferences made within GATES/GEB are based on firm, mathematical (algebraic) foundations. The proposed approach allows us to understand far better all physicochemical phenomena occurring in the system in question and improve some methods of analysis. All the facts testify very well about the potency of simulated calculations made, according to GATES, on the basis of all attainable physicochemical knowledge. Testing the complex redox and non-redox systems with use of iterative computer programs deserves wider popularization among physicochemists and chemists-analysts.

The GATES/GEB is put in context with constructivistic and deterministic principles, and GEB is perceived as the general law of nature, referred to as electrolytic (aqueous media) redox systems. It is proved that stoichiometry of reactions is not a primary concept in chemistry, and its application provides false results, for obvious reasons. From the GATES viewpoint, the stoichiometric reactions are only the basis to formulate the related equilibrium constants. GATES/GEB referred to modeling of redox titration curves in context with earlier approaches to this problem. The GATES/GEB is also presented in three other chapters issued in 2017 within InTech [68–70].

The dependency/independency criteria ascribed to $2{\cdot}f(O) - f(H)$ distinguishing between the relevant (non-redox and redox) systems are the properties of the equations obtained from the linear combination of the balances for H and O. Namely, the resulting equation is not independent of non-redox systems, since it is a linear combination of the remaining (charge and concentration) balances, whereas in the case of redox systems, this equation is linearly independent of those balances. This is a general property of nature, independent of the complexity of the system under consideration, which is the electrolytic system. GATES and GATES/GEB, in particular, are clear confirmation of the fact that the nature is mathematically designed and the true laws of nature are mathematical. In other words, the quantitative, mathematical method became the essence of science. To paraphrase a Chinese proverb, one can state that 'the lotus flower, lotus leaf and lotus seed come from the same root' [2]. Similarly, the three kinds of balances: GEB, charge and concentration balances come from the same family of fundamental laws of preservation. This compatibility is directly visible from the viewpoint of the approach II to GEB. The equivalent equations for GEB, based on a reliable law of the matter conservation, are equally robust as equations for charge and concentration balances. The complementarity of the GEB (approaches I and II) to other balances is regarded as the expression of harmony of nature and GATES/GEB as an example of excellent epistemological paradigm.

All earlier (dated from the 1960s) efforts made towards formulation of electrolytic redox systems were only clumsy attempts of resolution of the problem in question, as stated in review papers [2, 10–14]. These approaches were slavishly related to the stoichiometric reaction notations, involving only two pairs of indicated species participating in redox reaction; there were usually minor species of the system considered. The species different from those involved in the reaction notation were thus omitted in considerations. Moreover, the charge balance and concentration balances for accompanying substances were also omitted. Theoretical considerations were related to virtual cases, not to real, electrolytic redox systems.

9. Afterword: reductionism

The principles of science gaining are based on a conviction that complex phenomena occurred in nature, e.g. in electrolytic systems, that can be explained in terms of some general laws of the matter conservation; it is the basic assumption of reductionism [2]. These laws are expressed in terms of mathematical equations, valid for the systems of any degree of complexity. Reductionism is very similar to and has its roots in Occam's razor principle that gives precedence to simplicity, i.e. the explanation which requires the fewest assumptions. The power of reductionism lies in prediction and formulation; it is perceived as a good approximation of the macroscopic world. The knowledge thus obtained is verifiable and based on logical premises. This way

http://dx.doi.org/10.5772/intechopen.69341

the quantitative knowledge gained from the study of relatively simple systems can be synthesized in the knowledge obtainable from more complex systems. From this viewpoint, the knowledge obtained from physicochemical analysis involved, e.g. with determination of the stability constants of complex species formed in a particular system, can be perceived as a 'stone' used in construction of more complex systems. In this place, one can recall H. Poincaré who stated that 'Science is facts; just as houses are made of stones, so science is made of facts'. Any complex species must be equipped with its equilibrium constant value; a qualitative knowledge only (e.g. chemical formula) is insufficient in this respect. However, to construct the knowledge on more complex systems, these stones should be arranged according to a defined scheme (design), based on a set of compatible balances. Closely associated with reductionism is determinism—the philosophy that everything has a cause, and that a particular cause leads to a unique effect.

In any complex system, many particular reactions occur; the resultant reaction is the combination (superposition) of these elementary reactions which occur with different efficiencies that are known only after thorough physicochemical examination of the system in question. Similarly, the vibration in a polyatomic molecule is a superposition of normal vibrations. The physicochemical/thermodynamic knowledge on electrolytic systems is based on equilibrium constant values, referred to as the equilibrium system and all information about possible paths of particular reactions occurred in the system in question. Not all paths of chemical reactions are accessible, under defined conditions of analysis, involved with temperature and/or the presence of catalytic agents. This problem has been raised in Ref. [1], in context with GATES/GEB.

The approach II to GEB, preceded by the approach I to GEB, indicated new insight into redox systems and electrolytic systems in general. This way, the thermodynamic knowledge about redox systems was built in 1992 practically from scratch. The combination $2 \cdot f(O) - f(H)$ of elemental balances for H and O (approach II to GEB) is the quintessence of the generalized electron balance (GEB) that is the link needed for mathematical/algebraic description of electrolytic redox systems of any degree of complexity within the generalized approach to electrolytic systems as GATES/GEB.

Within GATES, a resolution of electrolytic systems is realized with use of iterative computer programs. Formally, the manner of resolution of this task is even easier than the one based on the formulation of some functional dependencies that require some simplifications, as a rule. The simplifications are not necessary in iterative methods.

Summarizing, the generalized electron balance (GEB), perceived within GATES as GATES/GEB, fulfils all the requirements imposed on reductionism. Its formulation is possible for any electrolytic redox systems—static and dynamic (titration), mono- and multiphase, equilibrium, metastable and non-equilibrium systems, of any degree of complexity, provided that all necessary physicochemical knowledge is attainable. Simply, it is the best possible thermodynamic approach to electrolytic redox systems.

GEB, within the context of GATES/GEB, confirms the validity of some statements expressed in the past:

- *A hidden connection is stronger than an obvious one* (Heraclitus)
- *All truths are easy to understand once they are discovered; the point is to discover them* (Galileo Galilei)

- *Everything should be made as simple as possible, but not simpler* (A. Einstein)
- *Minds are like parachutes: they function only when they are open* (J. Dewar)
- *A good theory is the best practical tool* (J.C. Maxwell)
- *Nothing is too wonderful to be true if it be consistent with the laws of nature* (M. Faraday).
- *A new scientific truth does not triumph by convincing its opponents (…), but rather because its opponents eventually die, and a new generation grows up that is familiar with it* (M. Planck).
- *Art of reasoning is nothing more than a well-ordered language*
- *Telling the truth in a time of universal deceit is a revolutionary act* (G. Orwell)
- *Man learns throughout his life, with the exception of school years*
- *Do not believe in any messages just because they are in force for a long time, in many countries. Do not believe in something just because a lot of people say/write it for a long time. Do not accept anything just because someone else said/wrote it. Do not believe in something just because it sounds plausible. Have confidence in what you have accepted as true after a long experience/ checking, what brings success to you and others* (own apprenticeship).

Notations

ChB	charge balance
D	titrand
EL	equilibrium law
G (≠ G)	Gibbs' function
G	number of balances
GEB	generalized electron balance
K	number of concentration balances
N_A	Avogadro's number
R	gas constant
T (≠ T)	temperature (in K)
T	titrant
^{T}	transposition sign for a vector
V	volume [mL] of T
V_0	volume [mL] of D
W	H_2O

Author details

Anna Maria Michałowska-Kaczmarczyk[1], Aneta Spórna-Kucab[2] and Tadeusz Michałowski[2]*

*Address all correspondence to: michalot@o2.pl

1 Department of Oncology, The University Hospital in Cracow, Cracow, Poland

2 Faculty of Chemical Engineering and Technology, Cracow University of Technology, Cracow, Poland

References

[1] Michałowski T. Application of GATES and MATLAB for resolution of equilibrium, metastable and non-equilibrium electrolytic systems. In: Michałowski T, editor. Applications of MATLAB in Science and Engineering. InTech-Open Access publisher in the fields of Science, Technology and Medicine; Rijeka, Croatia, 2011. pp. 1–34. Chapter 1. http://www.intechopen.com/books/show/title/applications-of-matlab-in-science-and-engineering

[2] Michałowski T, Toporek M, Michałowska-Kaczmarczyk AM, Asuero AG. New trends in studies on electrolytic redox systems. Electrochimica Acta. 2013;**109**:519–531

[3] Michałowski T, Michałowska-Kaczmarczyk AM, Toporek M. Formulation of general criterion distinguishing between non-redox and redox systems. Electrochimica Acta. 2013;**112**:199–211

[4] Michałowska-Kaczmarczyk AM, Toporek M, Michałowski T. Speciation diagrams in dynamic iodide + dichromate system. Electrochimica Acta. 2015;**155**:217–227

[5] Toporek M, Michałowska-Kaczmarczyk AM, Michałowski T. Symproportionation *versus* disproportionation in bromine redox systems. Electrochimica Acta. 2015;**171**: 176–187

[6] Michałowski T, Kupiec K, Rymanowski M. Numerical analysis of the Gran methods. A comparative study. Analytica Chimica Acta. 2008;**606**(2):172–183. http://www.sciencedirect.com/science/article/pii/S0003267007018673

[7] Michałowski T, Baterowicz A, Madej A, Kochana J. An extended Gran method and its applicability for simultaneous determination of Fe(II) and Fe(III). Analytica Chimica Acta. 2001;**442**(2):287–293

[8] Ponikvar M, Michałowski T, Kupiec K, Wybraniec S, Rymanowski M. Experimental verification of the modified Gran methods applicable to redox systems. Analytica Chimica Acta. 2008;**628**(2):181–189. http://pl.scribd.com/doc/173699262/ACA-2008-2

[9] Michałowski T, Toporek M, Rymanowski M. Overview on the Gran and other linearization methods applied in titrimetric analyses. Talanta. 2005;**65**(5):1241–1253. http://pl.scribd.com/doc/173699711/Talanta-65-2005-1241-1253

[10] Michałowski T. The generalized approach to electrolytic systems: I. Physicochemical and analytical implications. Critical Reviews in Analytical Chemistry. 2010;**40**(1):2–16

[11] Michałowski T, Pietrzyk A, Ponikvar-Svet M, Rymanowski M. The generalized approach to electrolytic systems: II. The generalized equivalent mass (GEM). Concept, Critical Reviews in Analytical Chemistry. 2010;**40**(1):17–29

[12] Michałowska-Kaczmarczyk AM, Asuero AG, Michałowski T. "Why not stoichiometry" versus "Stoichiometry – why not?" Part I. General context. Critical Reviews in Analytical Chemistry. 2015;**45**(2):166–188. http://www.ncbi.nlm.nih.gov/pubmed/25558777

[13] Michałowska-Kaczmarczyk AM, Asuero AG, Toporek M, Michałowski T. "Why not stoichiometry" versus "Stoichiometry – why not?" Part II. GATES in context with redox systems. Critical Reviews in Analytical Chemistry. 2015;**45**(3):241–269

[14] Michałowska-Kaczmarczyk AM, Michałowski T, Toporek M, Asuero AG. "Why not stoichiometry" versus "Stoichiometry – why not?" Part III, Extension of GATES/GEB on complex dynamic redox systems. Critical Reviews in Analytical Chemistry. 2015;**45**(4):348–366

[15] Michalowski T. Calculation of pH and potential E for bromine aqueous solution. Journal of Chemical Education. 1994;**71**:560–562. DOI: 10.1021/ed071p560

[16] Michałowski T, Lesiak A. Acid-Base titration curves in disproportionating redox ystems. Journal of Chemical Education. 1994;**71**:632–636. DOI: 10.1021/ed071p632

[17] Michałowski T, Rymanowski M, Pietrzyk A. Non-typical Brönsted's acids and bases. Journal of Chemical Education. 2005;**82**(3):470–472. http://pubs.acs.org/doi/abs/10.1021/ed082p470

[18] Michałowski T, Lesiak A. Formulation of generalized equations for redox titration curves. Chemia Analityczna (Warsaw). 1994;**39**:623–637. http://www.chem.uw.edu.pl/chemanal/

[19] Michałowski T, Wajda N, Janecki D. A unified quantitative approach to electrolytic systems. Chemia Analityczna (Warsaw). 1996;**41**:667–685. http://www.chem.uw.edu.pl/chemanal/PDFs/1996/CHAN1996V0041P00667.pdf

[20] Michałowski T, Ponikvar-Svet M, Asuero AG, Kupiec K. Thermodynamic and kinetic effects involved with pH titration of As(III) with iodine in a buffered malonate system. Journal of Solution Chemistry. 2012;**41**(3):436–446. http://www.springerlink.com/content/p2m73068h2q5u174/. DOI: 10.1007/s10953-012-9815-6

[21] Michałowska-Kaczmarczyk AM, Michałowski T, Toporek M. Formulation of dynamic redox systems according to GATES/GEB principles. International Journal of Electrochemical Science. 2016;**11**:2560–2578

[22] Michałowska-Kaczmarczyk AM, Michałowski T. Comparative balancing of non-redox and redox electrolytic systems and its consequences. American Journal of Analytical Chemistry. 2013;**4**(10):46–53. http://www.scirp.org/journal/PaperInformation.aspx?PaperID=38569

[23] Toporek M, Michałowska-Kaczmarczyk AM, Michałowski T. Disproportionation reactions of HIO and NaIO in static and dynamic systems. American Journal of Analytical Chemistry. 2014;**5**:1046–1056. http://www.scirp.org/journal/PaperInformation.aspx?PaperID=51637#.VHXKcWfpt74

[24] Michałowska-Kaczmarczyk AM, Rymanowski M, Asuero AG, Toporek M, Michałowski T. Formulation of titration curves for some redox systems. American Journal of Analytical Chemistry. 2014;**5**:861–878. http://www.scirp.org/journal/PaperInformation.aspx?paperID=50016#.VGOZ92dvHFw

[25] Michałowska-Kaczmarczyk AM, Michałowski T. Generalized electron balance for dynamic redox systems in mixed-solvent media. Journal of Analytical Sciences, Methods and Instrumentation. 2014;**4**(4):102–109. http://www.scirp.org/Journal/PaperInformation.aspx?PaperID=52018#.VH1N5Gfpt74

[26] Michałowska-Kaczmarczyk AM, Michałowski T, General relation valid for electrolytic systems. Journal of Analytical Sciences, Methods and Instrumentation. 2015;**5**:74–85. http://www.scirp.org/Journal/PaperInformation.aspx?PaperID=62292; http://file.scirp.org/pdf/JASMI_2015122910460421.pdf

[27] Michałowska-Kaczmarczyk AM, Michałowski T. Compact formulation of redox systems according to GATES/GEB principles. Journal of Analytical Sciences, Methods and Instrumentation. 2014;**4**(2):39–45. http://www.scirp.org/journal/PaperInformation.aspx?PaperID=46335

[28] Michałowski T, Pilarski B, Asuero AG, Michałowska-Kaczmarczyk AM. Modeling of acid-base properties in binary-solvent systems. In: Wypych G, editors. Handbook of Solvents, Vol. 1. Properties. Toronto: ChemTec Publishing; 2014. pp. 623–648. Chap. 9.4

[29] Michałowska-Kaczmarczyk AM, Michałowski T. GATES as the unique tool for simulation of electrolytic redox and Non-Redox systems. Journal of Analytical & Bioanalytical Techniques. http://omicsonline.org/open-access/gates-as-the-unique-tool-for-simulation-of-electrolytic-redox-and-non-redox-systems-2155-9872.1000204.pdf; http://www.speciation.net/Database/Journals/Journal-of-Analytical-amp-Bioanalytical-Techniques-;i3915

[30] Meija J, Michałowska-Kaczmarczyk AM, Michałowski T. Redox titration challenge. Analytical and Bioanalytical Chemistry. 2017;**409**(1):11–13

[31] Meija J, Michałowska-Kaczmarczyk AM, Michałowski T. Solution to redox titration challenge. Analytical and Bioanalytical Chemistry. 2017;**409**(17)

[32] Michałowski T, Pietrzyk A. Complementarity of physical and chemical laws of preservation in aspect of electrolytic systems (in Polish). Wiadomości Chemiczne. 2007;**61**:625–640

[33] Michałowski T. Calculations in Analytical Chemistry with Elements of Computer Programming [Internet]. Cracow, 2001. Available from: http://suw.biblos.pk.edu.pl/resourceDetails&rId=3974; http://www.biblos.pk.edu.pl/bcr&id=1762&ps=-12&dir=MD.MichalowskiT.ObliczeniaChemii.html

[34] Michałowski T, Pietrzyk A. The generalized electron balance concept. Derivation based on elementary rules of the matter conservation. In: Zuba D, Parczewski A, editors. Chemometrics: Methods and Applications. Publishing House of the Institute of Forensic Research, Cracow, 2006. pp. 415–422. https://pbn.nauka.gov.pl/works/210292

[35] Michałowski T. Electron balance as the basis of calculations in redox systems (in Polish). In: Use of Information Technology in Academic Teaching of Chemistry. Department of Chemistry, Jagiellonian University, Drukarnia Cyfrowa Eikon Plus, www.eikon.net.pl, 2007. pp. 162–169. http://www.chemia.uj.edu.pl/~ictchem/book.html

[36] Michałowski T. Generalized electron balance (GEB) as a law of preservation for electrolytic redox systems. In: 65th Annual Meeting of the International Society of Electrochemistry. Lausanne, Switzerland. 2014. http://myprogram14.ise-online.org/infiles/pdf/ise141331.pdf

[37] Marx D, Tuckerman ME, Hutter J, Parrinello M. The nature of the hydrated excess proton in water. Nature. 1999;**397**:601–604. DOI: 10.1038/17579

[38] Michałowski T, Pietrzyk A. A thermodynamic study of struvite+water system, Talanta. 2006;**68**:594–601. http://www.ncbi.nlm.nih.gov/pubmed/18970362

[39] Pietrzyk A, Michałowski T. Computer simulation of the process of struvite $MgNH_4PO_4$ $6H_2O$ dissolution in aqueous solutions (in Polish). In: Use of Information Technology in Academic Teaching of Chemistry. Department of Chemistry, Jagiellonian University, Drukarnia Cyfrowa Eikon Plus, www.eikon.net.pl, 2007. pp.137–146. http://www.chemia.uj.edu.pl/~ictchem/book.html

[40] Michałowski T, Pietrzyk A. Quasistatic processes in non-equilibrium two-phase systems with ternary salts: I. Struvite + aqueous solution (CO_2 + KOH). Chemia Analityczna (Warsaw). 2008;**53**:33–46. http://www.chem.uw.edu.pl/chemanal/PDFs/2008/CHAN2008V53P00033.pdf

[41] Pietrzyk A, Michałowski T. Conditions of struvite formation in aqueous systems. In: Zuba D, Parczewski A, editors. Chemometrics: Methods and Applications. Department of Chemistry, Jagiellonian University, Drukarnia Cyfrowa Eikon Plus, www.eikon.net.pl, 2006. Kraków; pp. 423–427. https://pbn.nauka.gov.pl/works/210292

[42] Michałowski T, Asuero AG. Thermodynamic modelling of dolomite behavior in aqueous media. Journal of Thermodynamics. 2012;**2012**:12. Article ID 723052. DOI: 10.1155/2012/723052. http://www.hindawi.com/journals/jtd/2012/723052/cta/

[43] Michałowski T. Effect of mutual solubility of solvents in multiple extraction. Journal of Chemical Education. 2002;**79**:1267–1268. DOI: 10.1021/ed079p1267. http://pubs.acs.org/doi/abs/10.1021/ed079p1267?journalCode=jceda8

[44] Michałowski T, Pilarski B, Dobkowska A, Młodzianowski J. Mathematical modeling and physicochemical studies on acid-base equilibria in binary-solvent systems. Wiadomości Chemiczne. 2010;**54**:124–154. http://suw.biblos.pk.edu.pl/resourceDetails&rId=4745%7D%7D%3C/opinion&rId=24741; http://www.ljemail.org/reference/ReferencesPapers.aspx?ReferenceID=1640416

[45] Pilarski B, Dobkowska A, Foks H, Michałowski T. Modelling of acid–base equilibria in binary-solvent systems: A comparative study. Talanta. 2010;**80**(3):1073–1080. http://www.sciencedirect.com/science/article/pii/S0039914009006110

[46] Asuero AG, Pilarski B, Dobkowska A, Michałowski T. On the isohydricity concept—Some comments. Talanta. 2013;**112**:49–54. http://www.ncbi.nlm.nih.gov/pubmed/23708536

[47] Michałowski T. Some remarks on acid-base titration curves. Chemia Analityczna. 1981;**26**: 799–813

[48] Inczédy J, Lengyel T, Ure AM. Compendium of Analytical Nomenclature, Definitive Rules, 1997. International Union of Pure and Applied Chemistry. Blackwell Scientific Publications, Oxford 1998 (the Orange Book). https://en.wikipedia.org/wiki/Compendium_of_Analytical_Nomenclature

[49] Segré E. Nuclei and Particles. 2nd ed. Reading, MA: Benjamin/Cummings; 1977

[50] Cetnar J. General solution of Bateman equations for nuclear transmutations. Annals of Nuclear Energy. 2006;**33**:640–645

[51] Massey F, Prentis J. Power law approximations for radioactive decay chains. Applied Mathematics and Computation. 2014;**245**:135–144

[52] Thibes R, de Oliveira SL. General solution to Bateman's differential equations with direct index notation. International Journal of Pure and Applied Mathematics. 2014;**93**(6):879–883

[53] Asuero AG, Michałowski T. Comprehensive formulation of titration curves referred to complex acid-base systems and its analytical implications. Critical Reviews in Analytical Chemistry. 2011;**41**:151–187. http://www.tandfonline.com/toc/batc20/41/2

[54] Michałowski T, Asuero AG. New approaches in modeling the carbonate alkalinity and total alkalinity. Critical Reviews in Analytical Chemistry. 2012;**42**:220–244. http://www.tandfonline.com/toc/batc20/42/3

[55] Michałowska-Kaczmarczyk AM, Michałowski T. Linear dependence of balances for non-redox electrolytic systems. American Journal of Analytical Chemistry. 2014;**5**:1285–1289. http://www.scirp.org/journal/PaperInformation.aspx?paperID=52615

[56] Yu L. Handbook of Analytical Chemistry. Moscow: Mir Publishers; 1975

[57] Wang H, Qiu Y, Schaefer 3rd HF. Pathways for the OH + Br_2 → HOBr + Br and HOBr + Br → HBr + BrO reactions. Journal of Physical Chemistry A. 2016;**120**(5):805–816. DOI: 10.1021/acs.jpca.5b11524

[58] Michałowska-Kaczmarczyk AM, Michałowski T. Dynamic buffer capacity in acid–base systems. Journal of Solution Chemistry. 2015;**44**(6):1256–1266. DOI: 10.1007/s10953-015-0342-0. https://www.ncbi.nlm.nih.gov/pmc/articles/PMC4495796/

[59] Michałowska-Kaczmarczyk AM, Michałowski T, Asuero AG. Formulation of dynamic buffer capacity for phytic acid. American Journal of Chemistry and Applications. 2015;**2** (1):5–9

[60] Michałowska-Kaczmarczyk AM, Michałowski T. Application of simms constants in modeling the titrimetric analyses of fulvic acids and their complexes with metal ions. Journal of Solution Chemistry. 2016;**45**:200–220. DOI: 10_1007-S10953-015-0430-1?locale=pl. https://www.infona.pl/resource/bwmeta1.element.springer

[61] Erdey L, Bodor A, Buzás H. Ascorbinometrische Bromatbestimmung. Fresenius' Zeitschrift für Analytische Chemie. 1952;**134**(6):412-419. DOI:10.1007/BF00433426

[62] Alajalov C. Commuter Card Game [Internet]. Available from: http://www.allposters.pl/-sp/Commuter-Card-Game-March-15-1947-plakaty_i8290703_.htm. [Accessed: 15 March 1947]

[63] Nemzer B, Pietrzkowski Z, Spórna A, Stalica P, Thresher W, Michałowski T, Wybraniec S. Betalainic and nutritional profiles of Pigment-Enriched root (*Beta vulgaris* L.) dried extracts. Food Chemistry. 2011;**127**(1):42–53. http://dx.doi.org/10.1016/j.foodchem.2010.12.081

[64] Wybraniec S, Michałowski T. New pathways of betanidin and betanin enzymatic oxidation. Journal of Agricultural and Food Chemistry. 2011;**59**(17):9612–9622. http://dx.doi.org/10.1021/jf2020107

[65] Wybraniec S, Stalica P, Spórna A, Nemzer B, Pietrzkowski Z, Michałowski T. Antioxidant activity of betanidin: Electrochemical study in aqueous media. Journal of Agricultural and Food Chemistry. 2011;**59**(22):12163–12170. http://www.ncbi.nlm.nih.gov/pubmed/21913685

[66] Wybraniec S, Starzak K, Skopińska A, Szaleniec M, Słupski J, Mitka K, Kowalski P, Michałowski T. Effects of metal cations on betanin stability in aqueous-organic solutions. Food Science and Biotechnology. 2013;**22**(2):353–363. http://dx.doi.org/10.1007/s10068-013-0088-7

[67] Wybraniec S, Starzak K, Skopińska A, Nemzer B, Pietrzkowski Z, Michalowski T. Studies on non-enzymatic oxidation mechanism in neobetanin, betanin and decarboxylated betanins. Journal of Agricultural and Food Chemistry. 2013;**61**(26):6465–6476. http://pubs.acs.org/toc/jafcau/61/26

[68] Michałowska-Kaczmarczyk AM, Spórna-Kucab A, Michałowski T. Solubility Products and Solubility Concepts. in: Descriptive Inorganic Chemistry Researches of Metal Compounds. InTech, Rijeka, Croatia; 2017

[69] Michałowska-Kaczmarczyk AM, Spórna-Kucab A, Michałowski T. A Distinguishing Feature of the Balance 2·*f*(O) – *f*(H) in Electrolytic Systems. The Reference to Titrimetric Methods of Analysis. In: Advances in Titration Techniques. InTech, Rijeka, Croatia, 2017

[70] Michałowska-Kaczmarczyk AM, Spórna-Kucab A, Michałowski T. Principles of Titrimetric Analyses According to Generalized Approach to Electrolytic Systems (GATES). In: Advances in Titration Techniques. InTech, Rijeka, Croatia, 2017

Chapter 3

Redox-Active Molecules for Novel Nonvolatile Memory Applications

Hao Zhu and Qiliang Li

Additional information is available at the end of the chapter

http://dx.doi.org/10.5772/intechopen.68726

Abstract

The continuous complementary metal-oxide-semiconductor (CMOS) scaling is reaching fundamental limits imposed by the heat dissipation and short-channel effects, which will finally stop the increase of integration density and the MOSFET performance predicted by Moore's law. Molecular technology has been aggressively pursued for decades due to its potential impact on future micro-/nanoelectronics. Molecules, especially redox-active molecules, have become attractive due to their intrinsic redox behavior, which provides an excellent basis for low-power, high-density, and high-reliability nonvolatile memory applications. This chapter briefly reviews the development of molecular electronics in the application of nonvolatile memory. From the mechanical motion of molecules in the Langmuir-Blodgett film to new families of redox-active molecules, memory devices involving hybrid molecular technology have shown advantageous potential in fast speed, low-power, and high-density nonvolatile memory and will lead to promising on-chip memory as well as future portable electronics applications.

Keywords: molecular electronics, redox-active molecules, self-assembled monolayer, nonvolatile memory, high-density, high-endurance, flash-like memory

1. Introduction

1.1. CMOS scaling challenges and impact on nonvolatile memory

The complementary metal-oxide-semiconductor (CMOS) scaling has deviated from the trends predicted by Moore and the scaling rules set forth by Dennard et al. due to both fundamental physical and technical limitations [1, 2]. This will inevitably slow down the pace of current CMOS scaling when approaching atomic dimension. Scaling limitations such as ultrathin gate oxide, channel length modulation, and off state leakage have become growing concern, and there are urgent needs to develop new semiconductor technology and solutions toward CMOS scaling challenges. For example, barriers such as doping, carrier transport, and series resistance scaling have been effectively avoided by innovations such as source/drain process upon silicon-on-insulator (SOI) structure, multigate field-effect transistor (FET), and SiGe BiCMOS technologies [3–6].

Currently, computing architectures and electronic systems built on CMOS components are still pursuing without signs of slowing down of requirements for low-power, fast speed, and high-density alternatives [7, 8]. Various scenarios of fundamentally new technology with advanced materials and structures have been proposed and investigated such as quantum computing, DNA computing, single electron device, spin transistor, and molecular electronics for both logic and memory applications. However, these new approaches still need in-depth studied before they can be considered for application in real-world device applications.

In modern electronic systems, the main functions lie in the data computing and data storage, which take up more than half of the semiconductor market. Solid-state mass storage occupies a large portion of this market; the demand is still growing explosively in areas such as portable electronic devices, due to their compatibility with CMOS scaling, suitability for harsh environment, and the fact that most types of memory are nonvolatile, which means that the data information can be maintained even without power supply. Currently, main types of nonvolatile memory technology that have been investigated include phase-change memory (PCM), ferroelectric random access memory (FeRAM), resistive random access memory (RRAM), magnetic random access memory (MRAM), and flash memory [9–13]. However, these nearer term new technologies have inherent disadvantages that may limit their implementations. For instance, even though FeRAM, PCM, and MRAM are currently in commercial production, they are still many years away from competing with dynamic RAM (DRAM) and NAND (Not-AND) flash for industry adoption [14, 15]. On the other hand, the retention time of floating-gate DRAM decreases with scaling, and the relatively poor endurance of the flash memory ($\sim 10^5$) is the critical limitation for its further application.

1.2. Molecular electronics

1.2.1. Introduction

Molecular technology has been aggressively pursued for its potential impact on nanoelectronics since the early 1970s due to its inherent scalability and intrinsic properties [16–18]. Molecular electronics has been considered to replace the conventional silicon-based computing even before hitting the fundamental limits [19–21]. Molecular electronic devices are typically fabricated by forming a self-assembled monolayer (SAM) or multiple layers on different surfaces using inexpensive and simple processing methods. Such device functions by the controlling of fewer electrons at a molecular scale, and therefore, has potential for fast speed, low-power, and ultrahigh-density device and circuit applications.

Molecular electronics are typically achieved through two fundamentally different approaches, which are graphically termed as "top-down" and "bottom-up". The top-down approach includes making nanoscale structures by machining and etching techniques. Molecular electronics relying on the bottom-up approach takes advantage of molecule self-assembly, building organic or inorganic structure by atom-by-atom or molecule-by-molecule techniques. In the past decades, the cross-disciplinary publications in the field of molecular electronics have dramatically increased by chemists, physicists, engineers, and other researchers. Various novel molecular device architectures and electronic systems have been introduced and explored. Nevertheless, most of the molecular electronics has been implemented by top-down approaches, and recently, the combination of top-down device fabrication (mainly lithography) with bottom-up molecule self-assembly has attracted more and more interest.

1.2.2. *Advantages and challenges of molecular electronics*

Molecular electronics competes to a large extent with conventional microelectronics based on traditional metal-oxide-semiconductor (MOS) structures. Molecular electronics has been expected to possess the following advantages [22]. First, the inherent scalability of molecules enables functional structures and ultrahigh device density with atomic control over a diversity of physical properties. Second, molecules can be self-assembled through intermolecular interaction to form nanostructures, and further for desired molecular devices and circuits. Third, molecular properties can be tailored with choice of composition and geometry, including electrical transport, binding, and structural properties. Fourth, a variety of molecules have multiple distinct stable electrical and geometric structures, therefore, molecular switching devices and circuits can be achieved through the transitions between different structures under electrical or chemical stimulus.

The major challenges molecular electronics are facing lie in the unknown reliability at high temperature, volatile environments, and electrical stress. The instability at high temperature makes most molecules incompatible with current CMOS process integration. The retention time is the biggest challenges for most of the molecular memory devices as at least >10-year retention time is necessary in order to be considered as a candidate technology for universal memory applications. Some specific condition and environment need to be taken care of when integrating molecular electronics in conventional CMOS devices and circuits. Furthermore, molecular technology development requires advancement in both molecular properties and device integration processes.

1.2.3. *Redox-active molecules*

Redox-active molecules have attracted more and more interest recently, due to their intrinsic and reliable redox behavior, which can be readily utilized for charge-storage memory applications. Physically, a redox-active molecule contains a redox component acting as the charge-storage center surrounded by insulators/barriers formed by the linkage and the surface group. The electrons tunnel through the barrier during the oxidation and reduction processes. Typically, the application of an oxidation voltage will cause electron loss in the redox molecules; reversely, the electrons will be driven back to the molecules by applying a reduction voltage. Generally, the redox molecules have multiple stable states. The switching between these states is dynamically reversible through the loss or capture of a charge, that is, oxidation and reduction of the redox centers. Distinct charged or discharge states can be deemed as logic on and off states, at different voltage with very fast write and erase speeds. It has been demonstrated that the redox molecules attached on silicon structures are stable and can endure more than 10^{12} program/erase cycles [23]. Such advantageous properties of redox molecules make them very attractive for future applications of fast speed, low-power, high-endurance, and high-density nonvolatile memory.

There have been great efforts to effectively integrate molecules as the active component for future micro-/nanoelectronic devices. The following sections will review the attachment and characterization of redox molecules on active surfaces, and the strategies to involve such attractive medium in low-power and high-density nonvolatile memory applications.

2. Redox molecules, attachment, and characterization techniques

2.1. Redox-active molecules

As described above, the generic structure of a redox-active molecule consists of redox-active components, linkage components, and surface attachment groups. **Figure 1** illustrates two examples of redox molecules with different components [24, 25]. α-Ferrocenylethanol (referred as ferrocene) molecule shown in **Figure 1a** has one Fe redox center and -OH linker component. $Ru_2(ap)_4(C_2C_6H_4P(O)(OH)_2)$ (referred as Ru_2) possesses two Ru redox centers, and the phosphonic acid component functions as the linker part. Such redox molecules have discrete energy states, which are accessible with relatively low, distinct, and quantized voltages. Ferrocene has a single cationic-accessible state and, therefore, can exhibit two states: neutral and monopositive. Ru_2 has two Ru metal atoms as redox centers; thus, it has two cationic states leading to three charge states: one neutral and two positive states. This is very attractive for multibit memory storage applications.

Despite the redox-active and linker components shown in **Figure 1**, a variety of different redox-active and linkage parts can be specifically designed and engineered for purposes including more redox states and attachment on desired surfaces via covalent bonds [26–29]. The molecules reviewer here are more focused on the attachment on Si and oxide structures with specific tethers and linkers, due to the prevalence of silicon in current microelectronic devices and the subsequent implementation in CMOS compatible nonvolatile memory devices.

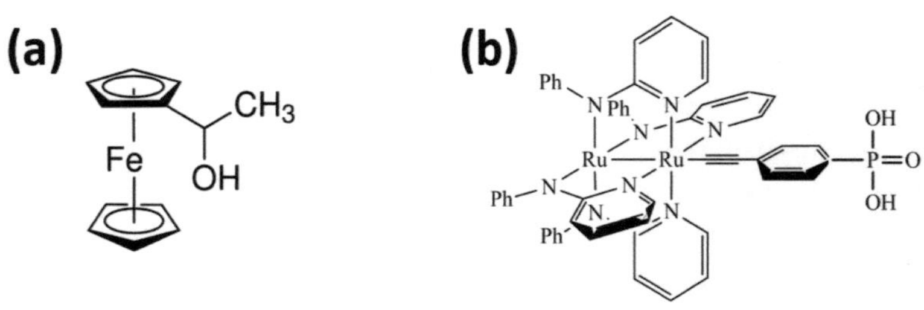

Figure 1. Structure of redox-active molecules (a) α-ferrocenylethanol and (b) $Ru_2(ap)_4(C_2C_6H_4P(O)(OH)_2)$ [24].

2.2. Molecule attachment and characterization techniques

2.2.1. *Attachment methods*

The attachment of well-ordered and tightly packed layers of molecules on active surfaces is important for the application of redox molecules in electronic devices. The attachment technique reviewed here is by using the self-assembled monolayers (SAMs) [24, 30–33]. SAMs are formed on active surfaces via covalent bonds to the atoms on the surface, and specific tether groups can be designed such that they can only attach to specific surfaces. The attachment of SAMs of redox molecules is via the use of chemical solution deposition [30]. A layer of 110 nm SiO_2 was first thermally grown on the Si substrate followed by the definition of square-shaped active areas (100 μm wide) using photolithography and wet etching. Then, a thin layer of SiO_2 was grown in the active area for the molecule attachment on the oxide surface, and it will also function as part of the tunnel barrier in the memory device. The solutions for deposition are prepared by dissolving redox molecules in organic solvents. The SAMs were then formed by placing droplets of the solution on the active areas with each drop kept in place for 3–4 min. The samples were held at elevated temperature in an N_2 environment during the attachment process. Saturated SAMs will be formed after ~30 min. An alternative approach to form SAMs is by immersing the substrate into the solution under same condition for a certain period of time. During the attachment process, redox molecules are covalently bonded to the desired surface through the linkage component. After the self-assembly process, the same organic solvent and the cleaning organic solvent were used to rinse the substrates in order to remove any residual molecules that are not bonded to the surfaces.

2.2.2. *Characterization techniques*

After the molecule attachment, cyclic voltammetry (CyV) was used to characterize the attachment quality and measure the molecule surface density. **Figure 2** shows the schematic of the CyV characterization setup [24]. The measurements were performed in a standard probe

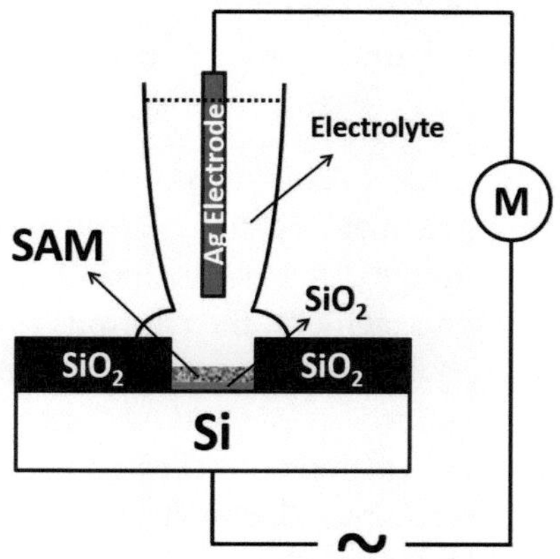

Figure 2. Schematic of the CyV characterization setup. "~" represents the voltage source and "M" represents the electrometer [24].

station, with the backside contact made via the probe station chuck. A solution of 1.0-M tetrabutylammonium hexafluorophosphate (TBAH) in propylene carbonate (PC) was used as the conducting gate electrolyte. Polypropylene micropipette tip containing the silver counter electrode and the electrolyte was lowered until only a small amount of electrolyte was spread across the active area. During the measurement, the voltage was applied on the substrate, and the CyV curves were obtained using a CHI600 electrochemical analyzer.

3. Redox molecules for nonvolatile memory applications

In this section, the early logic forms of molecular memory—molecular logic switching devices will be reviewed. The integration of redox molecules in a liquid electrolyte-involved nonvolatile memory device will be discussed where we also consider the device-related issues and limitations.

3.1. Introduction

3.1.1. CMOS memory technology

The continuous CMOS scaling and the impact on memory scaling has pushed for the investigation of alternative storage technologies, and many solutions have been proposed and studied. Solid-state mass-storage memory is getting a large part of the market, as introduced in the first section, due to its consolidated technology, better reliability, nonvolatility, and tolerance to harsh environment. Currently, the main solid-state reprogrammable nonvolatile memory is based on the floating-gate structure, which is also known as flash memory. Flash memory has fast read access times, good retention and reliability, and CMOS compatible fabrication process [34, 35]. However, physical limitations related to the difficulty in shrinking the tunnel and interpoly dielectric layer have further reduced the margins for the memory cell size reduction predicted by Moore's law. The floating-gate memory will suffer short-channel effects when the channel length is scaled below 100 nm. Leakage current will be significant during program/erase operations due to both drain-induced barrier lowering (DIBL) and subsurface punch-through effects.

In recent decades, charge-trapping nonvolatile memory has attracted intensive attention to replace the conventional floating-gate memory due to its advantages such as better scalability, lower power consumption, improved reliability, and simpler structure and fabrication process [36–38]. In a charge-trapping memory, the electrons are stored in a trapping layer, instead of the conducting floating gate in the conventional floating-gate memory. Different charge-storage media have been well studied, including conventional nitride material, various nanocrystals, and high-k dielectric materials. Incorporating redox-active molecules as the charge-storage medium in a Si-based nonvolatile memory is very interesting, as it will leverage the advantages afforded by a molecule-based active medium with the vast infrastructure of traditional metal-oxide-semiconductor (MOS) technology.

3.1.2. *History of molecular memory*

Memory application of molecules has been widely investigated, and one of the most common approaches in the molecular memory devices has been the bistable conductance switching devices. There are two major device architectures based on these devices: molecular crossbar circuit and nanocell molecular circuit.

In a crossbar circuit, the switching element is a metal/molecule/metal sandwich junction, wherein the molecules are located at the cross-section of two nanoscale metal wires [39–41]. The early demonstration of such a junction-switching device utilized the bistable rotaxane molecule that consists of two mechanically interlocked rings [39]. The molecular monolayers were deposited as a Langmuir-Blodgett film. The mechanical motion of such a molecule is an activated redox process, and the two stable mechanical states can exhibit different tunneling currents, representing logic on and off states. Note that ~100 on/off current ratio has been reported, but with limited endurance cycles. By taking advantage of the fabrication simplicity and high nanowire density, large-scale crossbar circuit has been realized [42]. However, the controlling of metal-molecule interface is the major challenge, upon which a new theory was put forward with experimental evidence indicating that the conductance switching in the metal/molecule/metal junction is independent of molecules, but attributed to the electrode reactions with molecules [43, 44].

Nanocell molecular circuit is a two-dimensional network of self-assembled metallic particles connected by molecules [45, 46]. The active component in a nanocell is also metal/molecule/metal switch. The molecules in a nanocell can show reprogrammable negative differential resistance which was initially believed to be originated from the redox center in the molecules. However, experimental results suggested that the electron transport phenomena is more likely due to the nanofilamentary metal switching, which is correlated with the similar metal-molecule interface problems as the crossbar circuit [46].

In addition to the two-terminal metal/molecule/metal junction devices, molecules have also been tested as the channel material in a standard three-terminal metal-oxide-semiconductor (MOSFET) architecture [47–49]. Even though organic field-effect transistors (OFETs) with low-power and low leakage current have been demonstrated, such OFETs suffer from the slower switching speed and limited device lifetime as compared with conventional silicon transistors. Moreover, the poor stability in a harsh environment is another disadvantage of these OFETs. Thus, the direct integration of molecules in advanced silicon CMOS technology is still under question and needs further research efforts.

3.2. Electrolyte/molecule/Si capacitors for memory application

3.2.1. *Introduction*

Based on the molecular logic switching devices, appropriate modifications of the molecular structures and switching elements have been designed to change the switching kinetics with enhanced memory performance. Different device platforms have been engineered to interface with molecules such as oxides, dielectrics, nanowires, and so forth. The memory effect achieved by integrating redox-active molecules is very promising, due to the intrinsic redox behavior of the molecules giving birth to fast speed, low operation voltage, and high-reliability memory devices.

Hybrid CMOS/molecular memory devices were proposed and studied by incorporating redox-active charge-storage molecules into Si structures through the self-assembly process [32, 33, 50–52]. Different redox-active components can be designed or synthesized for multiple redox states, thus for complex and high memory density. The memory retention properties can be effectively tuned through the engineering of the linkage component. The redox component such as the Fe in 4-ferrocenylbenzyl alcohol (referred as Fc-BzOH) and Zn in 5-(4-hydroxymethylphenyl)-10,15,20-trimesitylporphinatozinc(II) (referred as Por-BzOH) as shown in **Figure 3a** and **b**, respectively, can be in neutral and positively charged states through losing electrons [32, 51]. The molecular components surrounding the redox center function as the barrier against charge loss. Such an alternative scenario replacing the conventional tunneling, charge-trapping, and blocking layers in a charge-trapping memory with molecular components provides a smooth transition from CMOS technology to molecular electronics technology.

3.2.2. *Characterization and performance discussion*

The SAM attachment of the two redox molecules to the silicon surface was carried out by using the similar chemical solution deposition as described in the previous section. Benzonitrile was used as the organic solvent in the attachment process [32, 51]. The sample was placed in the solution for 80 min at 100°C. **Figure 3c** shows the structure of a fabricated electrolyte/molecule/Si capacitor memory cell. **Figure 4a** and **b** shows the CyV curves of Fc-BzOH and Por-BzOH molecular structures, both exhibiting oxidation and reduction peaks. Fc-BzOH shows one pair of redox peaks because the redox center Fe has only one neutral and one

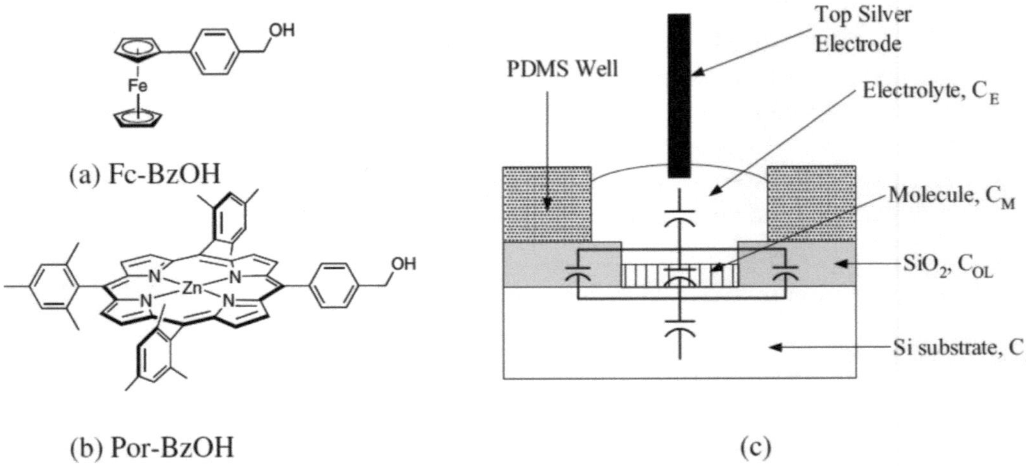

Figure 3. Molecular structure of redox molecules (a) Fc-BzOH and (b) Por-BzOH. (c) Schematic structure of the electrolyte/molecule/Si capacitor for memory application with a simplified equivalent circuit [32, 51].

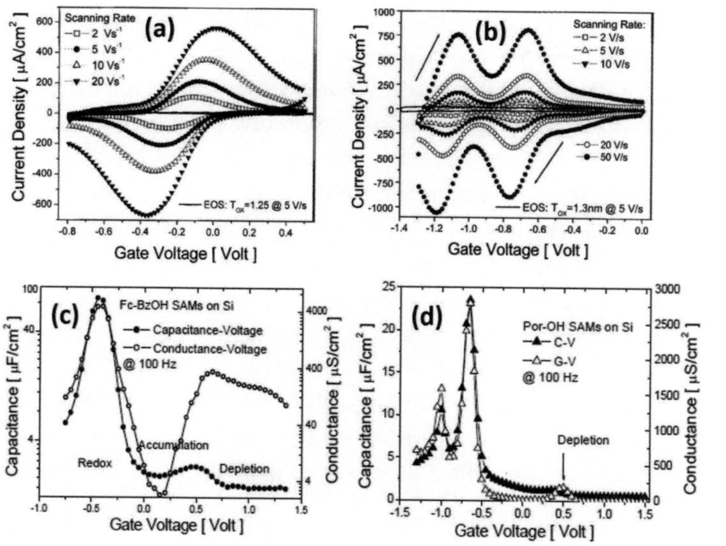

Figure 4. CyV of the electrolyte/molecule/Si structures with (a) Fc-BzOH and (b) Por-BzOH molecules at different voltage scan rates. *C-V* and *G-V* curves of the capacitor structures involving (c) Fc-BzOH and (d) Por-BzOH molecules at 100 Hz [31, 33, 52].

monopositive state. The Zn redox center in Por-BzOH exhibits one neutral and two (mono- and di-) positive states, thus, two pairs of redox peaks were observed.

Clear capacitance and conductance peaks related to the oxidation and reduction processes have been observed from the capacitance-voltage (*C-V*) and conductance-voltage (*G-V*) characteristics, as shown in **Figure 4c** and **d**, respectively [33, 52]. The *C-V* and *G-V* curves can be divided into three regions: depletion, accumulation, and redox regions. The depletion and accumulation are associated with the Si substrate, in a similar manner as MOS capacitors. The characteristics in the redox regions are due to the charging and discharging of molecules, resembling the CyV characteristics. Capacitance and conductance peaks were also observed at higher frequencies up to 5 kHz, but with lower capacitance peaks and higher conductance peaks. The lower capacitance peak at higher frequency is attributed to a lower effective capacitance of the electrolyte in series, while the higher conductance peak is due to the increasing frequency.

3.2.3. *Multibit molecular memory in one cell*

One of the simple yet effective approaches to increase the memory density is to employ charge-storage element containing molecules with multiple redox states. There are different approaches to realize multiple redox states, including mixed or stacked SAMs of different redox-active molecules with distinct voltage levels, and synthesis of molecule with multiple redox centers. Here, we review the first method by mixing the Fc-BzOH and Por-BzOH molecules in one memory cell [51]. The second approach will be reviewed in the following section.

Mixed SAMs of Fc-BzOH and Por-BzOH on silicon surfaces to achieve a four-state memory element were achieved by using the chemical deposition method of a mixture solution of Fc-BzOH and Por-BzOH in benzonitrile [51]. Molar ratios (Fc-BzOH/Por-BzOH) of 1:0.35, 1:1.4, and

1:3.5 were selected. The CyV curves of the mixed molecules are shown in **Figure 5a**. The results from the nonmixed pure molecules were also illustrated for comparison. One pair of Fc-BzOH peaks and two pairs of Por-BzOH peaks were clearly observed for the three mixed SAMs. With decreasing Fc-BzOH molar percentage, the corresponding peak height decreased substantially due to the changing surface coverage density. The coverage of Fc-BzOH was higher than that of Por-BzOH in the pure SAMs and even mixed SAMs with 1:1 molar ratio, because of the smaller size and the faster attachment of smaller molecule of Fc-BzOH. The redox peaks exhibited a scan rate-dependent peak separation, as shown in **Figure 5b**. Such phenomena could arise from either an increasing resistive drop in the electrolyte or limitations imposed by the intrinsic electron-transfer rate of redox centers on the Si substrate [51]. **Figure 5c–e** shows the *C-V* and *G-V* characteristics of the SAMs at 100 Hz. Each of the mixed SAMs shows three pairs of peaks corresponding to the charging and discharging behaviors of the Fc-BzOH and Por-BzOH molecules. As the molar ratio decreases, the capacitance and conductance peaks of Por-BzOH increase while those of Fc-BzOH decrease. This is also in agreement with the CyV measurement results.

Such a mixed SAM approach is very attractive owing to its simpler synthesis and more distinct potential of the redox states, as compared with a single molecule with multiple redox

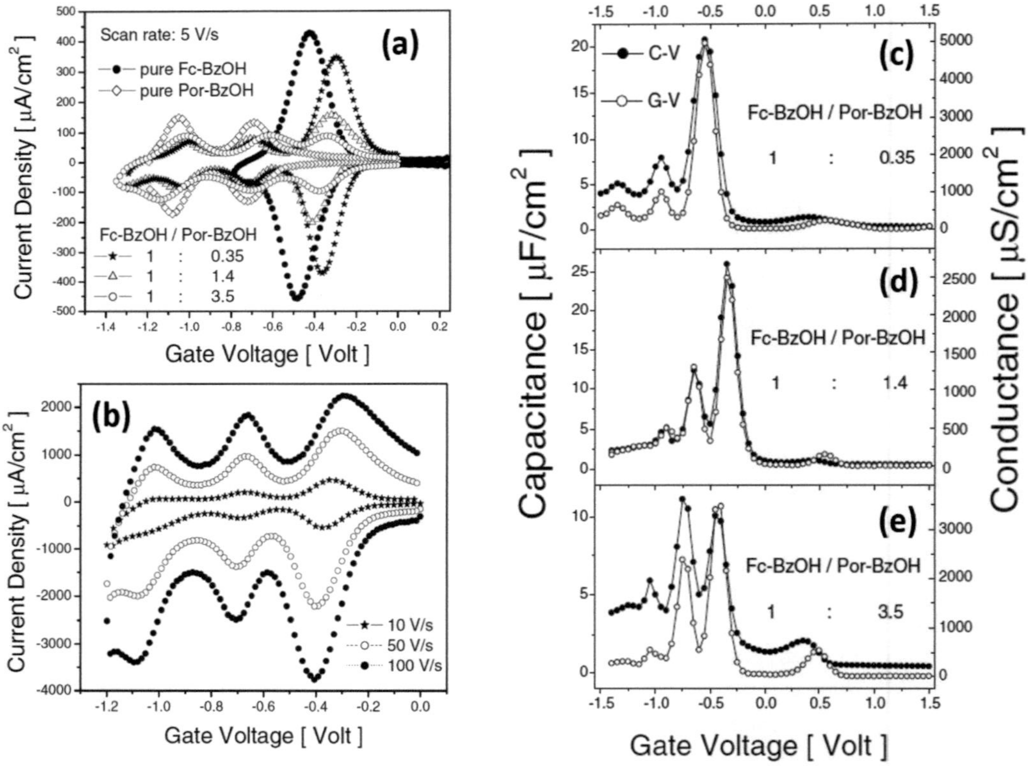

Figure 5. (a) CyV of SAMs of pure and mixed Fc-BzOH and Por-BzOH with different molar ratios. The scan rate was 5 V/s. (b) CyV at increasing scan rates of 10, 50, and 100 V/s. *C-V* and *G-V* characteristics with mixed SAMs with molar ratios of (c) 1:0.35, (d) 1:1.4, and (e) 1:3.5 [51].

states. However, the disadvantage of this method is that the density of a given peak goes down. Nevertheless, this approach still paves the way for constructing multibit information storage devices.

4. Solid-state molecular nonvolatile memory

4.1. Introduction

CMOS and semiconductor nonvolatile memory scaling have create huge demands for alternative memory technologies with higher scalability and better performance. The well-studied charge-trapping memory relies on hot electron injection from the channel into the charge-trapping medium through a tunneling layer. By attaching redox-active molecules onto silicon structures as the charge-storage medium can further enhance the memory density and enable further cell scaling [30, 53–55]. The previous section has introduced the implementation of redox molecules for nonvolatile memory applications. However, the liquid electrolyte-involved structure lacks effective protection for the molecules leading to deteriorated memory performance. By integrating redox molecules in a solid-state molecular memory cell with a solid-state insulating barrier deposited on both sides of the molecules, the possibility of orbital hybridization from the gate can be lessened. The structure of the solid-state molecular memory can be engineered with the material and layer thickness as well as the molecules, whose linker also works as the tunnel barrier and can be optimized by variation in structure and connectivity to obtain the desired tunneling probability, redox potential, and retention time. Such an integration of redox molecules in a solid-state memory provides an excellent platform to study the electrical behavior of the molecules and the devices with universal microelectronic characterization metrologies.

In this section, we first review a solid-state capacitor structure incorporating embedded redox molecules as the charge-trapping medium for high-endurance memory applications. Then, a Si nanowire FET-based molecular flash-like memory with faster operation speed, lower operation voltage, better reliability, and multibit charge storage will be introduced.

4.2. Metal/dielectric/molecule/oxide/Si capacitors for memory application

4.2.1. Device fabrication

Figure 6a shows the structure of the ferrocene molecule and the schematic of a metal/Al_2O_3/ferrocene/SiO_2/Si (MAFOS) solid-state capacitor structure [30]. The most important fabrication process steps are the molecule attachment on SiO_2 and the formation of Al_2O_3 encapsulating the molecules. After the 100 μm × 100 μm active areas were defined by photolithography and wet etching, a ~1.5 nm SiO_2 was grown in the active area by rapid thermal oxidation at 850°C for 2 min. The SAM was formed by placing the substrate in a solution of 3 mM ferrocene in dichloromethane at 100°C for 20 min. After the attachment, dichloromethane was used again to rinse the substrate to remove any physisorbed residuals on the surface.

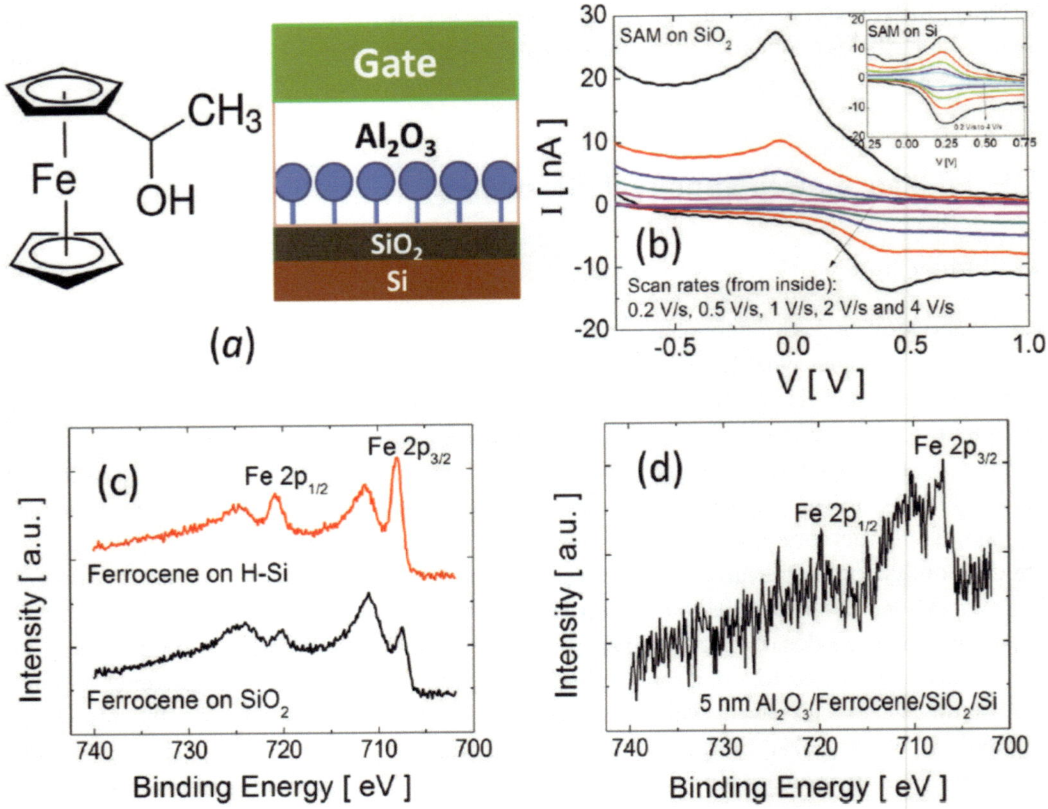

Figure 6. (a) Molecule structure of α-ferrocenylethanol (ferrocene) and the schematic of the MAFOS capacitor memory cell. (b) CyV curves of ferrocene-attached capacitor at different scan rates. Inset: CyV curves of the reference sample with ferrocene attached on the Si surface. (c) XPS spectra of the samples with ferrocene attached on Si and SiO_2 surfaces. (d) The XPS spectrum of the ferrocene-attached SiO_2 substrate with 5 nm Al_2O_3 covered on the top [30].

CyV was performed after the attachment of molecule. Other substrates were immediately loaded into the atomic layer deposition (ALD) vacuum chamber to deposit 20 nm Al_2O_3 at 100°C. Trimethyl aluminum (TMA) and H_2O were used as precursors. In the final step, a layer of 80 nm Pd was deposited and patterned on Al_2O_3 as the top gate. Three reference samples were fabricated for comparison: metal/Al_2O_3/ferrocene/Si (MAFS), metal/Al_2O_3/SiO_2/Si (MAOS), and metal/Al_2O_3/Si (MAS). For the ferrocene attachment on the Si surface in MAFS device, a hydrogen-terminated Si surface was obtained by dipping the substrate into 2% hydrofluoric acid for 30 s prior to the attachment [30].

4.2.2. Electrical characterization

Figure 6b shows the CyV curves of the electrolyte/ferrocene/SiO_2/Si structure at various scan rates. Oxidation and reduction peaks were observed at negative and positive gate voltage (V_G), respectively, through losing and restoring electrons from the SAM. By comparing the CyV curves with the results from the electrolyte/ferrocene/Si structure (inset in **Figure 6b**), the clear peak separation is due to the tunneling barrier (~1.5 nm SiO_2) and the molecular linker. The ferrocene surface coverage can be calculated from the oxidation peak, and the coverage was found to be 5.23×10^{13} and 3.14×10^{13} cm^{-2} for ferrocene attachment on Si and SiO_2 surfaces, respectively. X-ray photoelectron spectroscopy (XPS) was carried out before and after the ALD to examine the viability of ferrocene during the deposition. From the XPS spectra shown in **Figure 6c** and **d**, the ferrocene SAMs were well attached on both Si and SiO_2 surfaces, and the SAM survives the deposition of Al_2O_3.

The memory behavior was characterized by measuring the *C-V* hysteresis at 1 MHz. As shown in **Figure 7a**, large memory window was observed with the MAFOS device, and the counterclockwise hysteresis loop indicates the charge storage in ferrocene SAM. **Figure 7b** shows the flat-band voltage shift (ΔV_{FB}) of the MAFOS and three control samples as a function of applied programming/erasing (P/E) operations. The much smaller ΔV_{FB} observed from MAOS and MAS devices indicates the negligible charge storage at the dielectric interface traps. The asymmetric behavior observed from the MAFS device suggests the difficulty of

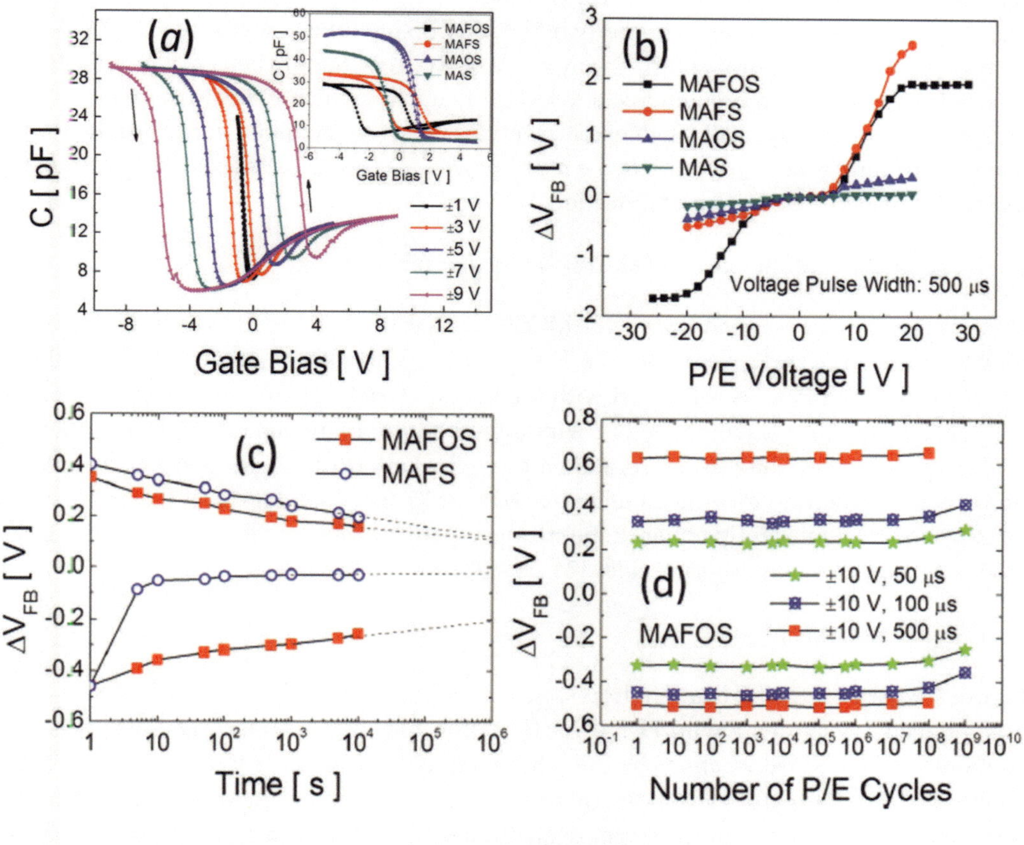

Figure 7. (a) *C-V* hysteresis curves of the MAFOS device at 1 MHz with different gate voltage scan ranges. Hysteresis of MAFOS and three control samples with ±5 V scan range were compared in the inset. (b) ΔV_{FB} of MAFOS and three control samples versus P/E voltage with 500 μs pulse width. (c) Retention properties of MAFOS and MAFS at room temperature. (d) Endurance properties of MAFOS with ±10 V P/E voltage but different pulse width [30].

maintaining molecules positively charged without the SiO_2 tunnel barrier. The stable and symmetric staircase ΔV_{FB} of MAFOS device originated from the reliable charging/discharging behavior of ferrocene and the effective charge separation by the SiO_2 tunnel barrier [30]. The charge density of the ferrocene SAM was calculated by using the following equation [30]:

$$\Delta V_{FB} = \frac{e \cdot n}{C} = e \cdot n \cdot \left(\frac{T_{Al_2O_3}}{\varepsilon_0 \, \varepsilon_{Al_2O_3}} \right) \tag{1}$$

where e is the elementary charge, n is the number of stored electrons, T_{Al2O3} and ε_{Al2O3} are the thickness and relative dielectric constant of Al_2O_3, respectively. The charge density was calculated to be 4.82×10^{12} cm^{-2}, which is only a small portion (~15%) of the coverage density obtained from the CyV results. Nevertheless, effective memory can still be realized and the charge density of the solid-state molecular memory is sufficiently high [30].

The room temperature retention of MAFOS and MAFS devices is shown in **Figure 7c** [30]. Note that ~60% charge loss was observed after 10^6 s retention time for MAFOS. Such fair retention is due to the relative thin SiO_2, and the oxide quality by rapid thermal oxidation is not satisfactory. The endurance property of the MAFOS device was measured with ±10 V P/E voltages and different pulse width. The memory device continues to behave well after 10^9 P/E cycle with the same memory window and a slightly upshift which is due to the accumulation of electron in deep traps [30]. Such excellent endurance characteristics are naturally derived from the intrinsic redox properties of the redox molecules, which have been well protected by the device structure design and fabrication.

4.3. Nanowire/nanotube-based flash-like molecular memory

Semiconductor nanowires and nanotube MOSFETs have been regarded as the building blocks for nanoelectronics and circuits [56, 57]. The quasi-one-dimensional nanowires have a larger surface-to-volume ratio as compared with the bulk materials. Therefore, less stored charges are needed to induce a same memory window. In addition, the nanowire or nanotube can enable a gate-surrounding structure, allowing excellent electrostatic gate control. The integration of redox molecules in semiconductor nanowire FET for solid-state flash-like memory can be expected to significantly enhance the memory performance by taking advantages of the inherent properties of redox molecules [24, 58–60].

4.3.1. Device fabrication

Figure 8 illustrates the fabrication process of a molecular flash memory based on a so-called "self-aligned" silicon nanowire FET [24, 61]. A thin film of Au catalyst (2–3 nm) was first sputtered at predefined locations by photolithography on a 300-nm SiO_2/Si substrate. The Si nanowires were then grown from the catalyst following the vapor-liquid-solid (VLS) mechanism in a low-pressure chemical vapor deposition (LPCVD) furnace at 440°C for 2 h with an ambient SiH_4 stream under a pressure of 500 mTorr. Immediately after the growth, the nanowires were oxidized at 750°C for 30 min by dry oxidation to form a 3-nm SiO_2 on which the SAM will be attached [24]. As compared with the SiO_2 grown by rapid thermal oxidation, the SiO_2 formed here is of much better quality by using the dry oxidation technique [30]. Therefore, enhanced memory retention can be expected. The next step is to pattern the source/drain

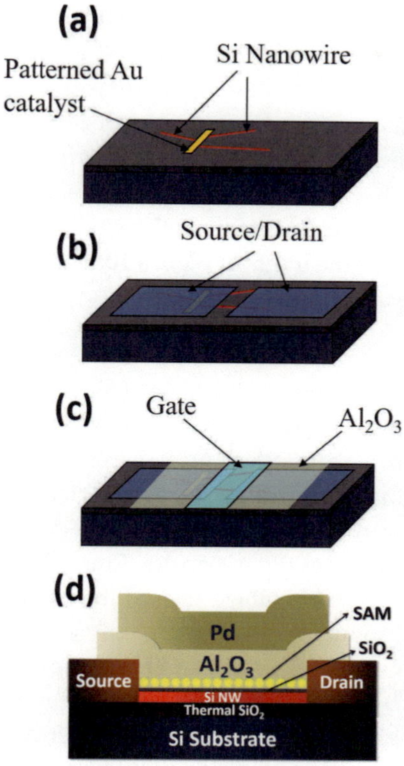

Figure 8. Schematic of the fabrication process of a molecular flash memory. (a) Au patterning on SiO_2, synthesis of Si nanowires and nanowire oxidation. (b) Formation of S/D electrodes and SAM attachment. (c) Deposition of Al_2O_3 by ALD and fabrication of top gate electrode. (d) Schematic structure of a completed molecular flash memory [24].

(S/D) electrodes with photolithography. A 2% HF wet etch was applied to remove the oxide from the nanowire at the patterned S/D area before 5/100 (unit:nm) Ti/Pt was deposited and lift-off to form S/D electrodes.

The SAM of ferrocene and Ru_2 redox molecules (**Figure 1**) were then formed on the SiO_2/Si nanowire channel by placing droplets of a solution of dichloromethane with 3-mM ferrocene and 2-mM Ru_2 on active areas separately [24]. Each drop was in place for 3–4 min and the substrates were held at 100°C in an N_2 environment during attachment. Saturated SAMs were formed after ~30 min. Dichloromethane solvent was used to rinse the substrates to remove the residual molecules. The samples were then immediately loaded into the ALD chamber for a layer of 25 nm Al_2O_3 deposition at 100°C with TMA and H_2O as precursors. The final step is the fabrication of 100-nm Pd top gate with the same photolithography and lift-off process as S/D electrodes. A reference sample without redox molecules was fabricated for comparison [24].

4.3.2. *Electrical characterization*

Figure 9 shows the transmission electron microscopy (TEM) image of the cross-section of a ferrocene-attached molecular flash memory device [24]. Clear gate-surrounding structure has been achieved, with a ~6-nm "intermixed" region observed.

Figure 10a and **b** shows the out characteristics, drain current (I_{DS}) versus drain voltage (V_{DS}) of the ferrocene molecular flash memory in linear and log-scale, respectively. Smooth and well-saturated I_{DS}-V_{DS} curves have been observed with negligible contact resistance. The leakage-affected and the weak, moderate, and strong inversion operation regions are shown in **Figure 10b**. From the transfer characteristics shown in **Figure 10c** and **d**, counterclockwise hysteresis loops have been observed at different gate voltage (V_{GS}) sweep ranges for both ferrocene and Ru_2 memory devices, suggesting the charge-trapping mechanism. From the log-scale I_{DS}-V_{GS} curves shown in the insets of **Figure 10c** and **d**, clear off states were achieved, and the on/off current ratio was as high as ~10^7 [24].

The P/E speed characteristics of the molecular memory (ferrocene) are shown in **Figure 11a** and **b**, demonstrated by the threshold voltage shift (ΔV_{TH}) under different P/E gate voltage pulses [24]. The P/E operations were performed by applying top gate voltage pulses while the substrate, S/D electrodes were all grounded. As shown in **Figure 11a**, with accumulative +10 V programming pulses, the threshold voltage showed a clear shift toward positive direction, indicating that the electrons were injected from the nanowire channel through the tunnel barrier and stored in the redox centers of the molecules. Erasing operations with −10 V gate voltage pulses back shifted the threshold voltage toward the negative direction, suggesting hole injection from the nanowire channel during erasing operations. The P/E speed characteristics of ferrocene and Ru_2 memory are summarized in **Figure 11c** and **d**, respectively. Both devices showed fast P/E speed, which arises from the intrinsic fast speed of the charging behavior of the redox molecules. The slightly faster erasing speed over programming speed was attributed to the more favorable hole injection in the Si/SiO_2/molecule/Al_2O_3/gate interface states, though the amount charge is very small even with ±10 V, 1 s stressing [24].

Figure 12a shows the ΔV_{TH} of the ferrocene molecular memory and the reference sample as a function of P/E voltages at a fixed pulse width of 500 μs. Negligible ΔV_{TH} was observed for

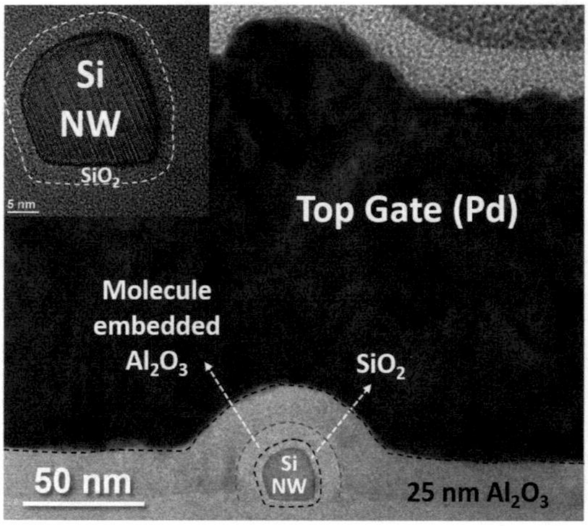

Figure 9. TEM image of the cross-section of a ferrocene-attached molecular flash memory device. The red-dashed line indicates the ferrocene-embedded region. Inset: cross-section of the nanowire channel, with the SiO_2 layer indicated by the dashed line [24].

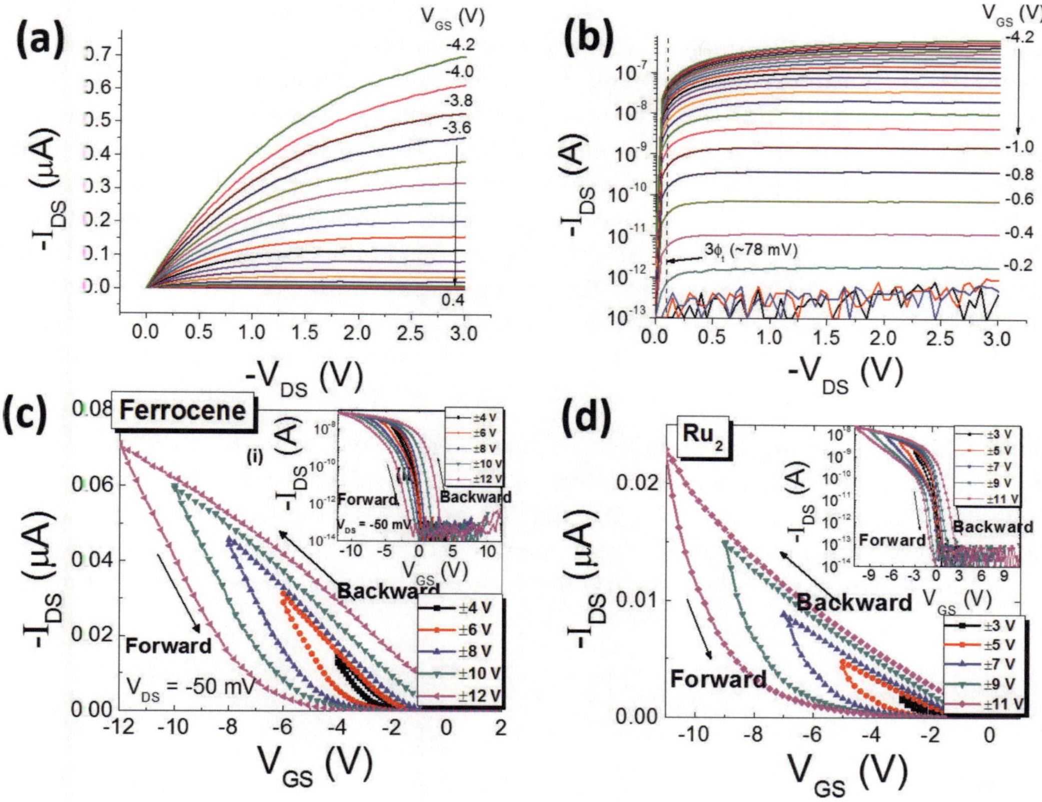

Figure 10. Output characteristics of ferrocene molecular memory in (a) linear and (b) log-scale. Transfer characteristics of (c) ferrocene and (d) Ru_2 molecular memory with different gate voltage sweep ranges. V_{DS} was set to −50 mV [24].

the reference sample, indicating the fact that the dominant charge-storage location lies in the molecules, rather than the solely traps within the gate dielectric interface or a dielectric interface. Clear staircase behavior was observed for the ferrocene memory, demonstrating discrete energy levels corresponding to various molecular orbitals. A saturated ΔV_{TH} was observed beyond ±26 V gate voltage, suggesting that all the available redox centers in the SAM have been occupied by injected charges. The charging density in the SAM can be thus calculated by

$$\Delta V_{TH} = \frac{qN}{C_{Redox}} \cong \frac{qN}{C_{AlO}} = qN\frac{\ln\left(\frac{t_{AlO\text{-}out}}{t_{AlO\text{-}in}}\right)}{2\pi\,\varepsilon_0\,\varepsilon_{AlO}\,L} \tag{2}$$

where q is the elementary charge, N is the total charge stored in the redox centers, C_{Redox} is the total capacitance arising between the redox centers and the metal gate, C_{AlO} and $\varepsilon;_{AlO}$ are the capacitance and relative dielectric constant of the Al_2O_3 layer, respectively, L is the channel length, $t_{AlO\text{-}out}$ and $t_{AlO\text{-}in}$ are the distances from the center of nanowire channel to the outside and inside surfaces of the Al_2O_3 layer. The charging density was found to be 6.96 × 10^{12} cm^{-2}, which is about 22% as compared with the coverage density from the CyV results [24]. This indicates that effective memory can be realized even with a partial (i.e., non-continuous) ferrocene SAM. Two-step charge-storage behavior was observed in the Ru_2 molecular memory due to the two redox centers in the Ru_2 molecule which can exhibit stable and distinct charged states at different voltage level. As shown in **Figure 12b**, with increasing programming gate voltage, the first step charged state was found at 10 V, with 0.8 V ΔV_{TH}. This means that the Ru_2 redox centers with lower voltage level have been occupied by electrons. Further increasing the programming voltage led to a second charged state, which was observed beyond 14 V V_{GS}, with 1.95 V ΔV_{TH}. Up to now, all the redox centers in Ru_2 SAM have been filled with injected electrons. Similarly, the overall charging density of Ru_2 SAM was calculated, and was found to be 1.12×10^{13} cm^{-2}, which is about 44% of the freshly attached SAM before Al_2O_3 deposition [24]. Such discrete charging behavior in Ru_2 molecules with multiple redox centers is very attractive for discrete multibit memory applications.

Figure 12c and **d** shows the room temperature retention properties of the ferrocene and Ru_2 molecular memory devices, respectively [24]. The devices were initially programmed/erased by ±10 V gate voltage with 500 and 100 μs pulse width. Only ~20% charge loss was observed with a projected 10-year memory window. As compared with the previous capacitor structure molecular memory, the improved retention shown here is attributed

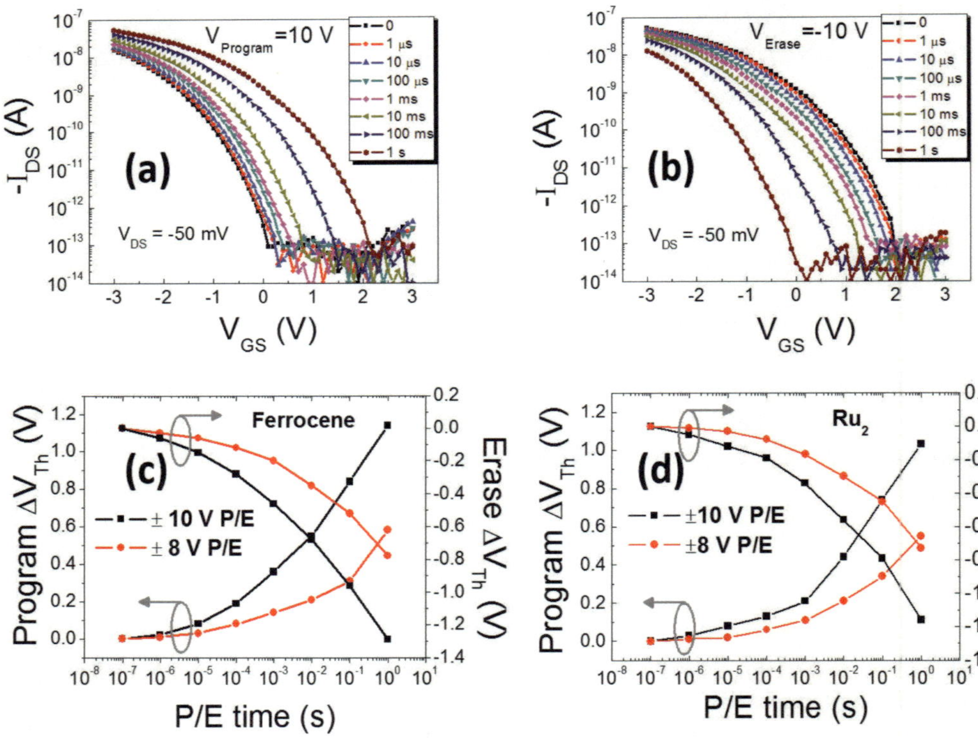

Figure 11. (a) Programming and (b) erasing operations of the ferrocene molecular memory under ±10 V P/E gate voltage pulses with accumulative time. P/E speed characterization of (c) ferrocene and (d) Ru_2 molecular memory devices, respectively [24].

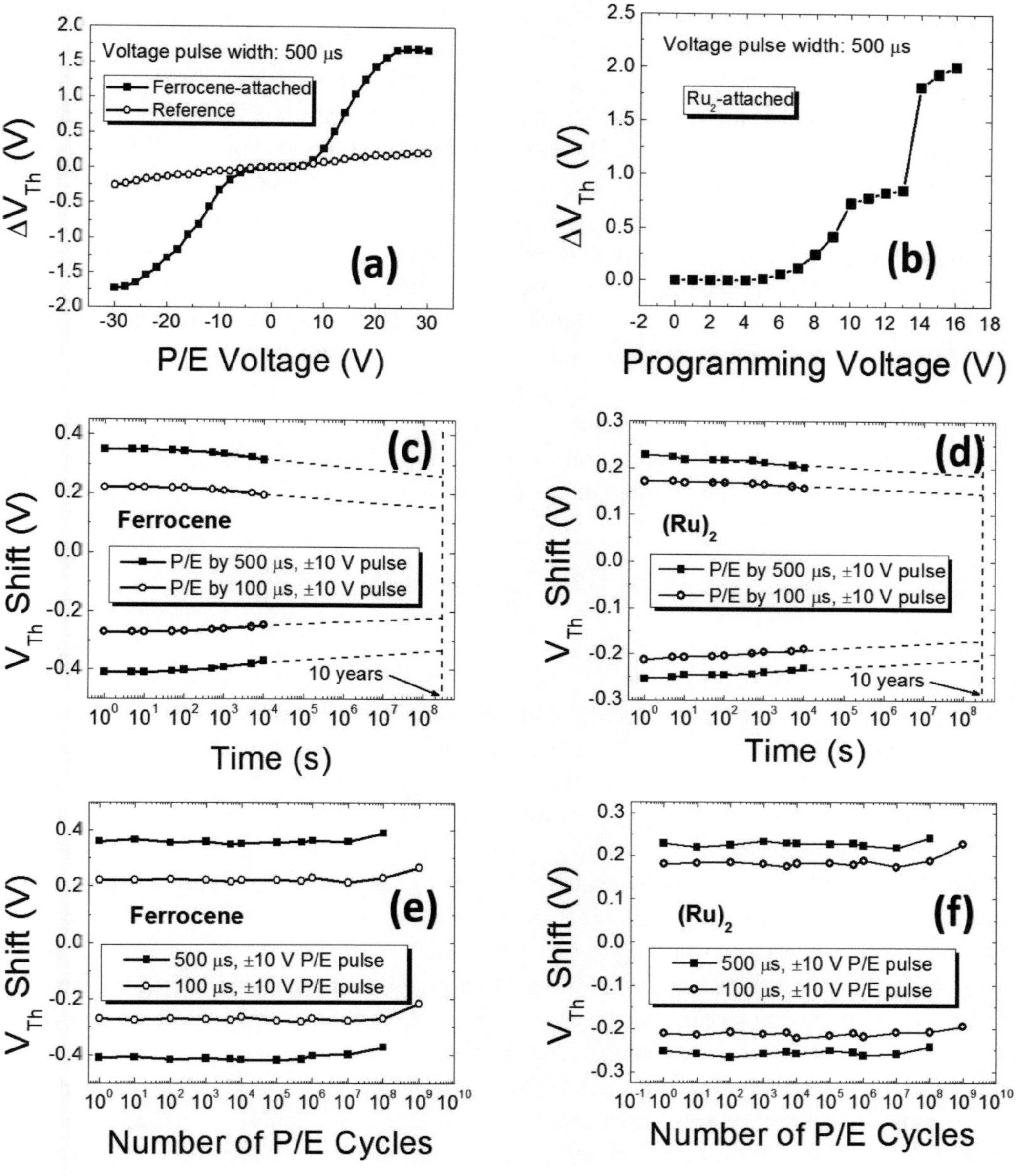

Figure 12. ΔV_{TH} of (a) ferrocene and reference sample and (b) Ru_2 molecular memory as a function of P/E gate voltage with 500 μs pulse width and increasing pulse height. Room temperature retention properties of (c) ferrocene and (d) Ru_2 molecular memory. The initial P/E pulses were ±10 V gate voltage with 500 and 100 μs pulse width, respectively. Endurance properties of (e) ferrocene and (f) Ru_2 molecular memory. Note that ±10 V gate voltage with 500 and 100 μs pulse width was applied [24].

to the high-quality tunnel oxide with clean solid/molecule and dielectric interfaces during the fabrication process. This further supports the mechanism of dominant charge storage located in the redox centers instead of interface states, since the recovery process of the interface states is quite fast leading to a poor retention. The endurance properties of the molecular memory devices are shown in **Figure 12e** and **f** [24]. Both devices exhibited excellent endurance characteristics with ±10 V P/E gate voltages and 500 and 100 μs pulse width. Negligible memory window degradation was observed after 10^8 P/E cycles by applying 500 μs P/E voltages. With shorter P/E voltages (100 μs), both devices still functioned perfectly after 10^9 P/E operation cycles. Such excellent endurance is 10,000 times better than that of the conventional floating-gate memory, and is resulted from the excellent reliability of the redox properties of the ferrocene and Ru_2 molecules [24]. However, less than 50% of the molecules in the SAM were effectively involved in the redox process, though the portion is slightly higher than that of the capacitor structure memory cell. Further research efforts are needed to increase the redox efficiency so as to lower the operation voltage and improve the operation speed. The memory density can be further increased with more carefully engineered molecules, and the demonstrated multibit memory concept is more reasonable and feasible than by just modulating the voltage level, as precise controlling of the charged states can be clearly defined and monitored with the physically discrete redox states.

5. Conclusions

The properties of redox-active molecules and the integration of molecular electronics in nonvolatile memory technology have been systematically discussed. So far, redox-active molecules have already shown their potential and advantageous properties for future low-power, high-density, and high-reliability nonvolatile memory. Solid-state integration of the redox molecules in flash-like nonvolatile memory devices enables the extension of the advantages afforded by the molecules to advanced electronic devices combined with universal semiconductor metrologies. Due to the intrinsic redox behavior of the molecules, large memory window with sufficient charge-trapping density can be achieved, and the $>10^9$ endurance cycles is about 10,000 times better than that of the conventional floating-gate memory. Furthermore, discrete multibit charge storage can be enabled by mixing various redox molecules or using molecules with multiple redox states. The current main barriers are the CMOS compatibility and the process issues introduced with the molecular integration. The realization of future molecular memory applications still requires a combination of empirical fabrication and rational designs for particular molecular electronics devices toward sophisticated application. Upon this, the molecular electronics will no doubt be shedding more new lights on the micro-/nanoelectronics society for creating next generation nonvolatile memory with enhanced performance.

Acknowledgements

Q. Li would like to acknowledge the support of Virginia Microelectronics Consortium (VMEC).

Author details

Hao Zhu[1]* and Qiliang Li[2]

*Address all correspondence to: hao_zhu@fudan.edu.cn

1 School of Microelectronics, Fudan University, Shanghai, PR China

2 Department of Electrical and Computer Engineering, George Mason University, Fairfax, VA, USA

References

[1] Dennard RH, Gaensslen FH, Yu HN, Rideout VL, Bassous E, Lebanc AR. Design of ion-implanted MOSFET's with very small physical dimensions. Proceedings of the IEEE. 1999;**87**:668-678

[2] Moore GE. Progress in digital integrated electronics. IEDM Technical Digest. 1975;11-13

[3] Sukegawa K, Yamaji M, Yoshie K, Furumochi K, Maruyama T, Morioka H, et al. High-performance 80-nm gate length SOI-CMOS technology with copper and very-low-k interconnects. Symposium on VLSI Technology, Digest of Technical Papers. 2000;186-189

[4] Chau R, Kavalieros J, Roberds B, Schenker R, Lionberger D, Barlage D, et al. 30 nm physical gate length CMOS transistors with 1.0 ps n-MOS and 1.7 ps p-MOS gate delays. IEDM Technical Digest. 2000;45-48

[5] Huang XJ, Lee WC, Kuo C, Hisamoto D, Chang LL, Kedzierski J, et al. Sub-50 nm p-channel FinFET. IEEE Transactions on Electron Devices. 2001;**48**:880-886

[6] Gan D, Hu C, Parker GE, Pao HH, Jolly G. n-p-n array yield improvement in a 0.18-μm deep trench SiGe BiCMOS process. IEEE Transactions on Electron Devices. 2012;**59**:590-595

[7] Horowitz M, Daily W. How scaling will change processor architecture. IEEE International Solid-State Circuits Conference, Digest of Technical Papers. 2004;**47**:132-133

[8] Chang L, Frank DJ, Montoye RK, Koester SJ, Ji BL, Coteus PW, et al. Practical strategies for power-efficient computing technologies. Proceedings of the IEEE. 2010;**98**:215-236

[9] Carley LR, Bain JA, Fedder GK, Greve DW, Guillou DF, Lu MSC, et al. Single-chip computers with microelectromechanical systems-based magnetic memory. Journal of Applied Physics. 2000;**87**:6680-6685

[10] Yoshida C, Tsunoda K, Noshiro H, Sugiyama Y. High speed resistive switching in Pt/TiO2/TiN film for nonvolatile memory application. Applied Physics Letters. 2007;**91**:223510

[11] Lankhorst MHR, Ketelaars B, Wolters RAM. Low-cost and nanoscale non-volatile memory concept for future silicon chips. Nature Materials. 2005;**4**:347-352

[12] Waser R, Dittmann R, Staikov G, Szot K. Redox-based resistive switching memories—nanoionic mechanisms, prospects, and challenges. Advanced Materials. 2009;**21**: 2632-2663

[13] Inoue N, Furutake N, Toda A, Tada M, Hayashi Y. PZT MIM capacitor with oxygen-doped Ru-electrodes for embedded FeRAM devices. IEEE Transactions on Electron Devices. 2005;**52**:2227-2235

[14] Meena JS, Sze SM, Chand U, Tseng T-Y. Overview of emerging nonvolatile memory technologies. Nanoscale Research Letters. 2014;**9**:526

[15] Baldi L, Bez R. The scaling challenges of CMOS and the impact on high-density non-volatile memories. Microsystem Technologies. 2007;**13**:133-138

[16] Mann B, Kuhn H. Tunneling through fatty acid salt monolayers. Journal of Applied Physics. 1971;**42**:4398-4405

[17] Aviram A, Ratner MA. Molecular rectifiers. Chemical Physics Letters. 1974;**29**:277-283

[18] Kuhn H, Mobius D. Systems of monomolecular layers-assembling and physico-chemical behavior. Angewandte Chemie International Edition in English. 1971;**10**:620-637

[19] Salvo B, Buckley J, Vuillaume D. Recent results on organic-based molecular memories. Current Applied Physics. 2011;**11**:e49-e57

[20] Heath JR. Molecular electronics. Annual Review of Materials Research. 2009;**39**:1-23

[21] Vuillaume D. Molecular nanoelectronics. Proceedings of the IEEE. 2010;**98**:2111-2123

[22] Heath J, Ratner M. Molecular electronics. Physics Today. 2003;**56**:43-49

[23] Liu Z, Yasseri A, Lindsey JS, Bocian D. Molecular memories that survive silicon device processing and real-world operation. Science. 2003;**302**:1543-1545

[24] Zhu H, Pookpanratana SJ, Bonevich JE, Natoli SN, Hacker CA, Ren T, Suehle JS, Richter CA, Li Q. Redox-active molecular nanowire flash memory for high-endurance and high-density nonvolatile memory applications. ACS Applied Materials & Interfaces. 2015;7:27306-27313

[25] Cummings SP, Savchenko J, Fanwick P. E, Kharlamova A, Ren T. Diruthenium alkynyl compounds with phosphonate capping groups. Organometallics. 2013;**32**:1129-1132

[26] Balakumar A, Lysenko A, Carcel C, Malinovskii V, Gryko D, Schweikart K, et al. Diverse redox-active molecules bearing o-, s-, or se-terminated tethers for attachment to silicon in studies of molecular information storage. Journal of Organic Chemistry. 2004;**69**:1435-1443

[27] Muthukumaran K, Loewe R, Ambroise A, Tamaru S, Li Q, Mathur G, et al. Porphyrins bearing arylphosphonic acid tethers for attachment to oxide surfaces. Journal of Organic Chemistry. 2004;**69**:1444-1452

[28] Schweikart K, Malinovskii V, Diers J, Yasseri A, Bocian D, Kuhr W. Design, synthesis, and characterization of prototypical multistate counters in three distinct architectures. Journal of Materials Chemistry. 2002;**12**:808-828

[29] Li J. Gryko D, Dabke R, Diers J, Bocian D, Kuhr W, Lindsey J. Synthesis of thiol-derivatized europium porphyrinic triple-decker sandwich complexes for multibit molecular information storage. Journal of Organic Chemistry. 2000;**65**:7379-7390

[30] Zhu H, Hacker CA, Pookpanratana SJ, Richter CA, Yuan H, Li H, et al. Non-volatile memory with self-assembled ferrocene charge trapping layer. Applied Physics Letters. 2013;**103**:053102

[31] Zhu H, Li Q. Novel molecular non-volatile memory: Application of redox-active molecules. Applied Sciences. 2016;**6**:7

[32] Li Q, Mathur G, Homsi M, Surthi S, Misra V, Malinovskii V, et al. Capacitance and conductance characterization of ferrocene-containing self-assembled monolayers on silicon surfaces for memory applications. Applied Physics Letters. 2002;**81**:1494-1496

[33] Li Q, Surthi S, Mathur G, Gowda S, Misra V, Sorenson TA. Electrical characterization of redox-active molecular monolayers on SiO_2 for memory applications. Applied Physics Letters. 2003;**83**:198-200

[34] Pavan P, Bez R, Olivo P, Zanoni E. Flash memory cells—An overview. Proceedings of the IEEE. 1997;**85**:1248-1271

[35] Atwood G. Future directions and challenges for ETox flash memory scaling. IEEE Transactions on Device and Materials Reliability. 2004;**4**:301-305

[36] Wu KH, Chien HC, Chan CC, Chen TS, Kao CH. SONOS device with tapered bandgap nitride layer. IEEE Transactions on Electron Devices. 2005;**52**:987-992

[37] Chen TS, Wu KH, Chung H, Kao CH. Performance improvement of SONOS memory by bandgap engineering of charge-trapping layer. IEEE Electron Device Letters. 2004;**25**:205-207

[38] Wrazien SJ, Zhao YJ, Krayer JD, White MH. Characterization of SONOS oxynitride nonvolatile semiconductor memory devices. Solid-State Electronics. 2003;**47**:885-891

[39] Chen Y, Ohlberg DAA, Li X, Stewart DR, Williams RS, Jeppesen JO, et al. Nanoscale molecular-switch devices fabricated by imprint lithography. Applied Physics Letters. 2003;**82**:1610-1612

[40] Collier CP, Wong EW, Belohradsky M, Raymo FM, Stoddart JF, Kuekes PJ, et al. Electronically configurable molecular-based logic gates. Science. 1999;**285**:391-394

[41] Collier CP, Mattersteig G, Wong EW, Yi L, Berverly K, Sampaio J, et al. A [2] catenane-based solid state electronically reconfigurable switch. Science. 2000;**289**:1172-1175

[42] Chen Y, Jung G-Y, Ohlberg DAA, Li X, Stewart DR, Jeppesen JO, et al. Nanoscale molecular-switch crossbar circuits. Nanotechnology. 2003;**14**:462-468

[43] Stewart DR, Ohlberg DAA, Beck PA, Chen Y, Williams RS, Jeppesen JO, et al. Molecule-independent electrical switching in Pt/organic monolayer/Ti devices. Nano Letters. 2004;**4**:133-136

[44] Lau CN, Stewart DR, Williams RS, Bockrath M. Direct observation of nanoscale switching centers in metal/molecule/metal structures. Nano Letters. 2004;**4**:569-572

[45] Tour JM, Zandt WL, Husband CP, Husband SM, Wilson LS, Franzon PD, et al. Nanocell logic gates for molecular computing. IEEE Transactions on Nanotechnology. 2002;**1**:100-109

[46] Tour JM, Cheng L, Nackashi DP, Yao YX, Flatt AK, St Angelo SK, et al. Nanocell electronic memories. Journal of the American Chemical Society. 2003;**125**:13279-13283

[47] Collet J, Vuillaume D. Nano-field effect transistor with an organic self-assembled monolayer as gate insulator. Applied Physics Letters. 1998;**73**:2681-2683

[48] Halik M, Klauk H, Zschieschang U, Schmid G, Dehm C, Schutz M, et al. Low-voltage organic transistors with an amorphous molecular gate dielectric. Nature. 2004;**431**:963-966

[49] Chen XL, Bao Z, Schon JH, Lovinger AJ, Lin YY, Crone B, et al. Ion-modulated ambipolar electrical conduction in thin-film transistors based on amorphous conjugated polymers. Applied Physics Letters. 2001;**78**:228-230

[50] Zhao Q, Luo Y, Surthi S, Li Q, Mathur G, Gowda S, et al. Redox-active monolayers on nano-scale silicon electrodes. Nanotechnology. 2005;**16**:257-261

[51] Li Q, Mathur G, Gowda S, Surthi S, Zhao Q, Yu L, et al. Multibit memory using self-assembly of mixed ferrocene/porphyrin monolayers on silicon. Advanced Materials. 2004;**16**:133-137

[52] Li Q, Surthi S, Mathur G, Gowda S, Zhao Q, Sorenson TA, et al. Multiple-bit storage properties of porphyrin monolayers on SiO_2. Applied Physics Letters. 2004;**85**:1829-1831

[53] Preiner MJ, Melosh NA. Creating large area molecular electronic junctions using atomic layer deposition. Applied Physics Letters. 2008;**92**:213301

[54] Shaw J, Zhong Y.-W, Hughes KJ, Hou T.-H, Raza H, Rajwade S, et al. Integration of self-assembled redox molecules in flash memory devices. IEEE Transactions on Electron Devices. 2011;**58**:826-834

[55] Shaw J, Xu Q, Rajwade S, Hou T.-H, Kan EC. Redox molecules for a resonant tunneling barrier in nonvolatile memory. IEEE Transactions on Electron Devices. 2012;**59**:1189-1198

[56] Xiang J, Liu W, Hu YJ, Wu Y, Yan H, Lieber CM. Ge/Si nanowire heterostructures as high-performance field-effect transistors. Nature. 2006;**441**:489-493

[57] Cui Y, Zhong ZH, Wang DL, Wang WU, Lieber CM. High performance silicon nanowire field effect transistors. Nano Letters. 2003;**3**:149-152

[58] Duan X, Huang Y, Lieber CM. Nonvolatile memory and programmable logic from molecule-gated nanowires. Nano Letters. 2002;**2**:487-490

[59] Li C, Fan W, Straus DA, Lei B, Asano S, Zhang D, et al. Charge storage behavior of nanowire transistors functionalized with Bis(terpyridine)-Fe(II) molecules: Dependence on molecular structure. Journal of the American Chemical Society. 2004;**126**:7750-7751

[60] Li C, Fan W, Lei B, Zhang D, Han S, Tang T, et al. Multilevel memory based on molecular devices. Applied Physics Letters. 2004;**84**:1949-1951

[61] Zhu H, Li Q, Yuan H, Baumgart H, Ioannou DE, Richter CA. Self-aligned multi-channel silicon nanowire field-effect transistors. Solid-State Electronics. 2012;**78**:92-96

Chapter 4

A New Control Strategy to Integrate Flow Batteries into AC Micro-Grids with High Wind Power Penetration

Leonardo Javier Ontiveros,
Gastón Orlando Suvire and Pedro Enrique Mercado

Additional information is available at the end of the chapter

http://dx.doi.org/10.5772/intechopen.69340

Abstract

The penetration of wind generation into AC micro-grids (MGs) has been increasing in recent years. Wind generation is uncontrollable, variable in nature, and uncertain. If the penetration level is high, the random variations of the wind power generation could cause problems for MGs to maintain the nominal system frequency. A typical solution is to employ energy storage systems (ESS) into the MG in order to compensate the wind power fluctuations. In this chapter, the use of a vanadium redox flow battery (VRFB) coupled with a power conditioning system (PCS) is suggested to enhance the frequency stability of a MG with high wind power penetration. A new control system is developed for the PCS/VRFB. The control system performs the load leveling of the wind generation and carries out the primary and secondary frequency control of the MG. Dynamic simulations of the proposed device are performed and demonstrate that the new control system improves the transient responses of the PCS/VRFB and the MG, during minor and/or severe disturbances.

Keywords: wind generation, AC micro-grids, energy storage, vanadium redox flow battery, power conditioning system, control system, frequency stability

1. Introduction

Wind generation is one of the most used renewable energy resources in AC micro-grids (MGs). Last year, wind power generation grew by 17% [1]. The incorporation of wind farms into the power system introduces new challenges to the operation, control, and planning of MGs, both for short-term intervals (seconds to minutes) and long-term intervals (minutes to hours). This is produced by the random variations of the wind resource. As a result, wind farms cannot be dispatched in the traditional sense. Conventional power plants struggle to meet variable demand, while wind turbines not. This leads to more problems into the MG. These problems increase when the penetration level is high, compromising the operation security and reliability of MGs [2, 3]. The traditional solution is to increase the reserve generation of conventional power plants, what causes a noneconomic dispatch. Other solutions are provided by the manufacturers of wind turbines, employing both the speed and the pitch control systems as methods of supplying power reserve for frequency support and power generation smoothing on variable-speed wind turbines [4–6]. However, these solutions depend on the operating point of the wind turbine and are less efficient than that provided by an energy storage system (ESS). The output power of wind generators is reduced in order to gain controllability, but sacrificing a part of the free wind energy and consequently increasing the more expensive output power of conventional generators to cover the same power demand.

ESS can be employed in order to improve the frequency stability of MGs with high wind power penetration. In this sense, several previous works focus on the problems caused by wind generation and how an ESS can effectively solve them [7–11]. However, these solutions are limited according to the amount of energy stored by the used ESS. Nowadays, a promising low-cost, large stationary advanced ESS for these applications is the vanadium redox flow battery (VRFB), which presents high-speed response and overload capacity characteristics [12–14].

The incorporation of the VRFB into MGs requires a power conditioning system (PCS) and appropriate control strategy to manage the power flow between the VRFB and the utility system [12, 13]. Many solutions using PCS for long-term ESS have been proposed and listed in Ref. [15], a bidirectional single-phase inverter with a DC/AC converter connects a battery energy storage system (BESS) to the AC grid. Additionally, an integrated nonlinear control strategy is proposed both to control converters and to manage the power flow direction between the BESS and a stiff grid. In Ref. [16], a DC-DC converter system is presented based on an input-series/output-parallel dual active bridge structure, in a full modular design. The proposed converter is dedicated to interface a DC voltage network with a battery-based energy storage device. In Ref. [17], a non-isolated online uninterruptible power supply (UPS) is presented. The proposed system consists of bridgeless boost rectifier, battery charger/discharger, and an inverter. For the inverter, a new control method is developed, which regulates the output voltage for both linear and nonlinear loads. In Ref. [18], a grid-connected hybrid PV-wind-battery system with a bidirectional DC/DC chopper is presented. The control strategy manages the power flow from different sources, provides generation reserve to the grid, charges the battery, and satisfies the load demand. The work presented in Ref. [19] proposes a PCS for zinc-bromine (Zn-Br) flow battery-based energy storage system. The PCS consists of four DC-DC converters, one DC-AC inverter, and a battery management system. The battery control strategy including the PCS and the stripping operation is described to perform the charging and the discharging of the flow battery in steady state.

The proposals mentioned above offer practical solutions to connect the VRFB to the AC network. However, much less has been done particularly on the utilization of the VRFB in emerging grid-interactive AC MGs, although major benefits apply [20]. Moreover, none of the aforementioned work has discussed strategies to stabilize the active power flow of wind farms employing long-term ESS and to contribute to the frequency control when faults occur in the electrical system.

Based on the previous discussion, this work employs the PCS presented by the same authors in Ref. [21] and develops a new control strategy, which performs the load leveling of the wind farm and the frequency regulation of the MG. The PCS consists of a 12-pulse thyristor converter with commutated capacitors (TCCC). A TCCC is a fast-response, solid-state power compensator that provides active power control at the point of connection to the MG for power quality improvements. This characteristic enables the TCCC/VRFB (1) to mitigate short-term and mid-term power fluctuations of wind turbines, (2) to correct frequency fluctuations of the MG, and (3) to provide the reserve generation required during minor and severe disturbances.

With the aim of demonstrating the validity of the three statements of the preceding paragraph, a model of a TCCC/VRFB device is presented with all its components represented in detail. Moreover, a new control design for this device is developed; this is the main contribution of this work. The control system implements a new approach based on multilevel control technique. To mitigate wind power fluctuations, the control includes a load-leveling compensator block. In addition, the control system incorporates a frequency regulator, which performs the primary and secondary frequency control of the MG (PFC and SFC, respectively). Validation of models and control schemes is carried out through simulations by using SimPowerSystems of Simulink/MATLAB™.

Considering the aspects mentioned above, this work is organized as follows: the VRFB and the proposed compensator are presented in Section 2. The new control system of the TCCC/VRFB is developed in Section 3. The model of the MG is presented in Section 4. Simulation tests are performed in Section 5. Finally, the conclusions are presented in Section 6.

2. The TCCC/VRFB compensator

The TCCC/VRFB compensator is composed of the vanadium redox flow battery and the thyristor converter with commutated capacitors; both of them are explained below.

2.1. Vanadium redox flow battery

The VRFB is an electrochemical ESS which consists of two electrolytes stored in two separate tanks. The electrolytes are formed by sulfuric acid with active vanadium species in different oxidation states: V4/V5 redox couple (catholyte) and V2/V3 redox couple (anolyte). These liquids circulate through the cell stack by pumps. The stack consists of several cells, each of which contains two half-cells that are separated by a proton exchange membrane (PEM). The electrochemical reactions take place in the half-cells, and then inert carbon felt polymer composite electrodes are employed in order to charge or discharge the battery through an external DC current. The general scheme of the VRFB is represented in **Figure 1(a)**.

The half-cell of the VRFB performs two simultaneous chemical reactions; these reactions occur on both sides of the membrane as shown in **Figure 1(b)**. During the charge process, the positive electrolyte (catholyte) delivers electrons to the negative electrolyte (anolyte) through

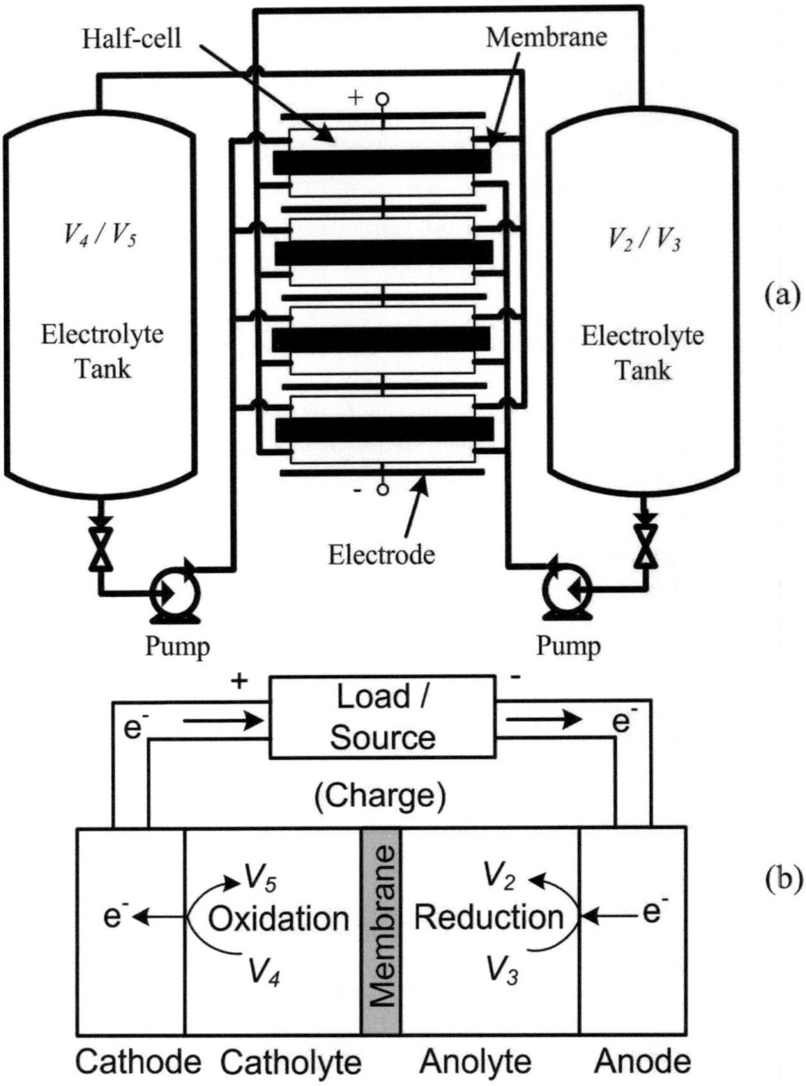

Figure 1. Vanadium redox flow battery: (a) general scheme and (b) chemical reactions.

the external circuit. Therefore, the oxidation occurs in the catholyte, and the reduction occurs in the anolyte. During the discharge process, the flow of electrons is reversed; the oxidation takes place in the anolyte and the reduction in the catholyte.

The battery produces a nominal cell potential of approximately 1.25 V. By connecting many cells into a "stack," the terminal voltage is obtained. The current density through the cell and the stack voltage establish the power available, while the supply of charged electrolyte to the stack establishes the energy available. So, the rated power and the energy stored can be upgraded by increasing or decreasing the stack and the electrolyte tank, respectively [22–24].

http://dx.doi.org/10.5772/intechopen.69340

2.2. Model of the VRFB

The VRFB model is composed by the stack model and the mechanical model, as shown in **Figure 2**. The complete model of the VRFB was developed in Ref. [22]. The stack model calculates the state of charge (*SOC*) of the electrolyte and the terminal stack voltage (U_{stack}). These values depend on the initial *SOC* and the stack current (I_{stack}). On the other hand, the mechanical model calculates the hydraulic losses caused by the electrolyte flow rate (Q). This flow rate is produced by two DC machines that drive the pumps. Therefore, the stack current is calculated from the difference between the terminal VRFB current (I_{VRFB}) and the pump current consumption (I_{pump}).

2.2.1. VRFB stack model

The proposed electrochemical model of the VRFB calculates two gains in order to obtain the stack voltage U_{stack} and the effective current I_{ef}. These parameters are the voltage and current gains (K_v, K_c), and they are obtained by solving (Eqs. (1) and (2)). Note that K_c and K_v gains depend on the stack current, the experimental efficiency curves (**Figure 3**) [23, 24], and the operating condition (charge or discharge):

$$K_{c,\,charge} = \eta_c K_{c,\,disch} = \eta_c \frac{(1+\eta_v)}{(1+\eta_e)} \tag{1}$$

$$K_{v,\,charge} = \frac{K_{v,\,disch}}{\eta_v} = \frac{(1+\eta_c)}{(1+\eta_e)} \tag{2}$$

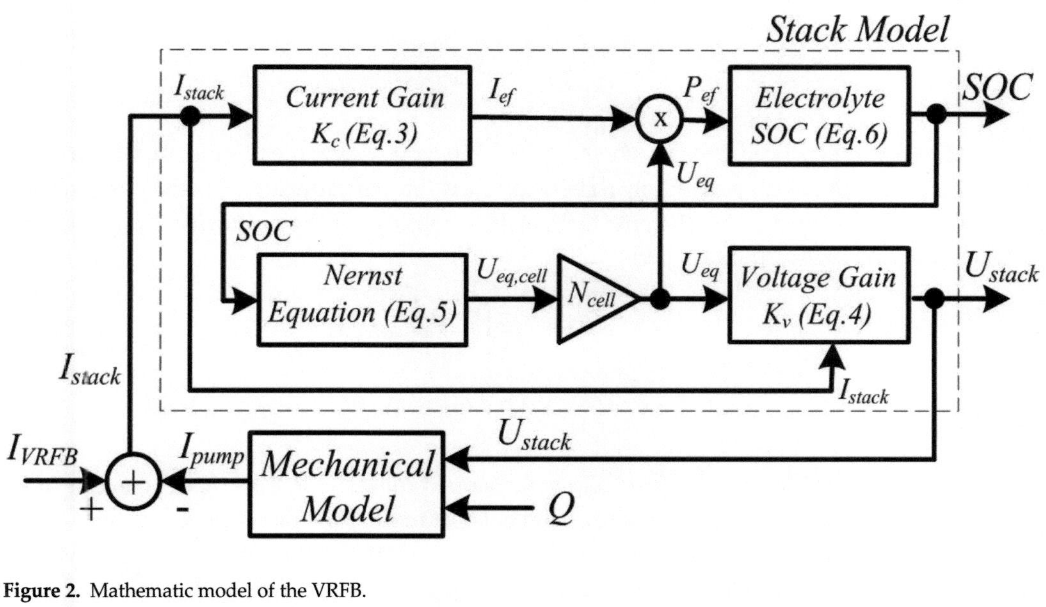

Figure 2. Mathematic model of the VRFB.

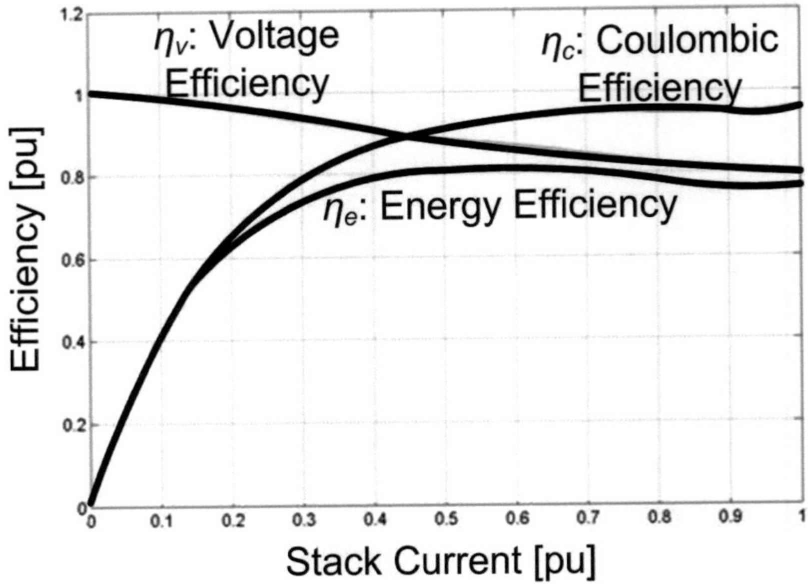

Figure 3. Voltage, coulombic, and energy efficiencies of the VRFB stack.

The effective stack current and the terminal stack voltage are obtained with (Eqs. (3) and (4)):

$$I_{ef} = K_c I_{stack} \tag{3}$$

$$U_{stack} = K_v U_{eq} \tag{4}$$

The equilibrium voltage (U_{eq}) can be calculated employing the Nernst equation (Eq. (5)) [25–28]:

$$U_{eq} = N_{cell}\left[U_0 + \frac{2RT}{F}\ln\left(\frac{SOC}{1-SOC}\right)\right] \tag{5}$$

where U_0 is the internal cell voltage when the SOC is 0.5 pu, N_{cell} is the number of cells in series connection, F is the Faraday constant, T is the absolute temperature, and R is the gas constant. The value of SOC depends on the effective stack power, the storage capacity (E_{max}), and the initial conditions (SOC_0) (Eq. (6)) [29]:

$$SOC = SOC_0 + \int \frac{-U_{eq} I_{ef}}{E_{max}} dt \tag{6}$$

2.2.2. *Mechanical model*

The mechanical model of the VRFB system is developed in Ref. [30]. This model consists of an analytical part and a numerical part. The analytical part models the pipes, bends, valves, tanks, and pumps. The numerical part describes the more complex stack hydraulic circuit. In addition to the mechanical model, in this work it is suggested to incorporate the equivalent DC

http://dx.doi.org/10.5772/intechopen.69340

current consumption of the pumps (I_{pump}). Therefore, the value of I_{pump} represents the mechanical losses of the VRFB system (Eq. (7)):

$$I_{pump} = \frac{P_{2pumps}}{U_{stack}} \quad (7)$$

2.3. Thyristor converter with commutated capacitors

A DC/AC bidirectional converter is required in order to connect the VRFB to the MG. In this regard, a 12-pulse thyristor converter with commutated capacitors on the AC side (TCCC) is proposed for controlling the active power flow of the VRFB [21]. The TCCC operates as a current source. Therefore, the polarization voltage of the VRFB stack determines the voltage at the DC bus (V_{conv}). As a result, the connection between the VRFB and the TCCC does not require a DC/DC chopper. **Figure 4** shows the proposed PCS unit for the VRFB system.

The most important aspects of the proposed PCS are:

- The converter can only draw an inductive current from the MG.
- A valve cannot be turned off actively. Instead, the current through one valve has to be brought to zero by turning on another valve.
- The reactive power consumption of the converter decreased by the additional voltage provided by the capacitors C_{con}. This leads to a reduction of the risk of commutation failures.

The TCCC contains tuned filters to smooth the AC current harmonics of order 12n ± 1. Additionally, the reactive power required by the thyristor converter is provided by these filters. From **Figures 2** and **4**, the following equalities are obtained:

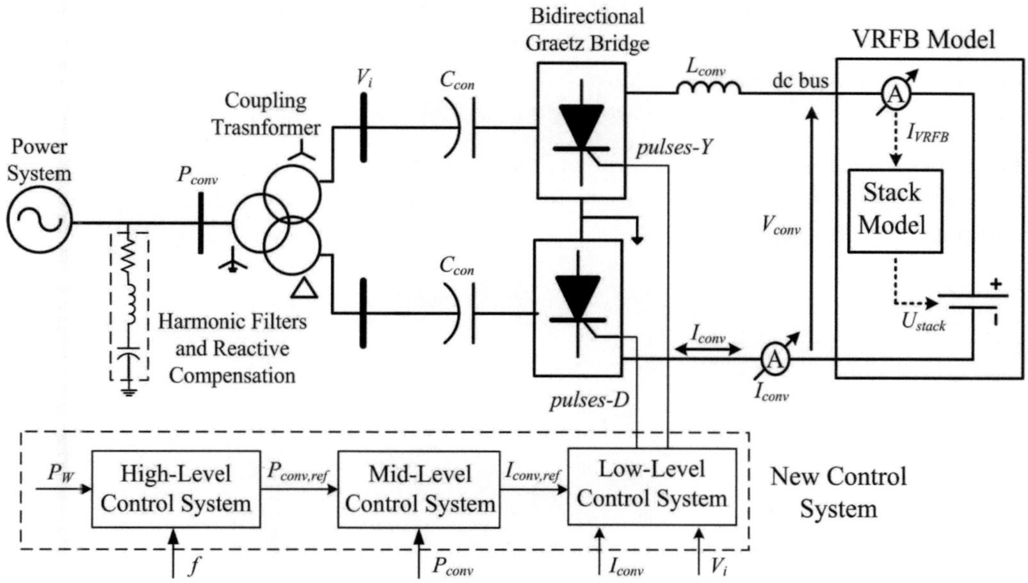

Figure 4. The proposed PCS unit (TCCC).

$$I_{VRFB} = I_{stack} + I_{pump} = I_{conv} \tag{8}$$

$$V_{conv} = U_{stack} \tag{9}$$

Eq. (8) shows that the VRFB terminal current is imposed by the converter current, whereas (Eq. (9)) demonstrates that the TCCC terminal voltage is the polarization voltage of the VRFB. The calculation of the TCCC current at the DC bus is obtained from Ref. [31]:

$$V_{conv} = b\frac{3}{\pi}\sqrt{2}V_{i,RMS}\cos\alpha - \frac{3}{\pi}(x_t - x_c)I_{conv} \tag{10}$$

where x_t and x_c are the transformer and capacitor (C_{con}) reactance, respectively; $V_{i,RMS}$ is the RMS voltage of the TCCC at the AC bus; α is the firing angle of the converter; and b is a sign factor which determines the operation mode of the converter: +1 for rectifier operation mode ($0 < \alpha < \pi/2$) and −1 for inverter operation mode ($\pi/2 < \alpha < \pi$). The DC current at VRFB terminals is obtained solving (Eqs. (8))–(10)):

$$I_{VRFB} = \frac{b}{(x_t - x_c)}\left(\sqrt{2}V_{i,RMS}\cos\alpha - \frac{\pi}{3}U_{stack}\right) \tag{11}$$

Considering constant values of $V_{i,RMS}$ and U_{stack}, I_{VRFB} decreases when α increases, while in rectifier operation mode, and I_{VRFB} increases when α increases during inverter operation mode.

3. Control system of the TCCC

In this section a new multilevel control system has been developed for the TCCC unit; this control system has its own control objectives for each hierarchical control level [9]. The proposed multilevel control system is shown in **Figure 5**, which is composed of three parts: a low-level control system, a mid-level control system, and a high-level control system.

The low-level and mid-level control systems are fully developed in Ref. [21], whereas the high-level control system (left side of **Figure 5**) is presented below which represents the main contribution of this work.

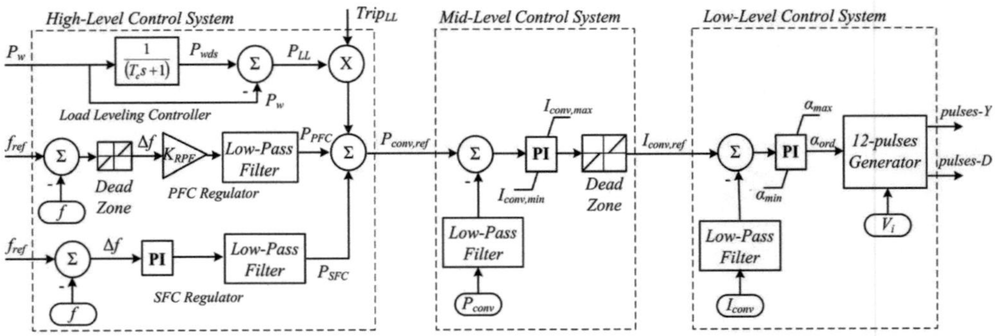

Figure 5. New control system of the TCCC.

http://dx.doi.org/10.5772/intechopen.69340

3.1. High-level control system

The high-level control system is responsible for determining the active power exchange between the TCCC and the utility system. The goal of the control algorithm is to perform the load leveling of the wind generation and to provide the generation reserve required for frequency regulation. This controller is composed of three parallel control loops: the load-leveling controller, the PFC regulator, and the SFC regulator. The aim of the load-leveling controller is to eliminate the turbulent component of the wind power (P_w). So, a first-order filter with a cutoff frequency at one cycle/hour is employed in order to obtain the diurnal and synoptic component of the wind power (P_{wds}) [32, 33]. The difference between P_{wds} and P_w is the compensation power P_{LL} that must be supplied or absorbed by the TCCC/VRFB unit. The load-leveling controller is always active, thus allowing for improvement in the power quality of the electric utility grid with wind generators.

On the contrary, the PFC and SFC regulators are only active when significant frequency deviations arise, contributing to recover the rated system frequency and thus improving the grid security. These controllers are in charge of minimizing the magnitude and duration of system disturbances by damping power oscillations. The purpose of this is to keep the system frequency between minimum and maximum levels during the transient dynamics. This is accomplished by measuring the frequency error Δf, which is proportional to the rate of change of the generator angle $d\delta/dt$ involved and consequently directly represents the power oscillation of the system. The value of Δf enters to proportional and proportional-integral (PI) control schemes, in order to calculate the generation reserve required to perform the PFC and SFC (P_{PFC}, P_{SFC}), respectively. In the PFC regulator, a dead-band block is incorporated within the control loop with the purpose of managing the participation of the TCCC/VRFB in the frequency control of the utility system. The low-pass filters are employed to eliminate the noise signal. Finally, the three components are added together to obtain the converter reference power $P_{conv,ref}$.

3.2. Mid-level and low-level control system

The mid-level control and the low-level control systems operate the TCCC as a controlled current source. The control algorithm is shown in **Figure 5**. The mid-level control system calculates the DC reference current ($I_{conv,ref}$) by comparing the reference power $P_{conv,ref}$ with the converter power (P_{conv}). The nonlinearity of the converter at low current is eliminated by a dead zone block. In the low-level control system, the reference firing angle (α_{ord}) for the thyristors is calculated from the reference current $I_{conv,ref}$ and the converter current (I_{conv}). Then, the value of α_{ord} enters to a 12-pulse generator in order to produce the firing pulses for the TCCC [31, 34].

4. Micro-grid model

A typical MG model shown in **Figure 6** is used for simulation studies; the SimPowerSystems® package of the MATLAB/Simulink software is employed to carry out the simulations. This model system is composed of a 132-kV-bulk power system (represented as an infinite AC bus)

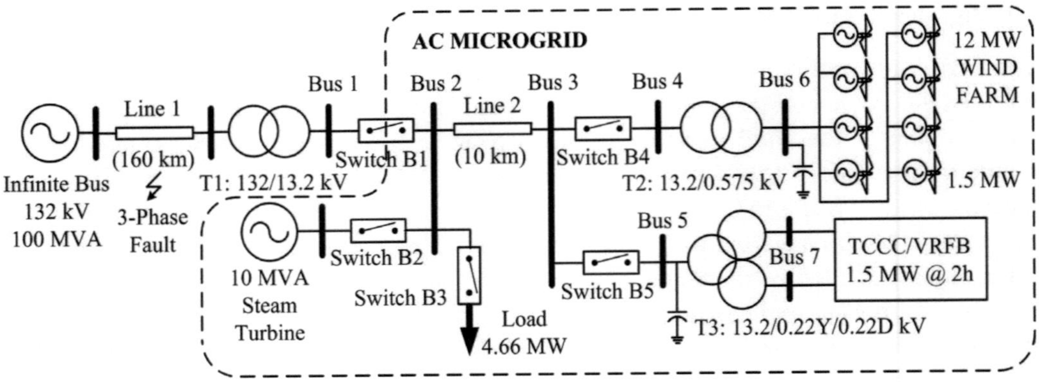

Figure 6. Single-line diagram of the test power system with an AC micro-grid including wind generation and the TCCC/VRFB system.

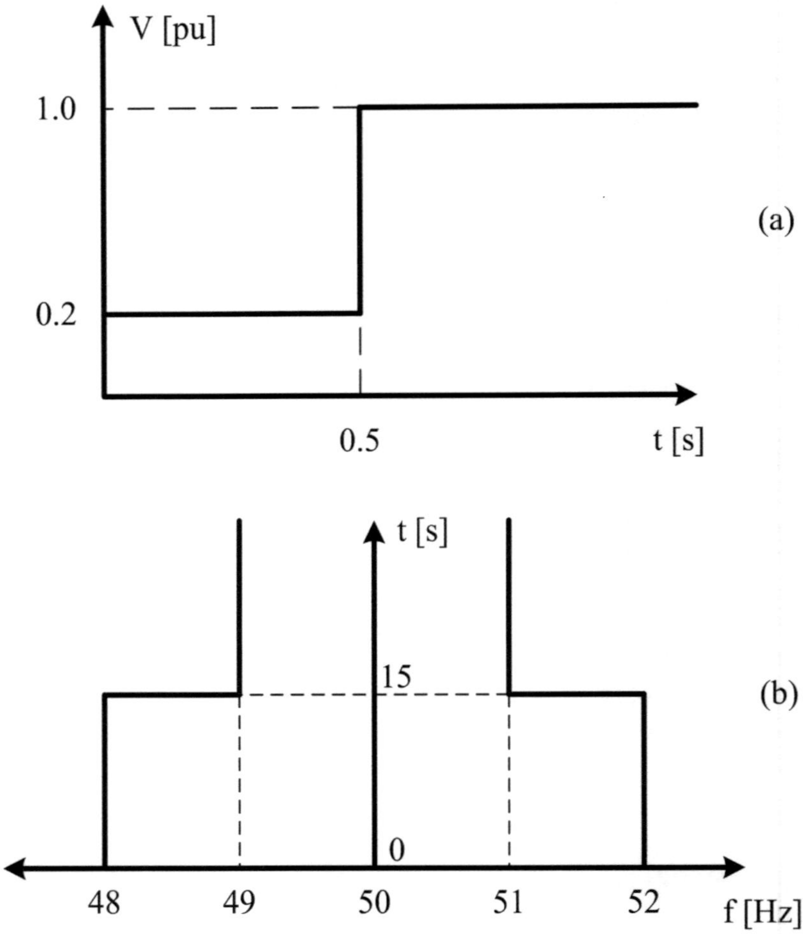

Figure 7. Capability curves of the wind farm: (a) voltage curve and (b) frequency curve.

connected to a 13.2 kV MG through a 160 km transmission line. The MG is composed of a 10 MVA steam turbine, a constant load, a 12 MW wind farm composed of eight double-fed induction generator (DFIG) wind turbines, and a 1.5 MW TCCC/VRFB unit. The energy and power ratings of the TCCC/VRFB unit are determined according to Refs. [35, 36] and proved through the simulations presented in Section V. The mathematical models of the TCCC and the steam turbine are detailed in the Appendix, and the wind farm model is developed in Refs. [37, 38].

The wind farm protection system is represented in **Figure 7**, which illustrates the voltage and frequency capability curves in **Figure 7(a) and (b)**, respectively. The protection system disconnects the wind farm when the terminal voltage is less than 0.2 pu during 0.5 s. On the other hand, frequency oscillations can also trip the wind farm. If the frequency of the MG is in the 49–48 or 51–52 Hz range during 15 s at least, the wind farm is switched off. If the frequency is lower than 48 Hz or higher than 52 Hz, the wind farm is instantaneously disconnected. The wind farm operates normally when the frequency of the MG is in the range of 49–51 Hz.

5. Simulation results

The frequency stability of the proposed MG is evaluated through simple events that impose high demands on dynamic response of the steam turbine and the TCCC/VRFB unit. In this regard, two case studies are considered. Case 1 evaluates the post-fault frequency stability of the MG when it is transmitting electric power to the 132 kV network. Case 2 examines the post-fault frequency stability of the MG when it is importing electric power from the bulk power system. In addition to the frequency stability analysis, Case 3 evaluates the quality power supply during the isolated operation of the MG. The under-frequency load shedding system, which could help in maintaining the frequency drop after contingencies within acceptable values, is not considered in all the cases aiming at highlighting the benefits of the proposed solution.

5.1. Case 1: unexpected disconnection of the MG during energy exportation

In this case, the frequency stability of the system is analyzed when an unexpected single-phase fault occurs at Line 1. In consequence, switch B1 opens 0.1 s after the occurrence of the event, causing the isolated operation mode of the MG. When the fault occurs at t = 150 s, the steam turbine and the wind farm are generating 9.5 and 3 MW, respectively. The load is 11 MW, so the power imbalance is 1.5 MW approximately.

Figure 8(a) shows the frequency deviation of the MG. Since the power generation is higher than the load, the frequency increases after the fault. When the control strategy developed in Ref. [21] is employed (previous control strategy), the frequency reaches the value of 52 Hz at t = 152.9 s; at this instant the wind farm switches off due to over-frequency. After that, the load is higher than the power generation, hence the system frequency decreases. The steam turbine switches off at t = 158.8 s due to under-frequency. Therefore, the MG collapses due to the lack of generation. A similar situation occurs when the new control strategy is applied and the rated power of the TCCC/VRFB is 1.0 MW. If the rated power of the TCCC/VRFB is 1.5 MW at least and the new control strategy is activated, the frequency reaches the value of 50.89 Hz at $t = 153.4$ s; after that the MG recovers the nominal value of the frequency (50 Hz).

Owing to the proportional characteristic of the PFC controller, the system frequency presents a steady-state error after the fault. In order to eliminate this error, the SFC controller participates in the frequency regulation. The steam turbine and the TCCC/VRFB compensator operate under this control scheme. These units restore the frequency after 70 s (**Figure 8(a)**).

The dynamic performance of the MG is represented in **Figure 8(b)–(d)**. After the fault at $t = 150$ s, the short-term reserve generation is activated in order to compensate the power unbalance of the MG. Since the power generation is higher than the load, the active power outputs of the steam turbine and the TCCC/VRFB unit decrease according to **Figure 8(b)** and **(d)**, respectively. According to **Figure 8(d)**, when the 1.0 MW TCCC/VRFB is operating in the saturation region, there is no frequency control. Hence, the wind farm and the steam turbine switch off at $t = 169$ and 175 s, respectively. The incorporation of the 1.5 MW TCCC/VRFB compensator and the proposed control strategy avoid the collapse of the MG, in accordance with **Figure 8(a)**.

5.2. Case 2: unexpected disconnection of the MG during energy importation

In this simulation, the frequency stability of the power system is analyzed when a fault occurs at line 1, whereas the MG is importing energy from the bulk power system. An unexpected single-phase fault occurs at $t = 150$ s; in a consequence switch B1 opens 0.1 s after the occurrence of the fault, causing the MG to operate in isolated mode. When the fault occurs, the

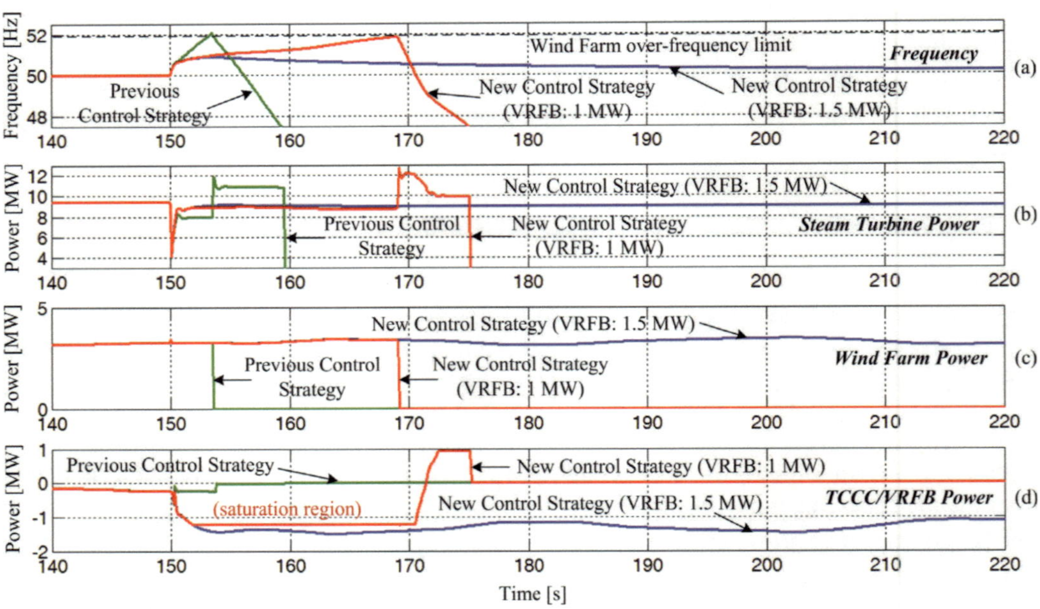

Figure 8. Dynamic response of the MG in Case 1: (a) frequency deviation, (b) steam turbine power, (c) wind farm active power, and (d) TCCC/VRFB active power.

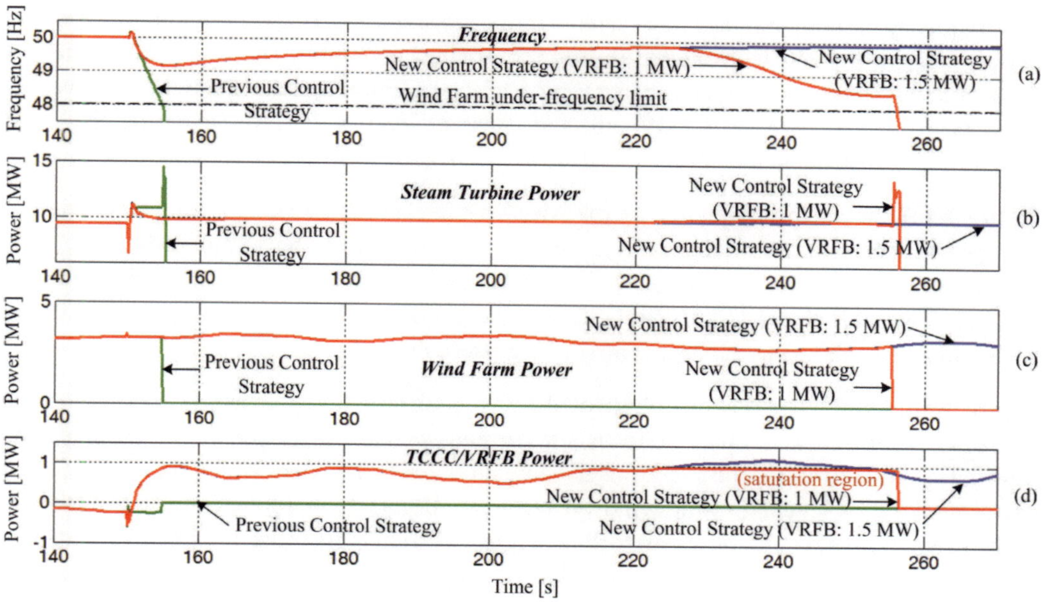

Figure 9. Dynamic response of the MG in Case 2: (a) frequency deviation, (b) steam turbine power, (c) wind farm active power, and (d) TCCC/VRFB active power.

steam turbine and the wind farm are generating 9.5 and 3 MW, respectively. The load is 14 MW, so the power unbalance is −1.5 MW approximately.

The frequency deviation of the MG is shown in **Figure 9(a)**. In this case, the load is higher than the power generation, so the frequency decreases after the fault. When the previous control strategy is employed, the frequency reaches the value of 48 Hz at $t = 155.6$ s; at this instant the wind farm switches off due to under-frequency. In a consequence, the system collapses owing to the lack of generation at $t = 156$ s. If the new control strategy is activated and the rated power of the TCCC/VRFB is at least 1.5 MW, the frequency reaches the value of 49.15 Hz at $t = 155.4$ s; after that the frequency asymptotically recovers its nominal value.

The dynamic response of the MG is shown in **Figure 9(b)–(d)**. Because the load is higher than the power generation, the short-term generation reserve is activated in order to compensate the power unbalance. Therefore, the steam turbine and the TCCC/VRFB increase the power output according to **Figure 9(b)** and **(d)**, respectively. Similarly in the previous case, when the 1.0 MW TCCC/VRFB is operating in the saturation region, there is no frequency control. Hence, the wind farm and the steam turbine switch off at $t = 256.2$ s. The incorporation of the 1.5 MW TCCC/VRFB unit and the proposed control strategy avoid the collapse of the MG as in the previous case (**Figure 9(a)**).

5.3. Case 3: isolated operation of the MG

Maintaining the voltage and frequency within certain limits in a MG is a basic operational requirement as many loads may be very sensitive to voltage and frequency deviations. Increasing wind penetration in the MG may lead to stability problems or produce unwanted voltage and frequency oscillation in isolated systems [2–4]. Bearing these aspects in mind, the dynamic performance of the MG is evaluated during the isolated operation. The operation conditions are as follows: the steam turbine is generating 4 MW (minimum operating power), the average wind farm active power is 3 MW, and the constant load is 7 MW. Therefore, the wind penetration is approximately 43%.

Figure 10(a) shows the frequency of the MG by considering two cases: with and without energy storage. When the TCCC/VRFB unit is not employed, the steam turbine presents problems in establishing the nominal value of the frequency as a result of random variations in wind power. When the TCCC/VRFB is used, the frequency deviations are less than 0.01 Hz approximately, enhancing the power quality supply of the MG.

Figure 10(b) shows the power output of the steam turbine and the wind farm. When the TCCC/VRFB unit is not employed, the steam turbine operates below the minimum operating power (4 MW) during several intervals. This situation is avoided when the TCCC/VRFB performs the load leveling of the wind power generation (**Figure 10(c)**), causing a reduction of the mechanical stress of the steam turbine. **Figure 10(d)** demonstrates that the steam turbine and the wind farm can control effectively the voltage of buses 2 and 3, respectively.

These simulations show that the TCCC/VRFB unit and the new control system enhance the dynamic response of MGs which incorporate wind generation, by performing the load leveling of wind turbines and carrying out the frequency regulation of the MG. It is important to notice that the DFIG wind farm and the TCCC/VRFB unit complement each other; the reactive power fluctuations generated by the TCCC are compensated by the wind farm, whereas the fluctuations of the wind power generation are smoothed by the TCCC/VRFB unit.

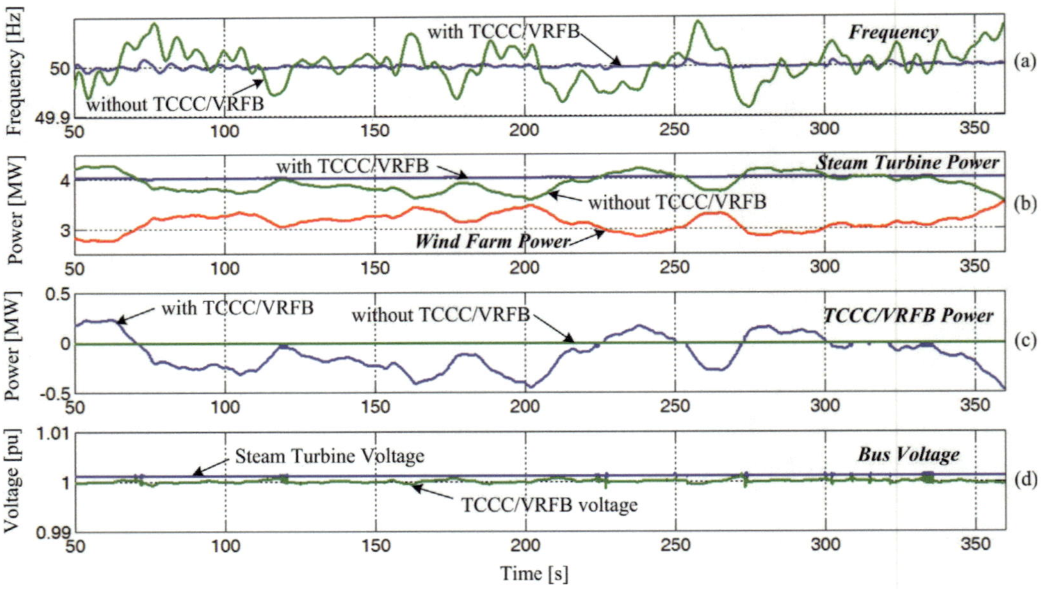

Figure 10. Dynamic response of the MG in Case 3: (a) frequency deviation, (b) steam turbine/wind farm power, (c) TCCC/VRFB active power, and (d) bus voltage.

http://dx.doi.org/10.5772/intechopen.69340

6. Conclusions

This work proposes a new control strategy for the TCCC/VRFB unit for enhancing the transient response of a MG with high wind power penetration. The control strategy has been implemented with a multilevel control topology. The high-level control system, which is the core of the contribution, sets the power that exchanges the TCCC/VRFB with the MG. The aims of this power flow are (1) to smooth the active power generated by a DFIG wind farm and (2) to provide the generation reserve required for PFC and SFC. The model aspects of the VRFB and the TCCC unit are also presented.

From the results obtained, it can be concluded that the developed control algorithms work satisfactorily. The TCCC/VRFB unit, with the operation of the load-leveling controller, effectively compensates the active power fluctuations from a wind farm. The complete system (wind farm plus TCCC/VRFB) generates a smoother power response than that of the system without the TCCC/VRFB. Moreover, the complete control system, now with the operation of the PFC and SFC controllers, also contributes to the recovery of the frequency when severe disturbances occur in the MG. Therefore, the incorporation of the TCCC/VRFB compensator and its new control system improves the integration of wind turbines into AC MGs.

Acknowledgements

This work has been supported by ELAC2014/ESE0034 from the European Union and its linked Argentinean national project 018/16, Res. MINCYT 542/15. We also appreciate the support from the Argentinean National Agency for the Promotion of Science and Technology (ANPCyT) through project PICT-2012-1733, the National Scientific and Technical Research Council (CONICET), and the National University of San Juan (UNSJ).

Appendix

TCCC model

The TCCC is composed of two bidirectional Graetz bridges, a coupling transformer with three windings, harmonic filters, and passive elements, as shown in **Figure 4**. The model of the thyristor is composed of an inductor L_{tyr}, a resistor R_{tyr}, and a DC voltage source that represents the forward voltage V_{ft}. These components are connected in series. A logical signal that depends on the gate signal g, the current I_{AK}, and the voltage V_{AK} controls the switch. A series R_{snub}-C_{snub} snubber circuit is connected in parallel with the thyristor model. **Figure 11(b)** shows the static VI characteristic of the model. The thyristor and TCCC parameters are shown in **Tables 1** and **2**.

Steam turbine model

According to **Figure 12**, the steam turbine model is composed of a four-stage steam turbine, a shaft with up to four masses, a speed governing system, and an automatic generation control (AGC) system. The speed governing system is composed of a servomotor that regulates the

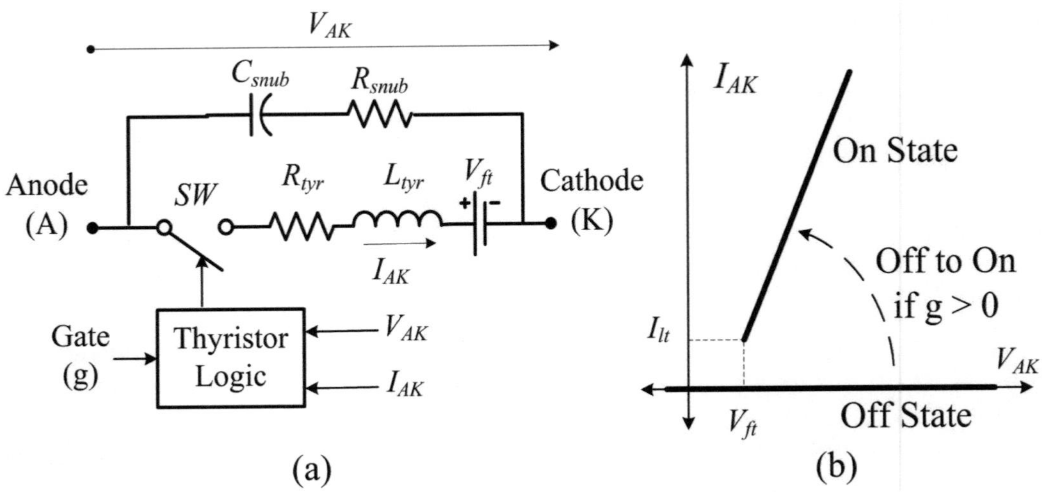

Figure 11. (a) Thyristor model and (b) static VI characteristic of the thyristor.

R_{tyr} (mΩ)	L_{tyr} (mH)	V_{ft} (V)	C_{snub} (nF)	R_{snub} (Ω)
1	0	2	100	2000

Table 1. Thyristor parameters.

Harmonic	Harmonic filters			Converter	
	L_{filt} (mH)	R_{filt} (Ω)	C_{filt} (nF)	C_{con} (mF)	L_{conv} (mH)
11th	117.52	4.06	712.4	30	7
13th	84.14	3.43	712.4	–	–
23–25th	24.68	558.46	712.4	–	–

Table 2. TCCC parameters.

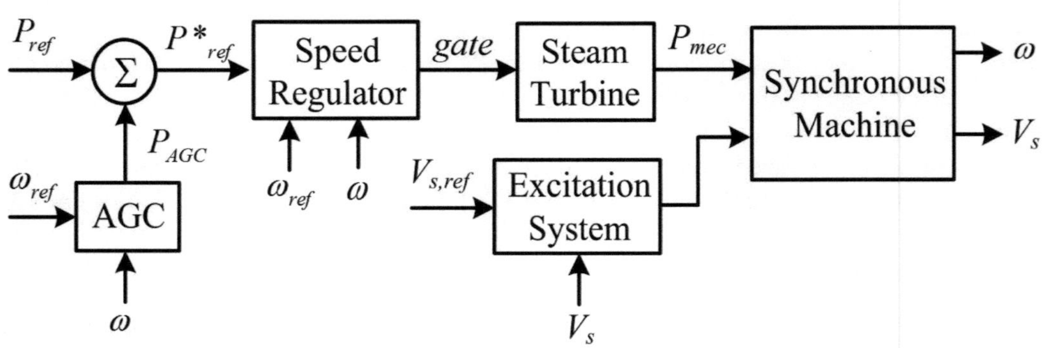

Figure 12. Steam turbine model.

turbine valve, a proportional regulator, and a speed relay. A PI regulator performs the secondary frequency control at the AGC subsystem. In this work, the load variations of the steam turbine are less than 4%; therefore, the boiler pressure is assumed constant during the simulation time [39].

Author details

Leonardo Javier Ontiveros*, Gastón Orlando Suvire and Pedro Enrique Mercado

*Address all correspondence to: ontiveros@iee.unsj.edu.ar

National Scientific and Technical Research Council (CONICET), San Juan, Argentina

References

[1] World Wind Energy Association Website. www.wwindea.org [Internet]. June 2016. Available from: http://www.wwindea.org/the-world-sets-new-wind-installations-record-637-gw-new-capacity-in-2015/ [Accessed: June 2016]

[2] Chandra D, Grimaccia F, Kumari M, Sydulu M, Mussetta M. Transient stability analysis of power system with grid integration of wind generation. International Review of Electrical Engineering (IREE). 2015;**10**(3):442-448. DOI: 10.15866/iree.v10i3.6037

[3] Daneshi H, Srivastava A. Impact of battery energy storage on power system with high wind penetration. In: IEEE PES, editor. Transmition and Distribution Conference and Exposition T&D 2012; May 2012; Orlando, FL, USA. Orlando, FL, USA: IEEE PES; 2012. pp. 1-8

[4] Chakib R, Essadki A, Cherkaoui M. Participation of DFIG wind turbine controlled by active disturbance rejection control in primary frequency control. International Review of Electrical Engineering (IREE). 2016;**11**(2):183-192. DOI: 10.15866/iree.v11i2.8150

[5] Vestas Wind Systems Website. www.vestas.com [Internet]. June 2016. Available from: http://www.vestas.com [Accessed: June 2016]

[6] GE Enegy Website. www.ge-energy.com [Internet]. June 2016. Available from: http://www.ge-energy.com/products_and_services/products/ [Accessed: June 2016]

[7] Xu Q, Zang H, Shi L, Du P. Researches on power system low-frequency oscillations damping with FESS. International Review of Electrical Engineering (IREE). 2011;**6**(5):2537-2544. DOI: 10.1109/ICPST.2006.321886

[8] Gan C, Prayun W, Tan C, Shamshiri M. Design of a hybrid diesel/PV/wind/battery system in remote areas. International Review of Electrical Engineering (IREE). 2014;**9**(2):420-430

[9] Suvire G, Mercado P. Active power control of a flywheel energy storage system for wind energy applications. IET Renewable Power Generation. 2012;**6**(1):9-16. DOI: 10.1049/iet-rpg.2010.0155

[10] Molina M, Mercado P. Power flow stabilization and control of microgrid with wind generation by superconducting magnetic energy storage. IEEE Transactions on Power Electronics. 2011;**26**(3):910-922. DOI: 10.1109/TPEL.2010.2097609

[11] Suvire G, Molina M, Mercado P. Improving the integration of wind power generation into AC microgrids using flywheel energy storage. IEEE Transactions on Smart Grid. 2012;**3**(4):1945-1954. DOI: 10.1109/TSG.2012.2208769

[12] Mohamed M, Ahmad H, Seman M. State of the art of all-vanadium redox flow battery: A review on research prospects. International Review of Electrical Engineering (IREE). 2012;**7**(5):5610-5622

[13] Molina M, Suvire G, Ontiveros L, Mercado P. Emerging energy storage technologies in utility power systems: A technical insight. In: Rose M, editor. Energy Storage. 1st ed. New York: Nova Science Publishers Press; 2011. pp. 235-312

[14] Gerami Moghaddam I, Fallahi F. Self scheduling program for VRB energy storage system in competitive electricity market. International Review of Electrical Engineering (IREE). 2010;**5**(4):1707-1714. DOI: 10.1109/POWERCON.2010.5666037

[15] Braier C, Torres M, Muñoz J, Rivera M, Espinosa E, Acuña P. Bidirectional power flow control of a single-phase current-source grid-tie battery energy storage system. In: IEEE Latin America, editor. 2015 IEEE 24th International Symposium on Industrial Electronics (ISIE); June 2015; Búzios, Brazil. Búzios, Brazil: IEEE Latin America; 2015. pp. 1372-1377

[16] Barrade P, Rufer A. Control of a modular DC-DC converter dedicated to energy storage. In: Power Electronics and Applications (EPE'15 ECCE-Europe); September 2015; Geneva, Switzerland: IEEE; 2015. pp. 1-9

[17] Aamir M, Mekhilef S. online transformerless uninterruptible power supply (UPS) system with a smaller battery bank for low power applications. IEEE Transactions on Power Electronics. 2017;**32**(1):233-247. DOI: 10.1109/TPEL.2016.2537834

[18] Mangu B, Akshatha S, Suryanarayana D, Fernandes B. Grid-connected PV-wind-battery based multi-input transformer coupled bidirectional DC-DC converter for household applications. IEEE Journal of Emerging and Selected Topics in Power Electronics. 2016;**4**(3):1086-1095. DOI: 10.1109/JESTPE.2016.2544789

[19] Lim J, Lee S, Kang K, Cho Y, Choe G. A modular power conversion system for zinc-bromine flow battery based energy storage system. In: IEEE, editor. Future Energy Electronics Conference (IFEEC); November 2015; Taipei, Taiwan. Taiwan: IEEE; 2015. pp. 1-5

[20] Darabi A, Hosseina M, Gholami H. Vanadium redox flow battery control in flexible microgrids. International Review of Electrical Engineering (IREE). 2013;**8**(4):1341-1348

[21] Ontiveros L, Mercado P. Thyristor-based flexible AC transmission system for controlling the vanadium redox flow battery. IET Renewable Power Generation. 2013;**7**(3):201-209. DOI: 10.1049/iet-rpg.2012.0361

[22] Ontiveros L, Mercado P. Modeling of a vanadium redox flow battery for power system dynamic studies. International Journal of Hydrogen Energy. 2014;**39**(16):8720-8727. DOI: 10.1016/j.ijhydene.2013.12.042

[23] Wang G. Minimising output power fluctuation of large photovoltaic plant using vanadium redox battery storage. In: IET, editor. Power Electronics, Machines and Drives (IET/PEMD); March 2012; Bristol, UK. UK: IET; 2012. pp. 1-6

[24] Wang G. PV power plant using hybrid energy storage system with improved efficiency. In: IEEE, editor. Power Electronics for Distributed Generation Systems (IEEE/PEDG); June 2012; Aalborg, Denmark. Denmark: IEEE; 2012. pp. 808-813

[25] Chahwan J, Abbey C, Joos G. VRB modelling for the study of output terminal voltages, internal losses and performance. In: Electric Power Conference (EPC); October 2007; Montreal, Canada. Canada: IEEE; 2007. pp. 387-392

[26] Skyllas-Kazacos M, Menictas C. The vanadium redox battery for emergency back-up applications. In: Int. Telecommunications Energy Conference (INTELEC); October 1997; Melbourne, Australia. Australia: IEEE; 1997. pp. 463-471

[27] Mousa A. Chemical and electrochemical studies of V(III) and V(II) solutions in sulfuric acid solution for vanadium battery applications [thesis]. New South Wales, Australia: University of South Wales (UNSW); 2003. 632 p. Available from: http://handle.unsw.edu.au/1959.4/54984

[28] Tsuda I, Kurokawa K, Nozaki K. Development of intermittent redox flow battery for PV system. In: World Conference on Photovoltaic Energy Conversion (WCPEC); December 1994; Waikoloa, HI, USA. USA: IEEE; 1994. pp. 946-949

[29] Wang W. Grid-connected wind farm power control using VRB-based energy storage system. In: Energy Conversion Congress and Exposition (IEEE/ECCE); August 2010; Delft, Netherlands. Netherlands: IEEE; 2010. pp. 3772-3777

[30] Blanc C, Ruffer A. Multiphysics and energetic modeling of a vanadium redox flow battery. In: International Conference on Sustainable Energy Technologies (ICSET); October 2008; Singapore. Singapore: IEEE; 2008. pp. 759-764

[31] Sood V. HVDC transmission. In: Rashid M, editor. Power Electronic Handbook. 1st ed. Cambridge, UK: Academic Press; 2001. pp. 575-596

[32] Söder L, Ackermann T. Wind power in power systems: An introduction. In: Ackermann T, editor. Wind Power and Power System. 1st ed. New Jersey, USA: John Wiley and Sons Press; 2005. pp. 33-39. DOI: 10.1002/9781119941842.ch4

[33] Van der Hoven I. Power spectrum of horizontal wind speed in the frequency range from 0.0007 to 900 cycles per hour. Journal of Meteorology. 1956;**14**(2):160-164. DOI: 10.1175/1520-0469(1957)014<0160:PSOHWS>2.0.CO;2

[34] Matlab and Simulink Website. www.mathworks.com [Internet]. April 2016. Available from: http://www.mathworks.com/help/physmod/powersys/ref/synchronized12pulsegenerator.html [Accessed: April 2016]

[35] Kerdphol T, Qudaih Y, Hongesombut K, Watanabe M, Mitani Y. Intelligent determination of a battery energy storage system size and location based on rbf neural networks for microgrids. International Review of Electrical Engineering (IREE). 2016;**11**(1):78-87. DOI: 10.15866/iree.v11i1.7718

[36] Martinez M, Mercado P, Molina M. Determinación del Tamaño y Ubicación Óptimo de un Sistema de Almacenamiento de Energía en un Sistema de Potencia para Regulación Secundaria de Frecuencia. In: ERIAC/CIGRE, editor. Décimo Quinto Encuentro Regional Ibero-americano del CIGRÉ; May 2013; Foz de Iguazú, Brasil. Brasil: ERIAC/CIGRE; 2013. pp. 1-8

[37] Ontiveros L, Mercado P, Suvire G. A new model of the double-fed induction generator wind turbine. In: IEEE/PES, editor. Transmission and Distribution Conference and Exposition: Latin America (T&D-LA); November 2010; Sao Paulo, Brazil. Brazil: IEEE/PES; 2010. pp. 263-269

[38] Ontiveros L, Mercado P. Modeling of variable-speed wind farms for power systems dynamic studies. In: CIGRE/ERIAC, editor. Décimo Cuarto Encuentro Regional Ibero-americano de Cigré; May 2011; Ciudad del Este, Paraguay. Paraguay: CIGRE/ERIAC; 2011. pp. 1-8

[39] Silvestri G. Steam turbines. In: Elliot T, editor. Standard Handbook of Powerplant Engineering. 1st ed. New York, USA: McGraw-Hill Press; 1989. pp. 2.25-2.61. DOI: 10.1036/0070194351

Chapter 5

Redox Flow Batteries: Fundamentals and Applications

Ruiyong Chen, Sangwon Kim and Zhenjun Chang

Additional information is available at the end of the chapter

http://dx.doi.org/10.5772/intechopen.68752

Abstract

A redox flow battery is an electrochemical energy storage device that converts chemical energy into electrical energy through reversible oxidation and reduction of working fluids. The concept was initially conceived in 1970s. Clean and sustainable energy supplied from renewable sources in future requires efficient, reliable and cost-effective energy storage systems. Due to the flexibility in system design and competence in scaling cost, redox flow batteries are promising in stationary storage of energy from intermittent sources such as solar and wind. This chapter covers basic principles of electrochemistry in redox flow batteries and provides an overview of status and future challenges. Recent progress in redox couples, membranes and electrode materials will be discussed. New demonstration and commercial development will be addressed.

Keywords: electrochemical energy storage, redox couples, electrolytes, electrodes, membranes

1. Introduction

Global economic growth with the increasing release of carbon dioxide disrupts our ecosphere and causes significant impacts on climate change. An environmentally friendly route to generate electricity from renewable sources such as wind and solar is desirable. To promote the utilization of renewable and sustainable energy and to enhance the stability of grid networks, energy storage systems are needed to store surplus electricity. The stored energy can be then delivered to end customers or to power grids upon need. It is becoming clear that the electrochemical energy storage using rechargeable batteries based on redox chemistry can provide a central solution to tackle such an issue. Through storing energy in recirculating liquid electrolytes, redox flow batteries have merits of decoupled energy density (tank size, electrolyte concentration, cell voltage and number dependent) and power generation capability (electrode size and reaction kinetics dependent). In terms of cost, system flexibility, quick response and safety concerns for large-scale applications, redox flow batteries show great advantages over other types of batteries such as lead-acid and lithium-ion batteries and are expected to have increasing commercial space through technological development in future. Therefore, the redox chemistry and technical fundamentals of flow batteries, which determine the technological success and market penetration, need to be well understood.

2. Classic vanadium redox flow batteries

Among various flow batteries, vanadium redox flow battery is the most developed one [1]. Large commercial-scale vanadium redox flow batteries are currently in construction. The structure and charge-discharge reactions of vanadium redox flow batteries are schematically shown in **Figure 1**. During discharging, reduction occurs at the cathode and oxidation occurs at the anode as shown in Eqs. (1)–(3) (discharge: →, charge: ←). While these redox reactions occur, proton ions diffuse across the membrane and electrons transfer through an external circuit.

$$\text{Negative electrode}: V^{2+} \leftrightarrow V^{3+} + e^- \tag{1}$$

$$\text{Positive electrode}: VO_2{}^+ + 2H^+ + e^- \leftrightarrow VO^{2+} + H_2O \tag{2}$$

$$\text{Overall reaction}: V^{2+} + VO_2{}^+ + 2H^+ \leftrightarrow VO^{2+} + V^{3+} + H_2O \tag{3}$$

The standard cell voltage for the all-vanadium redox flow batteries is 1.26 V. At a given temperature, pH value and given concentrations of vanadium species, the cell voltage can be calculated based on the Nernst equation:

$$E = 1.26\text{ V} - RT/F \ln\left(\left[VO^{2+}\right]\cdot\left[V^{3+}\right]\right)/\left(\left[VO_2{}^+\right]\cdot\left[H^+\right]^2\cdot\left[V^{2+}\right]\right) \tag{4}$$

where R, T and F are the universal gas constant, absolute temperature and Faraday constant, respectively. The crossover of vanadium ions through the membrane may occur, resulting in self-discharge with the unwanted mixing of vanadium species at both sides of the cell, as following [2]:

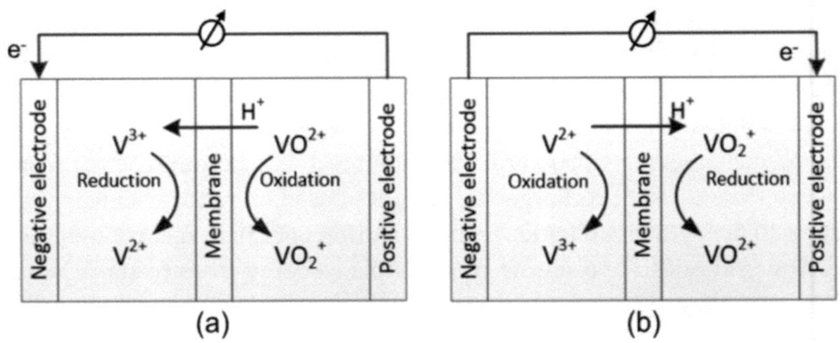

Figure 1. A schematic of a vanadium redox flow battery: (a) charge reaction and (b) discharge reaction.

At the negative electrode:

$$V^{2+} + VO^{2+} + 2H^+ \rightarrow 2V^{3+} + H_2O \tag{5}$$

$$2V^{2+} + VO_2{}^+ + 4H^+ \rightarrow 3V^{3+} + 2H_2O \tag{6}$$

$$V^{3+} + VO_2{}^+ \rightarrow 2VO^{2+} \tag{7}$$

At the positive electrode:

$$V^{2+} + 2VO_2{}^+ + 2H^+ \rightarrow 3VO^{2+} + H_2O \tag{8}$$

$$V^{3+} + VO_2{}^+ \rightarrow 2VO^{2+} \tag{9}$$

$$V^{2+} + VO^{2+} + 2H^+ \rightarrow 2V^{3+} + H_2O \tag{10}$$

Side reactions such as hydrogen evolution due to water decomposition and CO_2 evolution due to the oxidation of carbon-based electrode may occur during operation [3]. The battery performance is generally evaluated with three efficiencies: coulombic efficiency (CE), voltage efficiency (VE) and energy efficiency (EE), which are defined as following:

$$CE = \frac{\text{discharge capacity}}{\text{charge capacity}} \times 100\% \tag{11}$$

$$VE = \frac{\text{average discharge voltage}}{\text{average charge voltage}} \times 100\% \tag{12}$$

$$EE = CE \times VE \tag{13}$$

The CE reduces because of crossover of vanadium ions during cell operation and side reactions. The VE is related to the operation current density, ionic conductivity of membrane, electrode materials, flow rate and mass transport of electrolyte. As current density increases, the VE reduces due to the increase in polarization.

3. Types and configurations of redox flow batteries

Conventional redox flow batteries have two divided electrolyte reservoirs (**Figure 2a**). Catholyte and anolyte are separated by a membrane, which permits ions to pass through it. The most common working ions in aqueous media, H^+ (349.8 S cm^2 mol^{-1}) and OH^- (198.0 S cm^2 mol^{-1}), have the highest limiting molar conductivity among all known cations and anions, respectively [4]. All-vanadium redox flow batteries, for instance, have V^{3+}/V^{2+} redox reactions on the negative side (anolyte) and $VO_2{}^+/VO^{2+}$ on the positive side (catholyte). Such battery uses the same metal ions on both sides. Crossover of metal ions through the membrane will then not cause contamination of the electrolyte. In contrast, for redox flow batteries with different metal ions such as Fe^{3+}/Fe^{2+} and Cr^{3+}/Cr^{2+} in an iron-chromium flow battery, the cross-contamination via ion penetration may cause irreversible performance loss.

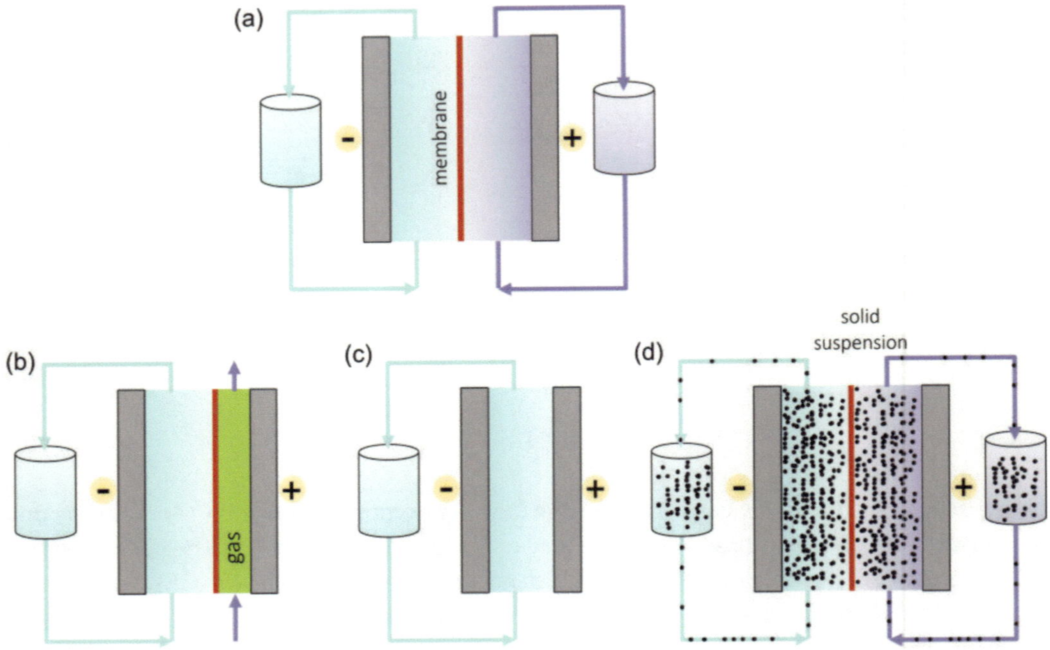

Figure 2. Configurations of (a) a conventional redox flow battery with two divided compartments containing dissolved active species, (b) a hybrid redox flow battery with gas supply at one electrode, (c) a redox flow battery with membrane-less structure and (d) a redox flow battery with solid particle suspension as flowing media.

Hybrid redox flow batteries such as zinc-bromine and zinc-cerium systems use metal stripping/plating reactions (Zn^{2+}/Zn, −0.76 V vs. [standard hydrogen electrode] SHE) on one of the electrodes inside the cell and the other side with normal soluble flowing electrolyte. Similarly, redox flow lithium batteries in non-aqueous electrolytes have been explored to make use of the low redox potential of Li^+/Li couple (−3.04 V vs. SHE). During charging, zinc or lithium is deposited from the electrolyte and during discharging, Zn^{2+} or Li^+ dissolves into the solution again. A shortcoming of such hybrid redox flow battery is that the energy storage capability is limited by the free space inside the cell accommodating the metal deposits.

A second-type hybrid redox flow batteries use gas such as Cl_2, O_2 and H_2 as the reaction medium or with gas evolution reaction at the cathode or anode (**Figure 2b**). For instance, oxygen reduction reaction ($O_2 + 4H^+ + 4e^- \rightleftarrows 2H_2O$) with high positive potential can be used as the cathode. The cell capacity is then only determined by the capacity of anolyte. Oxygen reduction reaction in non-aqueous electrolytes with the presence of lithium ions can proceed through: $O_2 + 2Li^+ + 2e^- \rightleftarrows Li_2O_2$. Moreover, oxygen reduction and oxidation during discharging and charging can be catalysed chemically with redox mediators [5]. Interestingly, the use of electrocatalysts for the oxygen reduction and oxidation as in a conventional system can be avoided. Note that the formation and deposition of Li_2O_2 proceed at porous matrix, which can be held statically in a gas diffusion tank over charging/discharging. Such a concept may maintain the character of decoupled energy and power for flow batteries.

For aqueous electrolytes, oxygen and hydrogen gas evolution reactions by electrolysis of water take place during charging at very positive and negative electrode potentials, respectively.

Hydrogen evolution reaction has been observed as a parasitic side reaction at the anode for some flow battery systems. Such behaviour has been used to store electricity and to generate hydrogen simultaneously ($2V^{2+} + 2H^{+} \rightarrow H_2 + 2V^{3+}$) as demonstrated in a vanadium-cerium flow battery [6]. Hydrogen generated can be then used to produce electricity in fuel cells.

The ionic conductivity and selectivity of membranes often significantly affect the overall cell performance for many redox flow batteries. High area resistance of membrane restricts the practical operation only at low current densities. Crossover of active species through membrane leads to performance loss over cycling. Redox chemistry of active species with formation of electrodeposits leads to another type of cell configuration without membranes and with only one electrolyte reservoir [7] (**Figure 2c**). Some selected membrane-free redox flow batteries are listed in **Table 1** [8–14]. Reasonable energy efficiencies and cycling stability have been observed. Considering the high cost of most commercial ion exchange membranes, such membrane-free cell configuration could enable simple operation and cost-effective applications.

Deposited anodic species should have slow dissolution rate in the presence of oxidized catholyte species as a self-discharge reaction. A direct reaction between the deposited metal and the other electroactive species in the electrolyte should be negligible or inhibited. Self-discharge effects must be minimized compared to a targeted rapid charging/discharging reaction. Acidic-supporting electrolyte is not suitable for anodic metal deposition. Solid-phase reactions in general have poor kinetics, in comparison with those in liquid electrolytes. The voltage efficiencies in most of the membrane-free flow batteries are relatively low (60–80%) restricted by mass transport and charge transfer kinetics. Compared to the flow-by configuration, an undivided battery with flow-through electrodes may assure enhanced mass transport. However, the flow rate will be largely limited.

A laminar flow battery using two-liquid flowing media, pumped through a slim channel without lateral mixing or with very little mixing, enables membrane-free operation. H_2 (flowing across anode with pumped liquid hydrobromic acid) aqueous bromine laminar flow

Flow batteries	Energy efficiencies	Cycling stability	Ref.
Pb/PbO_2	65%	Limited by the dendrite growth of Pb and formation of unwanted phase of β-PbO_2	[8]
Zn-NiOOH	86%	Stable over 1000 cycles with ~600% Zn excess	[9]
Cu-PbO_2	About 83% at 20.8 mA cm^{-2}	Stable over 450 cycles	[10]
Zn-Ce	About 75% at 20 mA cm^{-2}	Limited by Zn negative electrode, Zn residual on electrode after discharge	[11]
Zn-Quinone	About 40–70% at 30 mA cm^{-2}	Stable for low concentration quinone (50 mM)	[12]
H_2-Br_2	High round-trip efficiency at high current density up to 1 A cm^{-2}	Not given	[13]
Symmetric $Ru(acac)_3$	About 20% at about 2 mA cm^{-2}	Low coulombic and voltage efficiencies	[14]

Table 1. Membrane-free redox flow batteries.

battery demonstrated herewith allows high concentration reactants, fast reaction rates and a high peak power density (0.795 W cm^{-2}) [13].

Among various electrical energy storage technologies, redox flow batteries generally have relatively low energy density (for instance about 30 Wh L^{-1} for all-vanadium redox flow batteries). Thus, although recharging the electrolyte can be done by replacing the depleted one within a few minutes of transportation applications, redox flow batteries are only considered to be used in stationary energy storage. To increase the energy density, highly water-soluble species for instance LiI (solubility up to 8.2 M) and ZnI (7 M) can potentially enhance the volumetric energy density. The use of concentrated ZnI electrolyte leads to a high theoretical energy density of 322 Wh L^{-1} [15], which may even rival batteries based on lithium-ion chemistry ($LiFePO_4$ cathode, 223 Wh L^{-1}).

Another successful development is the redox flow lithium batteries. Pulverized energy-dense solid electrode materials such as $LiCoO_2$ and $LiFePO_4$ can be suspended in a flowable slurry, which is then circulated like a liquid-soluble electrolyte (**Figure 2d**). Due to the high molar concentration of lithium in the solid materials (for instance about 51.2 M for $LiCoO_2$ and 22.8 M for $LiFePO_4$, compared to about 1.6 M for vanadium species in conventional vanadium redox flow batteries), such flow batteries allow high volumetric energy density (about 580 Wh L^{-1} have been achieved [16]). Thus, redox flow batteries may find applications even in portable electronics and electric vehicles.

4. Redox electrochemistry of flow batteries

The overall system performance and cost for redox flow batteries depend largely on the flow cell redox electrochemistry. Great efforts have been made in search of alternative battery chemistry from electrolytes to electrodes [4, 17, 18]. The possible cell voltage depends on the selected redox couples (**Table 2**) and is limited by the electrochemical window of a given solvent-electrode system, stability of the supporting cation or anion and stability of the bipolar plate materials (**Figure 3**).

Table 2 summarises the electrochemical redox reactions at cathode and anode and cell open circuit voltage (OCV) for various reported redox flow batteries. For aqueous electrolytes, the typical cell voltage is below 1.5 V. To achieve high cell voltage, organic solvents with a broad electrochemical window such as acetonitrile (6.1 V) and propylene carbonate (6.6 V) are needed [4]. However, most of the used active species have poor solubility in organic solvents. High cell voltage in this case comes at the expense of low concentration of active species. A compromise among the solubility, cell voltage, reaction kinetics and suitable working temperature should be reached for selecting a suitable electrolyte. Ce^{4+}/Ce^{3+} redox reaction (from 1.44 to 1.70 V vs. SHE, depending on the type of supporting acidic electrolyte), occurring at potential beyond the stability limit of bipolar plate (**Figure 3**), needs special electrodes such as catalyst-coated titanium plate or mesh.

Many anodic reactions have low negative potential; the applications in aqueous batteries can be hindered by H_2 evolution due to the electrolysis of water with unwanted energy loss and an

Cathode redox reactions	Anode redox reactions	Cell OCV/V
$Br_2 + 2e^- \rightleftarrows 2Br^-$ or $Br_3^- + 2e^- \rightleftarrows 3Br^-$	$AQDSH \rightleftarrows AQDS^- + H^+ + e^-$	0.86
$[Mn(acac)_3]^+ + e^- \rightleftarrows Mn(acac)_3$	$[Mn(acac)_3]^- \rightleftarrows Mn(acac)_3 + e^-$	1.10
$Fe^{3+} + e^- \rightleftarrows Fe^{2+}$	$Cr^{2+} \rightleftarrows Cr^{3+} + e^-$	1.19
$VO_2^+ + 2H^+ + e^- \rightleftarrows VO^{2+} + H_2O$	$V^{2+} \rightleftarrows V^{3+} + e^-$	1.26
$I_3^- + 2e^- \rightleftarrows 3I^-$	$Zn \rightleftarrows Zn^{2+} + 2e^-$	1.30
$PbO_2 + 4H^+ + SO_4^- + 2e^- \rightleftarrows PbSO_4 + 2H_2O$	$Cu \rightleftarrows Cu^{2+} + 2e^-$	1.35
$ClBr_2^{2-} + e^- \rightleftarrows 2Br^- + Cl^-$	$VBr_2 + Br^- \rightleftarrows VBr_3 + e^-$	1.35
$Br_2 + 2e^- \rightleftarrows 2Br^-$ or $Br_3^- + 2e^- \rightleftarrows 3Br^-$	$2S_2^{2-} \rightleftarrows S_4^{2-} + 2e^-$	1.36
hydroquinone $\rightleftarrows$ *para*-benzoquinone $+ 2H^+ + 2e^-$	$Zn \rightleftarrows Zn^{2+} + 2e^-$	1.40
$O_2 + 4H^+ + 4e^- \rightleftarrows 2H_2O$	$V^{2+} \rightleftarrows V^{3+} + e^-$	1.49
$[Fe(CN)_6]^{3-} + e^- \rightleftarrows [Fe(CN)_6]^{4-}$	$Zn + 4OH^- \rightleftarrows [Zn(OH)_4]^{2-} + 2e^-$	1.58
$PbO_2 + 4H^+ + 2e^- \rightleftarrows Pb^{2+} + 2H_2O$	$Pb \rightleftarrows Pb^{2+} + 2e^-$	1.69
$2NiOOH + 2H_2O + 2e^- \rightleftarrows 2Ni(OH)_2$	$Zn + 4OH^- \rightleftarrows [Zn(OH)_4]^{2-} + 2e^-$	1.70
$TEMPO^+ + e^- \rightleftarrows TEMPO\bullet$	$Zn \rightleftarrows Zn^{2+} + 2e^-$	1.70
$VO_2^+ + 2H^+ + e^- \rightleftarrows VO^{2+} + H_2O$	$Zn \rightleftarrows Zn^{2+} + 2e^-$	1.76
$[Ru(acac)_3]^+ + e^- \rightleftarrows Ru(acac)_3$	$[Ru(acac)_3]^- \rightleftarrows Ru(acac)_3 + e^-$	1.77
$Br_2 + 2e^- \rightleftarrows 2Br^-$ or $Br_3^- + 2e^- \rightleftarrows 3Br^-$	$Zn \rightleftarrows Zn^{2+} + 2e^-$	1.82
$Cl_2 + 2e^- \rightleftarrows 2Cl^-$	$Zn \rightleftarrows Zn^{2+} + 2e^-$	2.12
$ClBr_2^- + 2e^- \rightleftarrows 2Br^- + Cl^-$	$Zn \rightleftarrows Zn^{2+} + 2e^-$	2.16
$[V(acac)_3]^+ + e^- \rightleftarrows V(acac)_3$	$[V(acac)_3]^- \rightleftarrows V(acac)_3 + e^-$	2.18
$Ce^{4+} + e^- \rightleftarrows Ce^{3+}$	$Zn \rightleftarrows Zn^{2+} + 2e^-$	2.20–2.46
$[Fe(bpy)_3]^{3+} + e^- \rightleftarrows [Fe(bpy)_3]^{2+}$	$[Ni(bpy)_3] \rightleftarrows [Ni(bpy)_3]^{2+} + e^-$	2.30
$Rubrene\bullet^+ + e^- \rightleftarrows Rubrene$	$Rubrene\bullet^- \rightleftarrows Rubrene + e^-$	2.30
$[Ru(bpy)_3]^{3+} + e^- \rightleftarrows [Ru(bpy)_3]^{2+}$	$[Ru(bpy)_3]^+ \rightleftarrows [Ru(bpy)_3]^{2+} + e^-$	2.60
$[Cr(acac)_3]^+ + e^- \rightleftarrows Cr(acac)_3$	$[Cr(acac)_3]^- \rightleftarrows Cr(acac)_3 + e^-$	3.40
$[Fe(CN)_6]^{3-} + e^- \rightleftarrows [Fe(CN)_6]^{4-}$	$Li \rightleftarrows Li^+ + e^-$	3.40
$OFN\bullet^+ + e^- \rightleftarrows OFN$	$BP\bullet^- \rightleftarrows BP + e^-$	4.52
$4Ce^{4+} + 2H_2O \rightarrow 4Ce^{3+} + 4H^+ + O_2$	$2V^{2+} + 2H^+ \rightarrow H_2 + 2V^{3+}$	Cell is chemically discharged [6]

Table 2. Selected redox reactions and cell OCV for redox flow batteries.

imbalance in the state of charge between two sides of the batteries. Through using concentrated electrolytes of water-in-ionic liquid (water in 1-butyl-3-methylimidazolium chloride, BMImCl) [19, 20] or water-in-salt (water in lithium, bis(trifluoromethylsulphonyl)imide, LiTFSI) [21], the onset of oxygen evolution and hydrogen reactions can be shifted to more positive and negative potentials, respectively. Broad electrochemical window of about 3 V has been achieved accordingly (**Figure 3**). It is considered that the amount of free water molecules reduces at such

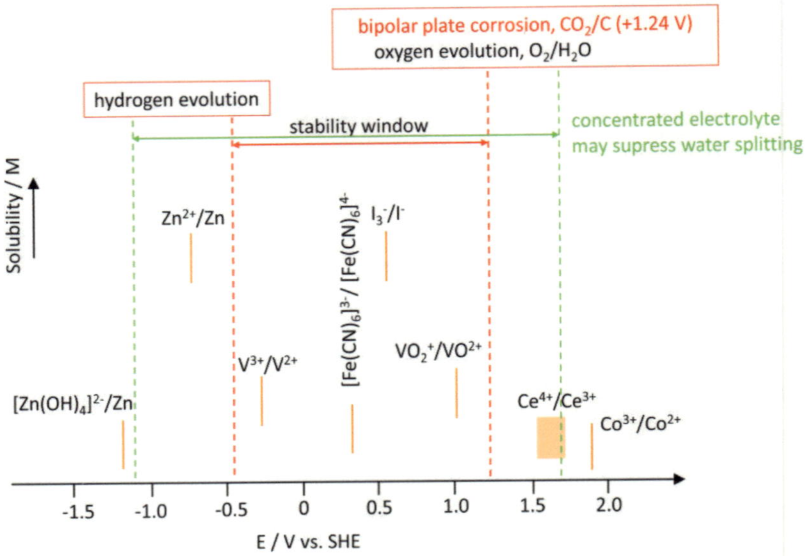

Figure 3. Potentials and relative solubility of selected inorganic and organic redox couples for redox flow batteries. Dotted lines show the electrochemical stability limit of typical aqueous electrolytes. Dashed lines show the possibility to extend the stability limit for aqueous electrolytes using concentrated electrolytes [19, 21].

concentrated mixtures. The inner Helmholtz layer close to the electrode surface is mostly occupied by the $[BMIm]^+$ cation or $TFSI^-$ anion, respectively. Water decomposition is then largely inhibited. The redox potentials for hydrogen and oxygen evolution reactions are pH dependent. Individual control in the pH values of the anolyte and catholyte with a multi-membrane system leads to high cell operation voltage of about 3 V [22].

In contrast to the electrochemical stability of the redox species and solvents, chemical stability of electroactive species and cell components is also critical for long-term operation. Vanadium electrolytes form solid precipitates at a temperature above 40 or below 10°C at concentrations above 1.6 M for all-vanadium redox flow batteries. Oxidizing V^{5+} and Ce^{4+} may cause degradation of membrane and the graphite electrode materials. Complexing agents are needed to store bromine, whereas phase separation (formation of water-insoluble emulsion) occurs for bromine complexes during charging for bromine-based flow batteries. Cross-contamination in bromine-polysulphide batteries may generate heat and release toxic Br_2 and H_2S.

High rate performance of redox flow batteries means high power generation capability. Ideally, two active species at both sides of the cell are expected to have close rate constants. However, mismatches in reaction rates are often observed. For many electrode reactions with sluggish kinetics, catalysts are needed to reduce the polarization (i.e. to improve the voltage efficiencies) and to improve the reaction rate (**Table 3**) [23]. Catalysts are generally applied onto a porous material, which offers high contact area for electrolytes. The supporting materials should have high electrical conductivity, mechanical stability, reasonable cost and high levels of oxygen and hydrogen evolution overpotential for aqueous system. Carbon-based materials are commonly used for this purpose [24].

Electrode reactions	Catalysts
$VO_2^{+} + 2H^{+} + e^{-} \rightleftarrows VO^{2+} + H_2O$	Mn_3O_4/carbon fibre
	ZrO_2
	Bi_2O_3
	Nanorod Nb_2O_5
	Ir-modification of carbon felt
	WO_3
	PbO_2
$V^{3+} + e^{-} \rightleftarrows V^{2+}$	Mn_3O_4/carbon fibre
	ZrO_2
	Bi_2O_3
	Nanorod Nb_2O_5
	TiC
$Cr^{3+} + e^{-} \rightleftarrows Cr^{2+}$	Noble catalysts
$Ce^{4+} + e^{-} \rightarrow Ce^{3+}$	Platinized titanium
$Cl_2 + 2e^{-} \rightleftarrows 2Cl^{-}$	RuO_2
$O_2 + 4H^{+} + 4e^{-} \rightleftarrows 2H_2O$	Pt/Ir mixed oxide

Table 3. Catalysts used for redox couple reactions.

5. Redox active organic electrolytes

Compared to the metal-based electrolytes for redox flow batteries with limited number and resource, organic molecules with unlimited chemical space allow low-cost (for instance, from \$5–10 kg^{-1} vs. \$27 kg^{-1} for vanadium) and high-performance operation. Fast reaction kinetics of organic compounds permit high power generation. High solubility can be realized by controlling the solubilizing functional groups. Redox potentials can be adjusted by varying the electron-donating ($-OH$, $-NH_2$) or -accepting ($-SO_3H$, $-NO_2$, $-PO_3H_2$) properties of the functional groups. By tuning the molecule size or grafting polymer chains, low membrane crossover can be obtained. High-performance organic-based aqueous redox flow batteries have been demonstrated recently (**Table 4**) [25–30].

Quinone-based organic compounds have received great attention, ranging from simple hydroquinone to large anthraquinone. These materials have merits of low cost and fast reaction rates. A peak power density of 1 W cm^{-2} has been observed for a 9,10-anthraquinone-2,7-disulfonic acid (AQDS)-bromide system [31], which is close to a reported peak power density of 1.34 W cm^{-2} for vanadium redox flow batteries. Compared to the relative small molecules such as hydroquinone and 2,2,6,6-tetramethylpiperidine 1-oxyl (TEMPO), large molecules such as AQDS and methyl viologen (MV) are expected to have low-membrane crossover rates. Functionalization of these active organic compounds with polymer backbone chains further enables the battery operation with low-cost size-exclusion membranes [32]. The development

Reaction mechanisms	Redox reactions	Redox potential/V vs. SHE	Ref.
$2e^-$, $2H^+$ redox reactions	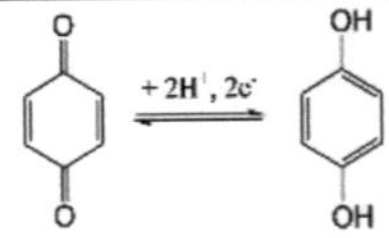	pH dependent, ranging from 0.56 to 0.75	[25]
$2e^-$, $2H^+$ redox reactions	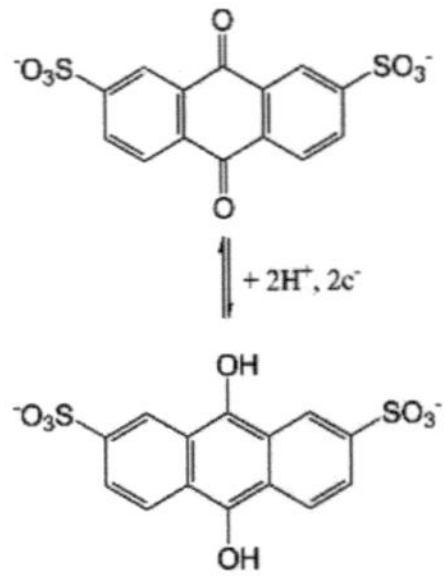	0.2	[26]
$2e^-$, $2H^+$ redox reactions	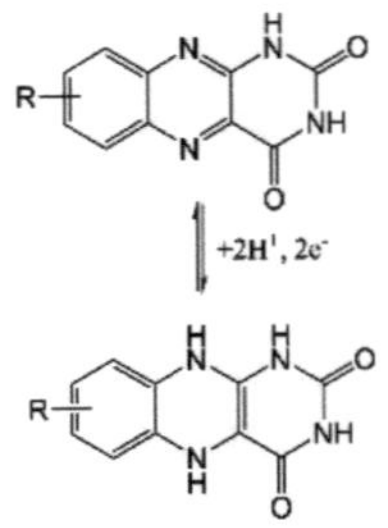	−0.73 for R = OH	[27]
Organic radicals	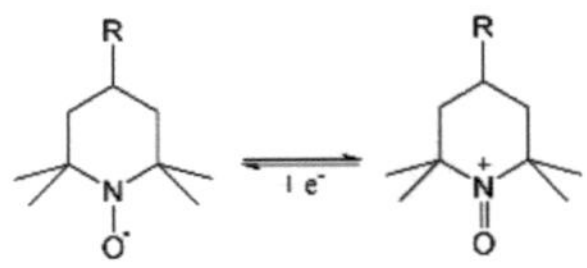	About 0.5 V in carbonate electrolyte	[28]
Organic radicals	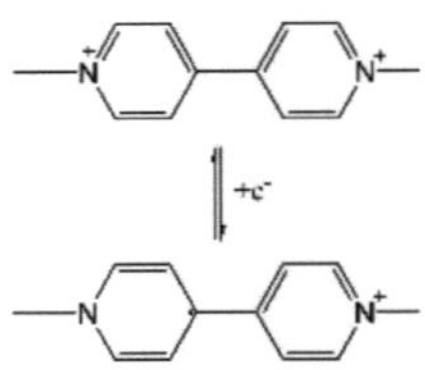	−0.45 V	[29]
Organodisulfide	$R{-}S{-}S{-}R \rightleftharpoons 2\,R{-}S^-$	About −1 V	[30]

Table 4. Selected redox active organic compounds used for redox flow batteries.

of organic active materials for redox flow batteries holds great promise for stationary electrochemical energy storage.

6. Semi-solid redox flow batteries

To overcome the restrictions in solubility of active species in liquid electrolytes, suspensions with energy-dense solid materials have been introduced for redox flow batteries. The concept was first demonstrated with intercalation materials by Chiang et al. [33], which are typically used for lithium ion batteries. Such semi-solid lithium redox flow batteries combine the merits of high energy density for lithium ion batteries and the decoupled character of conventional redox flow batteries. In order to form a percolation network for charge transfer, several strategies have been proposed: (i) dispersing conductive additive such as carbon into the electrolytes, (ii) adding redox mediators and (iii) inserting a metal wire as a current collector [34]. It has been found that the conductive electrolytes encounter the issue of shunt current between cells in a stack.

Energy-dense batteries, based on lithiation chemistry and intercalation chemistry of abundant elements (such as Na, Mg and Al etc.), contribute significantly to the transportable applications of various electronic devices and revolution of our modern societies. The successful development in these materials raises opportunities in new applications for flow batteries. Li-, Na- and organic molecule-based semi-solid redox flow batteries have been developed recently (**Table 5**) [33–37]. For a pumping system with solid suspension, the rheological properties of suspension need to be considered.

In contrast to the flow batteries with both (de)lithiation and electron transfer reactions occurring inside the electrochemical cells (**Figure 2d**), a new concept using redox shuttle molecules has been introduced [38], wherein solid active materials are kept statically in the tank and only the shuttle molecules are circulated in the electrochemical cell (**Figure 4**). Electrochemical redox reactions of the shuttle molecules go on at the electrode inside the cell, whereas chemical (de)lithiation of the active solid materials in the tank occurs through the reactions between the solid materials and the shuttle molecules. Since the active solid materials are not involved in the electrochemical reaction, conductive additives (such as carbon black) are not necessary in such a system. In addition, low concentration shuttle molecules of only several mM are sufficient to induce the (de)lithiation reaction of a large amount of solid materials.

Semi-solid flow batteries	Suspension	Remarks	Ref.
$LiCoO_2/Li_4Ti_5O_{12}$	26 vol% $LiCoO_2$, 0.8 vol% Ketjen; 25 vol% $Li_4Ti_5O_{12}$, 0.8 vol% Ketjen	C/3 to C/8 rate, high energy efficiency	[33]
$LiCoO_2/Li_4Ti_5O_{12}$	Carbon-free 0.5 vol% $LiCoO_2$, 1 vol% $Li_4Ti_5O_{12}$	Low current density from 0.002 to 0.008 mA cm^{-2}, low coulombic efficiency of about 11.5%	[34]
P2-type $Na_xNi_{0.22}Co_{0.11}Mn_{0.66}O_2/NaTi_2(PO_4)_3$	Active material with 1.3 wt% conductive additive	Current density below 0.5 mA cm^{-2}, low voltage efficiency of about 40%, energy density of about 9 Wh L^{-1}	[35]
Symmetric battery with polythiophene	Polythiophene (8.41 g L^{-1}), Ketjenblack (2 g L^{-1})	Low current density (<1 mA cm^{-2}), energy efficiency of 60.9% at 0.5 mA cm^{-2}	[36]
Zn/polyaniline	10 wt% polyaniline powder in suspension	0.28 V overpotential at 20 mA cm^{-2}	[37]

Table 5. Selected examples for semi-solid redox flow batteries.

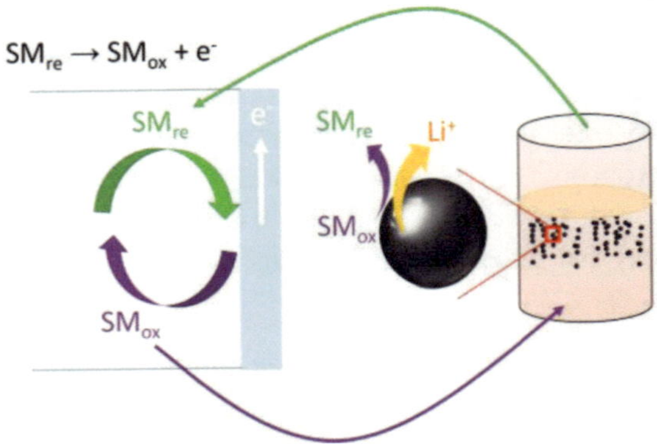

Figure 4. An illustration of a semi-solid redox flow battery with solid materials stored statically in the tank, and redox shuttle molecule (SM) circulated with electrolyte.

7. Conclusions and perspectives

Redox flow battery technology is relatively new and not yet well-developed. Rational electrolyte management and cell design can lead to the enhancement of energy storage capability and a reduction in construction cost. Novel electrolyte chemistry and development of a new configuration of flow batteries will create high system flexibility. Physiochemical and electrochemical redox properties of active couples, stability window of supporting electrolyte, selection of supporting ions, stability of electrode materials and cell components are key factors for successful applications. Future market penetration of flow batteries needs low cost, high energy density and high power density. The pace of recent development in the active organic molecules as electrolytes opens new strategies of cost-effective and sustainable solutions for large-scale stationary energy storage. The application of energy-dense solid materials in suspension for redox flow batteries may largely enhance the energy density of flow battery systems.

Acknowledgements

We thank the support from the basic research funding of KIST Europe ("Electrochemical energy transformation and energy storage"). Ruiyong Chen thanks Professor R. Hempelmann for his continuing support.

Abbreviations

AQDS	9,10-Anthraquinone-2,7-disulphonic acid
BMImCl	1-Butyl-3-methylimidazolium chloride
BP	Biphenyl
$BP^{\bullet -}$	Biphenyl radical anion
LiTFSI	Lithium bis(trifluoromethylsulphonyl)imide

MV	Methyl viologen dichloride
OCV	Open circuit voltage
OFN	Octafluoronaphthalene
$OFN^{\bullet+}$	Octafluoronaphthalene radical cation
SHE	Standard hydrogen electrode
TEMPO	2,2,6,6-Tetramethylpiperidine 1-oxyl

Author details

Ruiyong Chen[1,2]*, Sangwon Kim[1,2] and Zhenjun Chang[1,2,3]

*Address all correspondence to: r.chen@kist-europe.de

1 Transfercenter Sustainable Electrochemistry, Saarland University, Saarbrücken, Germany

2 Korea Institute of Science and Technology (KIST) Europe, Saarbrücken, Germany

3 College of Materials Science and Engineering, Jiangsu University of Science and Technology, Zhenjiang, China

References

[1] Skyllas-Kazacos M, Rychcik M, Robins RG, Fane AG, Green MA. New all-vanadium redox flow cell. Journal of the Electrochemical Society. 1986;**133**:1057–1058. DOI: 10.1149/1.2108706

[2] Tang A, Bao J, Skyllas-Kazacos M. Thermal modelling of battery configuration and self-discharge reactions in vanadium redox flow battery. Journal of Power Sources. 2012;**216**: 489–501. DOI: 10.1016/j.jpowsour.2012.06.052

[3] Liu H, Xu Q, Yan C, Qiao Y. Corrosion behavior of a positive graphite electrode in vanadium redox flow battery. Electrochimica Acta. 2011;**56**:8783–8790. DOI: 10.1016/j.electacta.2011.07.083

[4] Gong K, Fang Q, Gu S, Li SFY, Yan Y. Nonaqueous redox-flow batteries: Organic solvents, supporting electrolytes, and redox pairs. Energy & Environmental Science. 2015;**8**: 3515–3530. DOI: 10.1039/c5ee02341f

[5] Zhu YG, Jia C, Yang J, Pan F, Huang Q, Wang Q. Dual redox catalysts for oxygen reduction and evolution reactions: Towards a redox flow Li–O_2 battery. Chemical Communications. 2015;**51**:9451–9454. DOI: 10.1039/c5cc01616a

[6] Amstutz V, Toghill KE, Powlesland F, Vrubel H, Comninellis C, Hu X, Girault HH. Renewable hydrogen generation from a dual-circuit redox flow battery. Energy & Environmental Science. 2014;**7**:2350–2358. DOI: 10.1039/c4ee00098f

[7] Bamgbopa MO, Almheiri S, Sun H. Prospects of recently developed membraneless cell designs for redox flow batteries. Renewable and Sustainable Energy Reviews. 2017;**70**: 506–518. DOI: 10.1016/j.rser.2016.11.234

[8] Hazza A, Pletcher D, Wills R. A novel flow battery: A lead acid battery based on an electrolyte with soluble lead (II). Part I. Preliminary studies. Physical Chemistry Chemical Physics. 2004;**6**:1773–1778. DOI: 10.1039/b401115e

[9] Cheng J, Zhang L, Yang Y-S, Wen Y-H, Cao G-P, Wang X-D. Preliminary study of single flow zinc-nickel battery. Electrochemistry Communications. 2007;**9**:2639–2642. DOI: 10. 1016/j.elecom.2007.08.016

[10] Pan J, Sun Y, Cheng J, Wen Y, Yang Y, Wan P. Study on a new single flow acid Cu–PbO_2 battery. Electrochemistry Communications. 2008;**10**:1226–1229. DOI: 10.1016/j.elecom. 2008.06.008

[11] Leung PK, de León CP, Walsh FC. An undivided zinc-cerium redox flow battery operating at room temperature (295 K). Electrochemistry Communications. 2011;**13**:770–773. DOI: 10.1016/j.elecom.2011.04.011

[12] Leung PK, Martin T, Shah AA, Anderson MA, Palma J. Membrane-less organic–inorganic aqueous flow batteries with improved cell potential. Chemical Communications. 2016;**52**: 14270–14273. DOI: 10.1039/c6cc07692k

[13] Braff WA, Bazant MZ, Buie CR. Membrane-less hydrogen bromine flow battery. Nature Communications. 2013;**4**:2346. DOI: 10.1038/ncomms3346

[14] Chakrabarti MH, Roberts EPL, Bae C, Saleem M. Ruthenium based redox flow battery for solar energy storage. Energy Conversion and Management. 2011;**52**:2501–2508. DOI: 10.1016/j.enconman.2011.01.012

[15] Li B, Nie Z, Vijayakumar M, Li G, Liu J, Sprenkle V, Wang W. Ambipolar zinc-polyiodide electrolyte for a high-energy density aqueous redox flow battery. Nature Communications. 2015;**6**:6303. DOI: 10.1038/ncomms7303

[16] Chen H, Lu Y-C. A high-energy-density multiple redox semi-solid–liquid flow battery. Advanced Energy Materials. 2016;**6**:1502183. DOI: 10.1002/aenm.201502183

[17] Pan F, Wang Q. Redox species of redox flow batteries: A review. Molecules. 2015;**20**:20499–20517. DOI: 10.3390/molecules201119711

[18] Noack J, Roznyatovskaya N, Herr T, Fischer P. The chemistry of redox-flow batteries. Angewandte Chemie International Edition. 2015;**54**:9776–9809. DOI: 10.1002/anie.20141 0823

[19] Chen R, Hempelmann R. Ionic liquid-mediated aqueous redox flow batteries for high voltage applications. Electrochemistry Communications. 2016;**70**:56–59. DOI: 10.1016/j. elecom.2016.07.003

[20] Chen R, Hempelmann R. Ionic liquids-mediated aqueous electrolytes for redox flow batteries. In: 7th International Flow Battery Forum (IFBF). Swanbarton Limited, United Kingdom, 2016; 22-23. ISBN: 978-0-9571055-6-0

[21] Suo L, Borodin O, Gao T, Olguin M, Ho J, Fan X, Luo C, Wang C, Xu K. "Water-in-salt" electrolyte enables high-voltage aqueous lithium-ion chemistries. Science. 2015;**350**:938–943. DOI: 10.1126/science.aab1595

[22] Gu S, Gong K, Yan EZ, Yan Y. A multiple ion-exchange membrane design for redox flow batteries. Energy & Environmental Science. 2014;**7**:2986–2998. DOI: 10.1039/c4ee00165f

[23] Park M, Ryu J, Cho J. Nanostructured electrocatalysts for all-vanadium redox flow batteries. Chemistry—An Asian Journal. 2015;**10**:2096–2110. DOI: 10.1002/asia.201500238

[24] Chakrabarti MH, Brandon NP, Hajimolana SA, Tariq F, Yufit V, Hashim MA, Hussain MA, Low CTJ, Aravind PV. Application of carbon materials in redox flow batteries. Journal of Power Sources. 2014;**253**:150–166. DOI: 10.1016/j.jpowsour.2013.12.038

[25] Ding Y, Yu G. A bio-inspired, heavy-metal-free, dual-electrolyte liquid battery towards sustainable energy storage. Angewandte Chemie International Edition. 2016;**55**:4772–4776. DOI: 10.1002/anie.201600705

[26] Huskinson B, Marshak MP, Suh C, Er S, Gerhardt MR, Galvin CJ, Chen X, Aspuru-Guzik A, Gordon RG, Aziz MJ. A metal-free organic–inorganic aqueous flow battery. Nature. 2014;**505**:195–198. DOI: 10.1038/nature12909

[27] Lin K, Gómez-Bombarelli R, Beh ES, Tong L, Chen Q, Valle A, Aspuru-Guzik A, Aziz MJ, Gordon RG. A redox-flow battery with an alloxazine-based organic electrolyte. Nature Energy. 2016;**1**:16102. DOI: 10.1038/nenergy.2016.102

[28] Wei X, Xu W, Vijayakumar M, Cosimbescu L, Liu T, Sprenkle V, Wang W. TEMPO-based catholyte for high-energy density nonaqueous redox flow batteries. Advanced Materials. 2014;**26**:7649–7653. DOI: 10.1002/adma.201403746

[29] Liu T, Wei X, Nie Z, Sprenkle V, Wang W. A total organic aqueous redox flow battery employing a low cost and sustainable methyl viologen anolyte and 4-HO-TEMPO catholyte. Advanced Energy Materials. 2016;**6**:1501449. DOI: 10.1002/aenm.201501449

[30] Li B, Liu J. Progress and directions in low-cost redox flow batteries for large-scale energy storage. National Science Review. 2017; 4: 91–105. DOI: 10.1093/nsr/nww098

[31] Chen Q, Eisenach L, Aziz MJ. Cycling analysis of a quinone-bromide redox flow battery. Journal of the Electrochemical Society. 2016;**163**:A5057-A5063. DOI: 10.1149/2.0081601jes

[32] Winsberg J, Janoschka T, Morgenstern S, Hagemann T, Muench S, Hauffman G, Gohy J-F, Hager MD, Schubert US. Poly(TEMPO)/zinc hybrid-flow battery: A novel, "green," high voltage, and safe energy storage system. Advanced Materials. 2016;**28**:2238–2243. DOI: 10.1002/adma.201505000

[33] Duduta M, Ho B, Wood VC, Limthongkul P, Brunini VE, Carter WC, Chiang Y-M. Semi-solid lithium rechargeable flow battery. Advanced Energy Materials. 2011;**1**:511–516. DOI: 10.1002/aenm.201100152

[34] Qi Z, Liu AL, Koenig Jr GM. Carbon-free solid dispersion $LiCoO_2$ redox couple characterization and electrochemical evaluation for all solid dispersion redox flow batteries. Electrochimica Acta. 2017;**228**:91–99. DOI: 10.1016/j.electacta.2017.01.061

[35] Ventosa E, Buchholz D, Klink S, Flox C, Chagas LG, Vaalma C, Schuhmann W, Passerini S, Morante JR. Non-aqueous semi-solid flow battery based on Na-ion chemistry. P2-type $Na_xNi_{0.22}Co_{0.11}Mn_{0.66}O_2$-$NaTi_2(PO_4)_3$. Chemical Communications. 2015;**51**:7298–7301. DOI: 10.1039/c4cc09597a

[36] Oh SH, Lee C-W, Chun DH, Jeon J-D, Shim J, Shin KH, Yang JH. A metal-free and all-organic redox flow battery with polythiophene as the electroactive species. Journal of Materials Chemistry A. 2014;**2**:19994–19998. DOI: 10.1039/c4ta04730c

[37] Wu S, Zhao Y, Li D, Xia Y, Si S. An asymmetric Zn//Ag doped polyaniline microparticle suspension flow battery with high discharge capacity. Journal of Power Sources. 2015; **275**:305–311. DOI: 10.1016/j.jpowsour.2014.11.012

[38] Huang Q, Li H, Grätzel M, Wang Q. Reversible chemical delithiation/lithiation of $LiFePO_4$: Towards a redox flow lithium-ion battery. Physical Chemistry Chemical Physics. 2013;**15**:1793–1797. DOI: 10.1039/c2cp44466f

Chapter 6

On the Titration Curves and Titration Errors in Donor Acceptor Titrations of Displacement and Electronic Transference Reactions

Julia Martín, Laura Ortega Estévez and
Agustín G. Asuero

Additional information is available at the end of the chapter

http://dx.doi.org/10.5772/intechopen.68750

Abstract

An overview of the state of the art concerning with earlier approaches to titration in redox systems is given in this chapter, in which an overview on redox bibliography has also been undertaken. Titration error has been the subject of a variety of excellent papers, but the number of papers dealing with titration error in redox titrations is scarce. However, a single hyperbolic sine expression for the titration error in donor/acceptor titration of displacement and electronic transference reactions is derived in this chapter. The titration error expression is applicable to symmetrical redox reactions, that is to say, those in which no polynuclear species are involved in the equilibria. The donor versus acceptor particle notation is chosen to accentuate the analogy with that used in the description of acids and bases following the steps given by the French School and other recognized authors (Budevsky, Butler, Charlot, Gauguin, Inczedy, Monnier, Rosset). A diagram for the titration error in function of the difference between the end and equivalence point (pX) is drawn in order to facilitate the graphical calculation of titration error. A detailed error analysis concerning with the propagation of systematic and random error propagation in the titration error is given.

Keywords: titration curves, titration error, electronic transference reactions

1. Introduction

Titrimetry constitutes one of the former analytical methods, been applied since the late eighteenth century [1]. Based on reaction chemistry, it is still developing [2–5] and continues to be extensively employed in several analytical fields as well as in routine studies [6] due to its simplicity, with little sacrifice in accuracy and precision, and low cost [7–9].

Titrimetry is a fast technique easily automated and cheap in terms of equipment. It is together with gravimetry one of the two special existing methods to determine chemical composition on the basis of chemical reactions (primary method) being used for methodological and working levels [10, 11]. Independent values of chemical quantities expressed in SI units are obtained through gravimetry and titrimetry (classical analysis). In titrimetry the quantity of tested components of a sample is assessed through the use of a solution of known concentration added to the sample which reacts in a definite proportion. The reaction between the analyte and the reagent must be fast, complete (with an equilibrium constant very large), proceed according to a well-defined (known stoichiometry) chemical equation (without side-reactions), lead to a stable reaction product, and usually, and take place in solution, although in some cases precipitation reactions are involved. To identify the stoichiometric point, where equal amounts of titrant react with equal amounts of analyte, indicators are used in many cases to point out the end of the chemical reaction by a color change. The indicator reacts with either the analyte or the reagent to produce (in a clear and unambiguous way) a color change when the chemical reaction has been completed.

This is known as the "end point" of the titration, which should be as close as possible to the equivalence point. The difference between the end point and the equivalence point is the titration error. In order to reduce the uncertainty in the results, the end point can also be detected by instrumental rather than visual means, i.e., by potentiometry, where the potential difference is measured between a working electrode and a reference electrode; in photometry, where the light transmission is measured with a photometric sensor; or in coulometric, where the titrant is generated electrochemically and the amount of titrant is calculated from the current and time of reaction using Faraday's law.

Titrimetric methods are classified into four groups depending on the type of chemical reaction involved: acid-base, oxidation-reduction, precipitation, and complexometric. Among the most common applications of titrimetry are the measurement of anionic and cationic species, and neutral molecules of both organic and inorganic substances. Although it has been extensively used for the measurement of pure substances, it also performs well when trace constituents in complex matrices want to be evaluated.

Before the experience, the titrant is standardized by titration of the primary standard, which must be stable against the influence of light, air, and temperature. The main primary standards are potassium hydrogen phthalate (KHP), benzoic acid, tris(hydroxymethyl)aminomethane (TRIS, THAM), or Na_2CO_3 for acid-base titration, As_2O_3, $Na_2C_2O_4$, $K_2Cr_2O_7$, or KIO_3 for redox titration, NaCl for precipitation titrations, and $CaCO_3$, $Pb(NO_3)_2$, Zn, Ni or Cu for complexometry. During the experience, the temperature should be constant (for ordinary work $\pm$ 0.5°C). To reduce the overall uncertainty, titrimetry can also be performed on a mass/mass basis using a weight burette with a primary standard.

Titrimetric methods can also be classified based on the properties of the reaction: if the assay is performed with a known concentration of titrant or with an unknown concentration of titrant is called direct or indirect titration, respectively, whereas back titration involves an excess of titrant that reacts until the equivalence with a known concentration of sample.

2. Titration in redox systems: earlier approaches

Redox titrations were essential in volumetric analysis as well as in electrometric techniques. The first attempts to quantify titration curves, including redox systems, and data processing [12] were from 1920s to 1930s. As suggested by Goldman's review [13], the first theoretical developments were due to Kolthoff [14], who provided the link between the application of physical chemistry and titrimetry [15–17] (though it was criticized by his Dutch colleagues in his earlier career). The understanding of redox equilibria provided the platform to the development on electrode processes [13].

One can plot against pE (or pe) as variable either the logarithm of the concentration (or activity) of each species (maintaining the other concentration constant) or the logarithm of the ratio of the concentration (or activity) of the species [13] using as a reference one oxidation state. Then, the diagram, entirely composed of straight lines, is simple and easy to construct [18]. From log (concentration) versus E diagrams redox titrations may be drawn [19–22], even the titration error by an appropriate modification [23, 24]. In fact, logarithmic diagrams, introduced by Bjerrum [25], have found a wide use in the study of titration systems, mainly in the acid-base domain [23, 26, 27]. On the basis of logarithmic diagrams, Maccà and Bombi [28] discuss the symmetry properties of titration curves and the linearity of Gran plots. In the particular case of redox systems, distribution diagrams are not as commonly used as they are for acid-base and complex ion systems [18, 29–31].

In 1960s titration was revisited to improve formulation of titration curves, in the papers by Bard and Simonsen, Bishop and Goldman's papers as indicated in the review of Goldman [13], "formal" potentials, E'_{0i}, being used in those approaches instead of standard potentials. de Levie [32–34] pointed out the similarity between the profile of redox and acid-base titration curves, related to a polyprotic acid titrated with NaOH. Equivalence points were also discussed. Some further papers were related to redox indicators and titration errors.

Anfält and Jagner [35] reviewed and evaluated the accuracy and precision of some frequently potentiometric end-point indicator methods. Yongnian and Ling [36] used mathematical methods to assess the application and advancement of titration methods. Although it is common to find examples of applications to redox titrations [37, 38] their actual use in redox titrations has been scarce [39–42]. The Gran methods were modified later [43–45], without unnecessary simplifications.

The most common topics dealing redox titrations and published in the last two decades were: to calculate stoichiometric point potentials by different types of redox titration [46]; the use of linear regression analysis of potentiometric data through linearization of the titration function [47, 48]; approximated titration curves [49] by logarithmic function studying the prediction of random error in titration parameters [50]; as well as the factors affecting the accuracy [51] of redox titrations. Nonlinear regression analysis allows the simultaneous evaluation of several parameters from the data obtained [52] in a single redox titration. A general algorithm and a program for the calculation and construction of titration curves [53] have been proposed. Numerical and computer simulations [54, 55] and equations by nonlinear least squares with novel weighting functions [56] for redox potentiometric data were also evaluated.

The analogy between acid-base and redox-behavior, as particle exchange reactions (of protons or electrons, respectively) [57, 58], has been interpreted through the years [59–67].

Titration error has been the subject of excellent publications, but the number of papers dealing with redox titration error is relatively scarce. The focus has been mainly put in acid-base, precipitation, and complex formation reaction titrations. In this chapter, an attempt is given to devise a titration error theory applicable to donor-acceptor titration of displacement and electronic transference (redox) reactions. The error will be formulated as a function of the titration parameters, in a hyperbolic sine expression way. As a matter of fact a detailed treatment of the analysis error issue is also carried out. The error analysis requires differentiating with respect to given variables leading at first sight to complex expressions, which at the end finally appears to be compact.

3. On the titration error in donor/acceptor titrations of displacement and electronic transference reactions

Acid-base, complexation, and precipitation titrations have been extensively dealt with both in scientific and educational literature. Redox titrations, although being of primary importance, have received less attention [68]. Titration error has been the subject of several excellent papers, the emphasis being placed, however, mainly on acid-base [69–73] as well as on precipitation [74–79], and complex formation titration reactions [80, 81]. A number of papers dealing with the topic of titration error in redox titration [13, 24, 82, 83] have also been appeared. Redox titrations hold an important place in simple, fast, low-cost analysis of redox-active-species [56]. In this chapter, a single equation is presented for symmetrical reactions for which the reduced and oxidized species of each half-reaction should be the same. Therefore the treatment is not strictly valid for such couples as $Cr_2O_7^{2-}/Cr^{3+}$ or I_2/I^- [81].

Let *SD* be a particle donor (weak or strong), to be titrated with a particle acceptor *TA*. The analyst can choose the titrant and will always use a strong one to obtain better results. The systems implied in the titration reaction and their corresponding equilibrium constants will be given by

$$TA + bX \rightleftharpoons TD \qquad K_T = \frac{[TD]}{[TA][X]^b} \tag{1}$$

$$SA + aX \rightleftharpoons SD \qquad K_S = \frac{[SD]}{[SA][X]^a} \tag{2}$$

X being the particle transferred in the semi-reactions involved in the global titration reaction

$$bSD + aTA \rightleftharpoons bSA + aTD \tag{3}$$

to which corresponds the following equilibrium constant

$$K_{eq} = \frac{[SA]^b[TD]^a}{[SD]^b[TA]^a} = \frac{K_T^a}{K_S^b} \tag{4}$$

Proton free equilibria are assumed first in the presentation for the sake of clarity though redox equilibria that are independent of pH are relatively few. Effects of such factors as hydrogen ion concentration and complexing ligands may be easily incorporated [84] in the corresponding conditional constants. The particle X may be a proton, electron, cation, or an uncharged molecule. This notation is chosen to accentuate the analogy with that used in the description of acids and bases by the French School and other recognized authors [19, 29–31, 85–92]. Anyway, as TA is a strong particle acceptor, it must occur that $K_T >> K_S$.

From Eq. (3) it is readily seen that in any moment of the titration it holds that (through the entire concentration range)

$$b[TD] = a[SA] \tag{5}$$

On the other hand, we may define the (relative) titration error as

$$\Delta T = \left(\frac{bC_T - aC_S}{aC_S}\right)_{\text{end}} = T - 1 \tag{6}$$

where C_T and C_S are the analytical concentrations of titrant and analyte, respectively,

$$C_T = [TA] + [TD] \tag{7}$$

$$C_S = [SA] + [SD] \tag{8}$$

and T, the fraction titrated, is defined as the ratio between the amount of titrant added and the initial amount of analyte at any moment of the titration. Thus, depending on the nature of the titration, ΔT might be either positive or negative. When the titration is carried out in the reverse order the same result is obtained, but the equation now bears a minus sign.

Therefore, from Eqs. (6)–(8), it follows that

$$\Delta T = \left(\frac{b[TA] - a[SD]}{aC_S}\right)_{\text{end}} \tag{9}$$

By combining Eqs. (5) and (1), we have

$$[TA] = \frac{a}{b}\frac{[SA]}{K_T[X]^b} \tag{10}$$

and from Eq. (2) we obtain

$$[SD] = K_S[SA][X]^a \tag{11}$$

By substituting the values of $[TA]$ and $[SD]$ given by Eqs (10) and (11)

$$\Delta T = \frac{[SA]}{C_S}\left(\frac{1}{K_T[X]^b} - K_S[X]^a\right) \tag{12}$$

As the molarity fraction of the species $[SA]$ is given by

$$f_{SA} = \frac{[SA]}{C_S} = \frac{1}{1 + [X]^a K_S} \tag{13}$$

Eq. (12) may be transformed into

$$\Delta T = \frac{1}{1 + [X]^a K_S}\left(\frac{1}{K_T[X]^b} - K_S[X]^a\right) \tag{14}$$

By multiplying and dividing the right hand side of Eq. (14) by $\sqrt{(K_S/K_T)}$ we get

$$\Delta T = \frac{\sqrt{\frac{K_S}{K_T}}}{1 + [X]^a K_S}\left(\frac{1}{\sqrt{K_T K_S}[X]^b} - \sqrt{K_T K_S}[X]^a\right) \tag{15}$$

In the equivalence point, when the exact stoichiometric amount of titrant has been added, in addition to Eq. (15), the following condition is satisfied

$$a[SD] = b[TA] \tag{16}$$

and then, from Eqs. (1) and (2), it follows that when the exact stoichiometric amount of titrant has been added

$$\sqrt{K_T K_S} = \frac{1}{[X]_{eq}^{(a+b)/2}} \tag{17}$$

$$pK_{eq} = \frac{\log K_T + \log K_S}{a + b} \tag{18}$$

and then, the potential at the equivalence point is independent of the concentration of the reactants and thus unaffected by dilution. Note, however, that Eq. (18) is not perfectly general, because the simple relation of Eq. (16) for reactants and products is not always valid. The species involved in the equilibria may be polynuclear. The p^X in this instance varies with dilution.

By substituting Eq. (17) into Eq. (15)

$$\Delta T = \frac{\sqrt{\frac{K_S}{K_T}}}{1 + [X]^a K_S} [X]^{(a-b)/2} \left(\left(\frac{[X]_{eq}}{[X]} \right)^{(a+b)/2} - \left(\frac{[X]}{[X]_{eq}} \right)^{(a+b)/2} \right) \tag{19}$$

A chemical error will arise because of lack of agreement between the end point and equivalence point. The difference between the end point and the equivalence point of a titration is the source of systematic error of determination. Taking into account that

$$\Delta pX = pX_{end} - pX_{eq} \tag{20}$$

and

$$\sinh x = \frac{e^x - e^{-x}}{2} \tag{21}$$

after some manipulation, the following expression may be easily obtained

$$\Delta T = \frac{2\sqrt{\frac{K_S}{K_T}}}{1 + [X]^a K_S} [X]^{(a-b)/2} \sinh\left(\ln 10 (a+b) \frac{\Delta pX}{2} \right) \tag{22}$$

or

$$\Delta T = W \sinh\left(\ln 10 (a+b) \frac{\Delta pX}{2} \right) \tag{23}$$

where the shape coefficient W is depending on the particle concentration in the end point titration when asymmetrical titrations are being considered

$$W = \frac{2\sqrt{\frac{K_S}{K_T}}}{1 + [X]^a K_S} [X]^{(a-b)/2} \tag{24}$$

Low values of the stoichiometric coefficients a, b, as well as low difference a−b values, and large K_T values lead to lower errors.

In the vicinity of the equivalence point, $+1 >> [X]^a_{end} K_S$ and so

$$W = 2\sqrt{\frac{K_S}{K_T}} [X]^{(a-b)/2}_{end} \tag{25}$$

In those cases in which the titration reaction is symmetrical, $a = b$, and then

$$W = 2\sqrt{\frac{K_S}{K_T}} \tag{26}$$

and the following formula is obtained for the titration error

$$\Delta T = 2K_{eq}^{-1/(2a)} \sinh\left(\ln 10 a \Delta pX\right) \tag{27}$$

Note that the titration error may be formulated as a hyperbolic sine expression. Hyperbolic functions are of great worth in parameter estimation as shown by Asuero [93].

The methodology developed in this section of the chapter is going to be applied forward to some experimental situations characteristics of redox titration reactions. The calculations made by the hyperbolic sine method are checked with the procedure devised by de Levie [21, 26, 62]. A detailed treatment of systematic and random errors associated with the titration error is carried out in the following.

Note that the equations developed in this contribution can easily take into account the lateral reactions by using the corresponding lateral reaction coefficients and the conditional constants involved. A condition is however required, namely that the pH remains constant in the course of titration, which cannot always be achieved. Michalowski [4, 94–100] has given a general and definitive solution to the problem of redox equilibria, which does not require any restriction.

The beginnings of the rigorous GATES/Generalized Electron Balance (GEB) approach of Michalowski, which can be interpreted as a new natural law, dates back from 1992 to 1995. This approach has recently been shown repeatedly in the bibliography solving complex chemical problems and requires the use of nonlinear regression and a high level language such as MATLAB. The equations developed here, although very modest, have an obvious didactic interest and can be seen in the case of the redox equilibria as an alternative route to that given by de Levie [62].

4. Electronic transfer reactions

In the following, some titration curves of typical oxide-reduction reactions, involving Ce^{4+} and MnO_4^- as titrant, are the subject of study. The transferred particle, the electron, takes the place of [X] in Eq. (22). The numerical values obtained by applying the hyperbolic sine method proposed in this contribution are checked against the method devised by de Levie [62], thus verifying the identity of the results in all cases.

4.1. Fe^{2+} titration curve with Ce^{4+}

The equilibrium constant and *pe* of the semi-reaction $Ce^{4+} + e = Ce^{3+}$ ($E_{0T} = 1.44$ v) are given (Ce^{4+} is the acceptor) by

$$K_T = \frac{[Ce^{3+}]}{[Ce^{4+}][e]} \qquad pe = \log K_T + \log\frac{[Ce^{4+}]}{[Ce^{3+}]} \tag{28}$$

Also, for the half-reaction $Fe^{3+} + e = Fe^{2+}$ ($E_{0S} = 0.68$ v) (Fe^{2+} is the donor)

$$K_S = \frac{[Fe^{2+}]}{[Fe^{3+}][e]} \qquad pe = \log K_S + \log \frac{[Fe^{3+}]}{[Fe^{2+}]} \tag{29}$$

The equilibrium constant of the overall reaction is expressed as

$$Ce^{4+} + Fe^{2+} = Ce^{3+} + Fe^{3+}, \qquad K = \frac{[Ce^{3+}][Fe^{3+}]}{[Ce^{4+}][Fe^{2+}]} = \frac{K_T}{K_S} \tag{30}$$

From Eq. (22) taking into account that $[X] = e$, and that $a = b = 1$

$$\Delta T = \frac{2\sqrt{K_S/K_T}}{1+[e]K_S}[e]^{(1-1)/2}\sinh\left(\ln 10(1+1)\left(\frac{pe - pe_{eq}}{2}\right)\right) = \frac{2\sqrt{K_S/K_T}}{1+10^{-pe}K_S}\sinh\left(\ln 10(pe - pe_{eq})\right) \tag{31}$$

From the Nernst equations applied to the two half-directions involved

$$E = E_T^0 + 0.06\log\frac{[Ce^{4+}]}{[Ce^{3+}]} \qquad \frac{E}{0.06} = \frac{E_T^0}{0.06} + \log\frac{[Ce^{4+}]}{[Ce^{3+}]} \qquad pe = pe_T^0 + \log\frac{[Ce^{4+}]}{[Ce^{3+}]} \tag{32}$$

$$E = E_T^0 + 0.06\log\frac{[Fe^{3+}]}{[Fe^{2+}]} \qquad \frac{E}{0.06} = \frac{E_T^0}{0.06} + \log\frac{[Fe^{3+}]}{[Fe^{2+}]} \qquad pe = pe_S^0 + \log\frac{[Fe^{3+}]}{[Fe^{2+}]} \tag{33}$$

and taking into account Eqs. (28) and (29) are reached

$$pe = \frac{E}{0.06} \tag{34}$$

$$pe_T^0 = \frac{E_T^0}{0.06} = \log K_T \qquad \log K_T = \frac{1.44}{0.06} = 24 \quad K_T = 10^{24} \tag{35}$$

$$pe_S^0 = \frac{E_S^0}{0.06} = \log K_S \qquad \log K_S = \frac{0.68}{0.06} = 11.333 \quad K_S = 2.154 \cdot 10^{11} \tag{36}$$

The value of pe_{eq} is calculated, Eq. (32), from the expression

$$pe_{eq} = \frac{\log K_T + \log K_s}{a+b} = \frac{24 + 11.333}{1+1} = 17.667 \tag{37}$$

whereby the potential at the point of equivalence is given by the expressions

$$E_{eq} = pe_{eq} \cdot 0.06 = 17.667 \cdot 0.06 = 1.06; \qquad E_{eq} = \frac{E_T^0 + E_S^0}{2} = \frac{1.44 + 0.68}{2} = 1.06 \tag{38}$$

Note that

$$T = \Delta T + 1 \tag{39}$$

which allows us to calculate the titration curve (**Figure 1**)

$$pe = f(T) \qquad E = f(T) \tag{40}$$

or the graph of the titration error (**Figure 2**)

$$\Delta T = f(\Delta pe) \qquad \Delta T = f(E) \tag{41}$$

The required calculations are detailed (from 0.62 to 0.82 v) in **Table 1**. Note that when $T = 0.5$, the potential value, $E = 0.68$ v, coincides with the normal potential of the Fe^{3+}/Fe^{2+}. Continuing the calculations would prove that when $T = 2$, $E = 1.44$ v, normal potential value of the Ce^{4+}/Ce^{3+} pair.

4.2. Tl^{+} titration curve with Ce^{4+}

For the system $Tl^{3+} + 2e = Tl^{+}$ ($E_{0S} = 1.25$ v) the equilibrium constant K_s and *pe*

$$K_S = \frac{[Tl^{+}]}{[Tl^{3+}][e]^2} \qquad pe = \frac{\log K_S}{2} + \frac{1}{2}\log\frac{[Tl^{3+}]}{[Tl^{+}]} \tag{42}$$

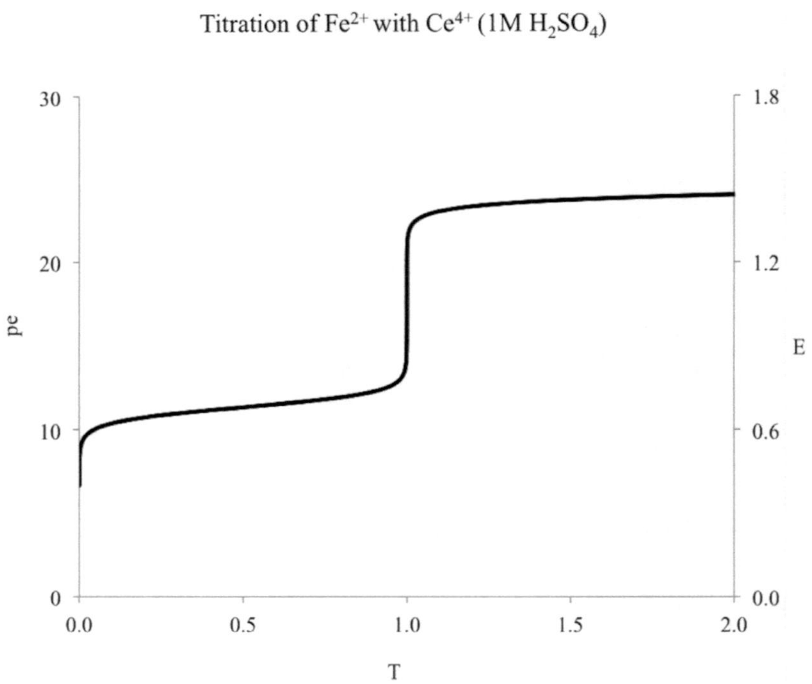

Figure 1. Titration curve of Fe^{2+} with Ce^{4+} in acid medium (H_2SO_4 1 M).

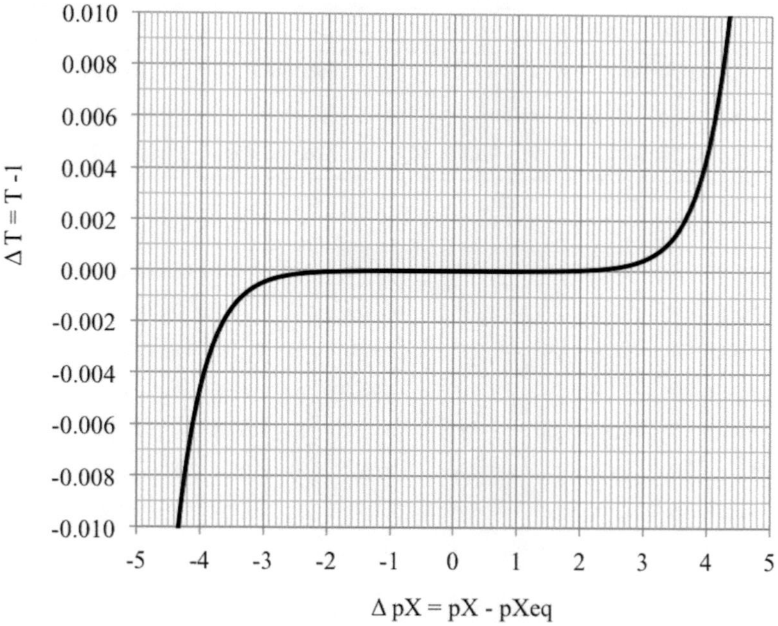

Figure 2. Titration error diagram $\Delta T = f(\Delta pX)$.

The overall reaction and its equilibrium constant are expressed as

$$2Ce^{4+} + Tl^{+} = 2Ce^{3+} + Tl^{3+} \qquad K = \frac{[Ce^{3+}]^2[Tl^{3+}]}{[Ce^{4+}]^2[Tl^{+}]} = \frac{K_T^2}{K_S} \tag{43}$$

Applying the Nernst equation to the half-reaction Tl^{3+}/Tl^{+}

$$E = E_S^0 + \frac{0.06}{2}\log\frac{[Tl^{3+}]}{[Tl^{+}]} \qquad \frac{E}{0.06} = \frac{E_S^0}{0.06} + \frac{1}{2}\log\frac{[Tl^{3+}]}{[Tl^{+}]} \qquad pe = pe_S^0 + \frac{1}{2}\log\frac{[Tl^{3+}]}{[Tl^{+}]} \tag{44}$$

$$pe_S^0 = \frac{E_S^0}{0.06} = \frac{\log K_S}{2} \qquad \log K_S = 2\frac{1.25}{0.06} = 41.666 \qquad K_S = 4.642 \cdot 10^{41} \tag{45}$$

$$pe_{eq} = \frac{\log K_T + \log K_s}{a+b} = \frac{24 + 41.666}{1+2} = 21.889 \tag{46}$$

$$E_{eq} = pe_{eq} \cdot 0.06 = 21.889 \cdot 0.06 = 1.313; \; E_{eq} = \frac{E_T^0 + 2E_S^0}{1+2} = \frac{1.44 + 21.25}{3} = 1.313 \tag{47}$$

The values of K_S (Eq. 45), K_T (Eq. 35), pe (Eq. 34) and pe_{eq} (Eq. 46) can be replaced in Eq. (22), taking into account that $a = 2$ and $b = 1$

Ce(IV) + Fe(II) = Ce(III) + Fe(III)

E0T=1.44	*neT*=1	*nHT*=0	*f*(*pHT*)= 0	*pH*=−0.3	1M H_2SO_4
E0S=0.68	*neS*=1	*nHS*=0	*f*(*pHS*)= 0	*E0T′*=1.44	
KT′=1.000E+24	*KS′*=2.154E+11	*Wc*=9.283E−07	*pXeq*=17.667	*E0S′*=0.68	

Ws	[*X*]^(*a*−*b*)/2	*W*	sine h	Δ*T*	*T*	*E*	*X*	*pX*	Δ*pX*	LOG (ABS(Δ*T*))
0.091	1	8.439E−08	−1.077E+07	−0.9091	0.0909	0.62	4.642E−11	10.333	−7.333	−0.041
0.128	1	1.188E−07	−7.339E+06	−0.8720	0.1280	0.63	3.162E−11	10.500	−7.167	−0.059
0.177	1	1.645E−07	−5.000E+06	−0.8227	0.1773	0.64	2.154E−11	10.667	−7.000	−0.085
0.240	1	2.230E−07	−3.406E+06	−0.7597	0.2403	0.65	1.468E−11	10.833	−6.833	−0.119
0.317	1	2.943E−07	−2.321E+06	−0.6830	0.3170	0.66	1.000E−11	11.000	−6.667	−0.166
0.405	1	3.762E−07	−1.581E+06	−0.5948	0.4052	0.67	6.813E−12	11.167	−6.500	−0.226
0.500	1	4.642E−07	−1.077E+06	−0.5000	0.5000	0.68	4.642E−12	11.333	−6.333	−0.301
0.595	1	5.521E−07	−7.339E+05	−0.4052	0.5948	0.69	3.162E−12	11.500	−6.167	−0.392
0.683	1	6.340E−07	−5.000E+05	−0.3170	0.6830	0.70	2.154E−12	11.667	−6.000	−0.499
0.760	1	7.053E−07	−3.406E+05	−0.2403	0.7597	0.71	1.468E−12	11.833	−5.833	−0.619
0.823	1	7.638E−07	−2.321E+05	−0.1773	0.8227	0.72	1.000E−12	12.000	−5.667	−0.751
0.872	1	8.095E−07	−1.581E+05	−0.1280	0.8720	0.73	6.813E−13	12.167	−5.500	−0.893
0.909	1	8.439E−07	−1.077E+05	−0.0909	0.9091	0.74	4.642E−13	12.333	−5.333	−1.041
0.936	1	8.691E−07	−7.339E+04	−6.378E−02	0.9362	0.75	3.162E−13	12.500	−5.167	−1.195
0.956	1	8.871E−07	−5.000E+04	−4.436E−02	0.9556	0.76	2.154E−13	12.667	−5.000	−1.353
0.969	1	8.999E−07	−3.406E+04	−3.065E−02	0.9693	0.77	1.468E−13	12.833	−4.833	−1.514
0.979	1	9.087E−07	−2.321E+04	−2.109E−02	0.9789	0.78	1.000E−13	13.000	−4.667	−1.676
0.986	1	9.149E−07	−1.581E+04	−1.447E−02	0.9855	0.79	6.813E−14	13.167	−4.500	−1.840
0.990	1	9.191E−07	−1.077E+04	−9.901E−03	0.9901	0.80	4.642E−14	13.333	−4.333	−2.004
0.993	1	9.220E−07	−7.339E+03	−6.767E−03	0.9932	0.81	3.162E−14	13.500	−4.167	−2.170
0.995	1	9.240E−07	−5.000E+03	−4.620E−03	0.9954	0.82	2.154E−14	13.667	−4.000	−2.335

Table 1. Titration curve of Fe(II) with Ce(IV): hyperbolic sine method.

$$\begin{aligned} \Delta T &= \frac{2\sqrt{K_S/K_T}}{1+[\mathrm{e}]^2 K_S}[\mathrm{e}]^{\frac{2-1}{2}}\sinh\left(\ln 10(2+1)\left(\frac{pe-pe_{eq}}{2}\right)\right) \\ &= \frac{2\sqrt{K_S/K_T}}{1+10^{-2pe}K_S}\sqrt{e}\,\sinh\left(\frac{3}{2}\ln 10(pe-21.889)\right) \end{aligned} \quad (48)$$

The titration curve $E = f(T)$ is shown in **Figure 3**, together with those corresponding to other half-reactions exchanging a single electron, VO^{2-}/VO_2^{+} ($E_0 = 1.001$ v), NO_3^{-}/NO_2 ($E_0 = 0.80$ v), and Fe^{3+}/Fe^{2+} ($E_0 = 0.68$ v).

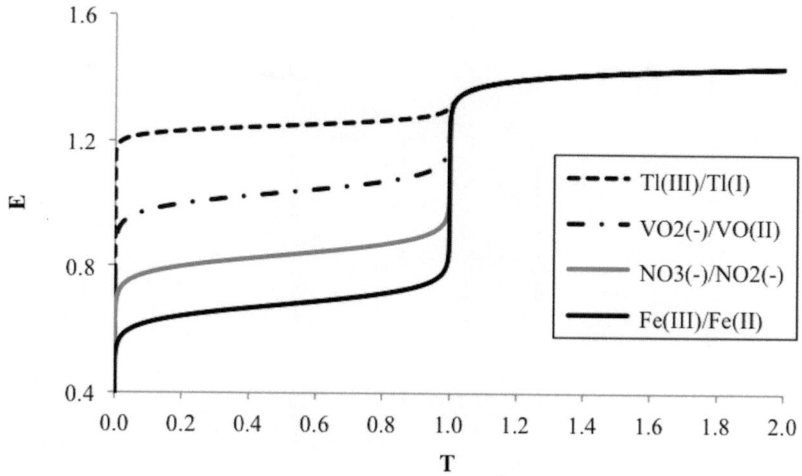

Figure 3. Titration curves of several redox systems with Ce^{4+} as a titrant.

4.3. $Fe^{2+} + Tl^{+}$ titration curve with Ce^{4+}

In this particular case, the total ΔT function (or total T) is additive, i.e., ΔT is the sum of the values of ΔT (T) corresponding to the individual titration of Fe^{+2} with Ce^{+4} and Tl^{+} with Ce^{+4}. The reactions and equations involved have been previously described in Sections 4.1. and 4.2.. Thus, in this case we should only sum the values given by Eqs. 31 (Fe^{+2} with Ce^{+4}) and 48 (Tl^{+} with Ce^{+4}). The corresponding titration curve of a mixture of Fe^{2+} and Tl^{+} with Ce^{4+} calculated in this way is shown in **Figure 4**.

4.4. Fe^{2+} titration curve with MnO_4^- as titrant

For the half-reaction $MnO^{4-} + 5\,e + 8\,H^+ = Mn^{2+} + 4\,H_2O$ ($E_{0T} = 1.51$ v)

$$K_T = \frac{[Mn^{2+}]}{[MnO_4^-][e]^5[H^+]^8} \qquad pe = \frac{\log K_T}{5} - \frac{8}{5}pH + \frac{1}{5}\log\frac{[MnO_4^-]}{[Mn^{2+}]} \tag{49}$$

$$E = E_0^T + \frac{0.06}{5}\log\frac{[MnO_4^-][H^+]^8}{[Mn^{2+}]} = E_0^T - \frac{0.06 \cdot 8}{5}pH + \frac{0.06}{5}\log\frac{[MnO_4^-]}{[Mn^{2+}]} \tag{50}$$

and thus following the previous procedure

$$\frac{E}{0.06} = \frac{E_0^T}{0.06} - \frac{8}{5}pH + \frac{1}{5}\log\frac{[MnO_4^-]}{[Mn^{2+}]} \qquad pe = pe_0^T - \frac{8}{5}pH + \frac{1}{5}\log\frac{[MnO_4^-]}{[Mn^{2+}]} \tag{51}$$

$$pe = \frac{E}{0.06} \tag{52}$$

$$pe_0^T = \frac{E_0^T}{0.06} = \frac{\log K_T}{5} \qquad \log K_T = 5\frac{E_0^T}{0.06} \qquad K_T = 10^{5\frac{E_0^T}{0.06}} \tag{53}$$

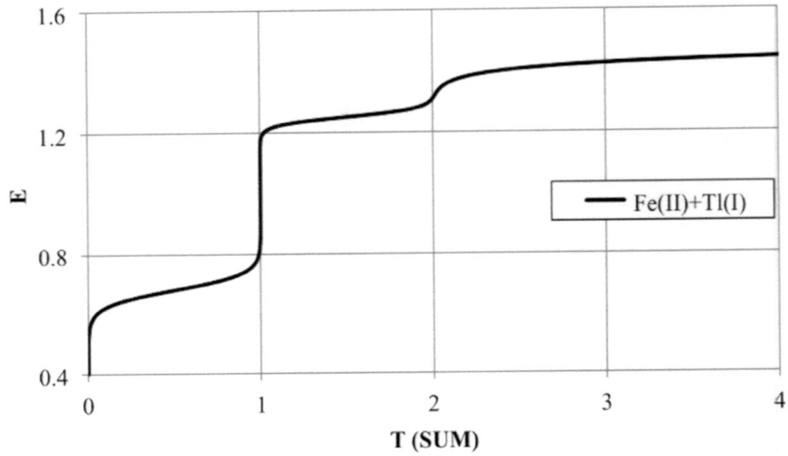

Figure 4. Titration curve of a mixture of Fe^{2+} and Tl^{+} with Ce^{4+}.

$$pe_0^{T'} = pe_0^T - \frac{8}{5}pH = \frac{\log K_T}{5} - \frac{8}{5}pH = \frac{\log K'_T}{5} \qquad \log K'_T = \log K_T - 8pH \tag{54}$$

The terms of Eq. (54) are conditional (pH-dependent). In the equivalence point

$$pe'_{eq} = \frac{\log K'_T + \log K'_S}{a+b} \tag{55}$$

The expression for the valuation error will be given by ($a = 1, b = 5$)

$$\Delta T = \frac{2\sqrt{K'_S/K'_T}}{1+[e]K'_S}[e]^{\frac{1-5}{2}}\sinh\left(\ln 10(5+1)\left(\frac{\Delta pe'}{2}\right)\right) = \frac{2\sqrt{K'_S/K'_T}}{1+[e]K'_S}[e]^{-2}\sinh(3\ln 10(pe - pe'_{eq})) \tag{56}$$

The curve and the titration error are shown in **Figures 5** and **6**, respectively, at different pH values. Part of the necessary calculations, at pH = 0, is shown in **Table 2**. When $T = 2$ and pH = 0, $E = 1.51$ v, normal MnO_4^-/Mn^{2+} system potential.

4.5. Titration curve of V^{2+} with MnO_4^-

Oxido-reductor systems involving the vanadium oxidation states 2+, 3+, 4+, and 5+ are shown in **Table 3**. The V^{2+} ion undergoes various successive ionizations at 3+, 4+, and 5+ when MnO_4^- is added, being appreciated in the curve (**Figure 7**), the three corresponding jumps. The total ΔT function (or total T) is additive. The titration curve is the sum of the contributions of each individual reaction

$$\Delta T_{SUM} = \Delta T_{32} + \Delta T_{43} + \Delta T_{54} \tag{57}$$

$$T_{SUM} = \Delta T_{SUM} - 3 = T_{32} + T_{43} + T_{54} \tag{58}$$

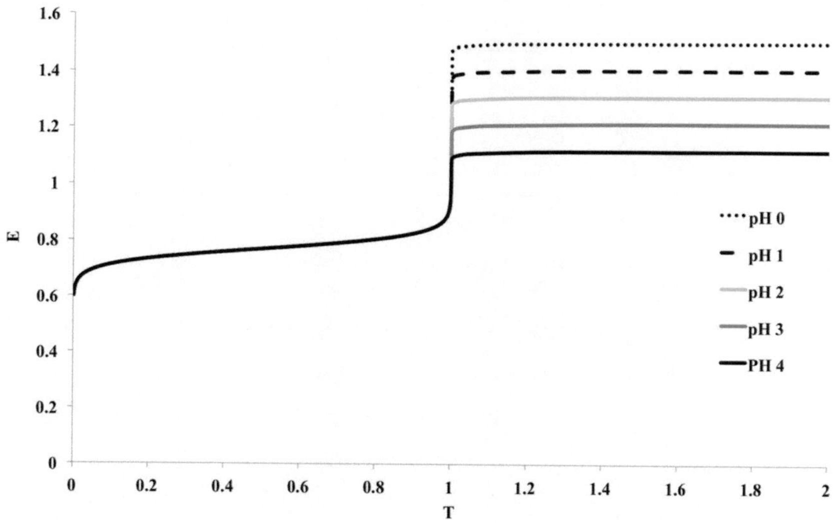

Figure 5. Titration curve of Fe^{2+} with MnO_4^-, at different *pH* values.

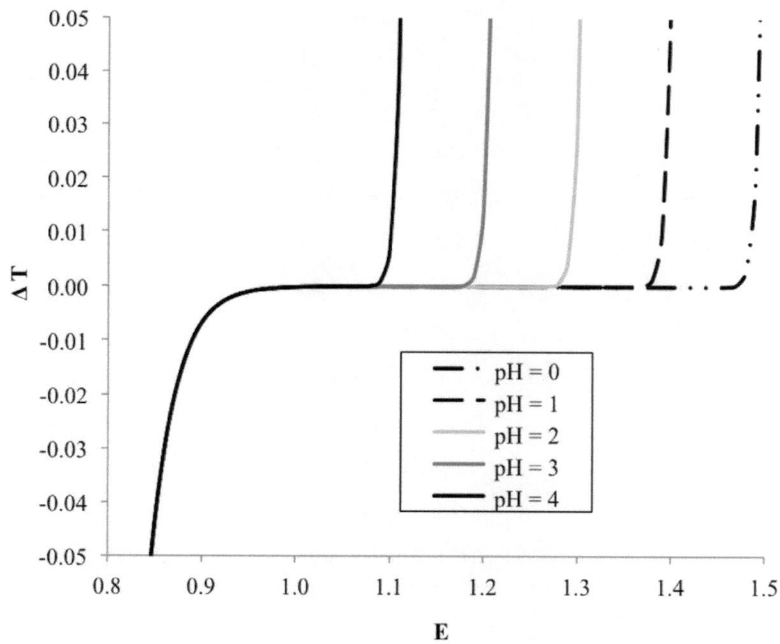

Figure 6. Titration error diagram $\Delta T = f(E)$ at different *pH* values.

The head of the spreadsheet in EXCEL with all the necessary elements to carry out the numerical calculations applying the equations of the type of Eq. (22) is shown in **Table 3**. From the corresponding EXCEL sheet we have extracted a few columns, those corresponding to the

			pH=0	**$MnO_4^- + 5\ Fe(II) + 8\ H^+ = Mn(II) + 5\ Fe(III) + 4\ H_2O$**
E0T=1.51	*neT*=5	*nHT*=8	*f(pHT)*= 0	*E0T'*=1.51
E0S=0.77	*neS*=1	*nHS*=0	*f(pHS)*= 0	*E0S'*=0.77
KT'=6.8129E+125	*KS'*=6.81292E+12	*Wc*=6.32456E−57	*pXeq*=23.111	*Eeq*=1.387

Ws	[X]^(*a*−*b*)/2	W	sine h	Δ*T*	*T*	E	X	*pX*	Δ*pX*	LOG(ABS (Δ*pX*))
0.0909	4.642E+23	2.669E−34	−3.406E+33	−0.9091	0.0909	0.71	1.468E−12	11.833	−11.278	−4.139E−02
0.1280	1.000E+24	8.095E−34	−1.077E+33	−0.8720	0.1280	0.72	1.000E−12	12.000	−11.111	−5.948E−02
0.1773	2.154E+24	2.415E−33	−3.406E+32	−0.8227	0.1773	0.73	6.813E−13	12.167	−10.944	−8.473E−02
0.2403	4.642E+24	7.053E−33	−1.077E+32	−0.7597	0.2403	0.74	4.642E−13	12.333	−10.778	−1.193E−01
0.3170	1.000E+25	2.005E−32	−3.406E+31	−0.6830	0.3170	0.75	3.162E−13	12.500	−10.611	−1.656E−01
0.4052	2.154E+25	5.521E−32	−1.077E+31	−0.5948	0.4052	0.76	2.154E−13	12.667	−10.444	−2.256E−01
0.5000	4.642E+25	1.468E−31	−3.406E+30	−0.5000	0.5000	0.77	1.468E−13	12.833	−10.278	−3.010E−01
0.5948	1.000E+26	3.762E−31	−1.077E+30	−0.4052	0.5948	0.78	1.000E−13	13.000	−10.111	−3.923E−01
0.6830	2.154E+26	9.306E−31	−3.406E+29	−0.3170	0.6830	0.79	6.813E−14	13.167	−9.944	−4.989E−01
0.7597	4.642E+26	2.230E−30	−1.077E+29	−0.2403	0.7597	0.80	4.642E−14	13.333	−9.778	−6.193E−01
0.8227	1.000E+27	5.203E−30	−3.406E+28	−0.1773	0.8227	0.81	3.162E−14	13.500	−9.611	−7.514E−01
0.8720	2.154E+27	1.188E−29	−1.077E+28	−0.1280	0.8720	0.82	2.154E−14	13.667	−9.444	−8.928E−01
0.9091	4.642E+27	2.669E−29	−3.406E+27	−0.0909	0.9091	0.83	1.468E−14	13.833	−9.278	−1.041E+00
0.9362	1.000E+28	5.921E−29	−1.077E+27	−0.0638	0.9362	0.84	1.000E−14	14.000	−9.111	−1.195E+00
0.9556	2.154E+28	1.302E−28	−3.406E+26	−0.0444	0.9556	0.85	6.813E−15	14.167	−8.944	−1.353E+00
0.9693	4.642E+28	2.846E−28	−1.077E+26	−0.0307	0.9693	0.86	4.642E−15	14.333	−8.778	−1.514E+00
0.9789	1.000E+29	6.191E−28	−3.406E+25	−0.0211	0.9789	0.87	3.162E−15	14.500	−8.611	−1.676E+00
0.9855	2.154E+29	1.343E−27	−1.077E+25	−0.0145	0.9855	0.88	2.154E−15	14.667	−8.444	−1.840E+00
0.9901	4.642E+29	2.907E−27	−3.406E+24	−0.0099	0.9901	0.89	1.468E−15	14.833	−8.278	−2.004E+00
0.9932	1.000E+30	6.282E−27	−1.077E+24	−0.0068	0.9932	0.90	1.000E−15	15.000	−8.111	−2.170E+00
0.9954	2.154E+30	1.356E−26	−3.406E+23	−0.0046	0.9954	0.91	6.813E−16	15.167	−7.944	−2.335E+00

Table 2. Titration curve of Fe(II) with $MnO_4(-)$.

values of *E*, *pX*, *T*1, *T*2, *T*3 and T_{SUM} (1, 2, 3 refer to 32, 43, and 54, respectively, in Eqs. 57 and 58, that is, the oxidations of V^{2+} to V^{3+}, from this to VO^{2+} and from this to VO_2^{+}), which are shown in **Table 4**.

Potential values varying from 0.05 in 0.05 units have been varied in this table in order to cover the entire valuation curve. **Figure 7** has been drawn instead with potential variations of only 0.01 v.

http://dx.doi.org/10.5772/intechopen.68750

	E0T	$MnO_4^- + 8\,H^+ + 5e = Mn(II) + 4\,H_2O$		
	E032	$V(III) + e = V(II)$		
	E043	$VO(II) + 2\,H^+ + e = V(III) + H_2O$		
	E054	$VO_2(I) + 2\,H^+ + e = VO(II) + H_2O$		*pH*=0
E0T=1.51	*neT*=5	*nHT*=8	*f*(*pHT*)= 0	E0T′=1.51
E0S32=−0.255	*neS32*=1	*nHS32*=0	*f*(*pHS32*)= 0	E0S32'=−0.255
E0S43=0.337	*neS43*=1	*nHS43*=2	*f*(*pHS43*)= 0	E0S43'=0.337
E0S54=1.001	*neS54*=1	*nH54*=2	*f*(*pH54*)= 0	E0S54'=1.001
KT'=6.813E+125				
KS32'=5.523E−05	*Wc32*=1.817E−65	*pXeq1*=20.264		
KS43'=4.137E+05	*Wc43*=1.558E−60	*pXeq2*=21.908		
KS54'=4.32318E+16	*Wc54*=5.321E−55	*pXeq3*=23.753		

Table 3. Head of the spreadsheet in EXCEL of the titration curve of V(II) with MnO_4.

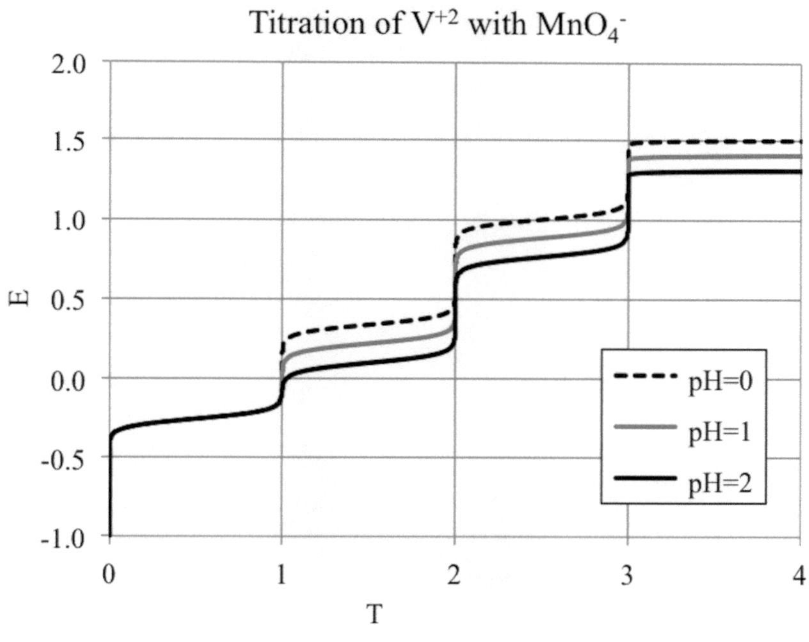

Figure 7. Titration curve of V^{2+} with MnO_4^-, at *pH* 0, 1 and 2.

5. Criterion for the quantitative titration and the influence of side-reactions

The principle of redox titrations is that the solution of a reducing agent is titrated with a solution of an oxidizing agent (or vice versa)

*T*1	*T*2	*T*3	*T* SUM	*E*	*pX*
0.0001	0	0	0.0001	−0.5	−8.333
0.0006	0	0	0.0006	−0.45	−7.5
0.0038	0	0	0.0038	−0.4	−6.667
0.0254	0	0	0.0254	−0.35	−5.833
0.151	0	0	0.151	−0.3	−5
0.5478	0	0	0.5478	−0.25	−4.167
0.8919	0	0	0.8919	−0.2	−3.333
0.9825	0	0	0.9825	−0.15	−2.5
0.9974	0	0	0.9974	−0.1	−1.667
0.9996	0	0	0.9996	−0.05	−0.833
0.9999	0	0	0.9999	0	0
1	0	0	1	0.05	0.833
1	0.0001	0	1.0001	0.1	1.667
1	0.0008	0	1.0008	0.15	2.5
1	0.0052	0	1.0052	0.2	3.333
1	0.0343	0	1.0343	0.25	4.167
1	0.1947	0	1.1947	0.3	5
1	0.6222	0	1.6222	0.35	5.833
1	0.9182	0	1.9182	0.4	6.667
1	0.9871	0	1.9871	0.45	7.5
1	0.9981	0	1.9981	0.5	8.333
1	0.9997	0	1.9997	0.55	9.167
1	1	0	2	0.6	10
1	1	0	2	0.65	10.833
1	1	0	2	0.7	11.667
1	1	0.0001	2.0001	0.75	12.5
1	1	0.0004	2.0004	0.8	13.333
1	1	0.003	2.003	0.85	14.167
1	1	0.0203	2.0203	0.9	15
1	1	0.1238	2.1238	0.95	15.833
1	1	0.4904	2.4904	1	16.667
1	1	0.8677	2.8677	1.05	17.5
1	1	0.9781	2.9781	1.1	18.333
1	1	0.9967	2.9967	1.15	19.167
1	1	0.9995	2.9995	1.2	20
1	1	0.9999	2.9999	1.25	20.833

T1	*T2*	*T3*	*T SUM*	*E*	*pX*
1	1	1	3	1.3	21.667
1	1	1	3	1.35	22.5
1	1	1	3	1.4	23.333
1	1	1	3	1.45	24.167
1.1468	1.1468	1.1468	3.4403	1.5	25
2	2	2	6	1.51	25.167

Table 4. Titration curves of V(II) with MnO_4^-.

$$yOx_T + z\mathrm{Red}_S \leftrightarrow y\mathrm{Red}_T + zOx_S \tag{59}$$

At the equivalence point

$$pe_{\mathrm{eq}} = \frac{1}{z+y}(\log K_S + \log K_T) \tag{60}$$

The criterion of the quantitative titration can be deduced if we consider that the substance to be determined must be oxidized (or reduced) during the titration to an extent of 99.9% [23]. This means that the amount of determinant remaining untitrated at the equivalence point should not exceed 0.1% of that originally present, i.e.,

$$\log \frac{[Ox_S]}{[\mathrm{Red}_S]} \geq 3 \tag{61}$$

and

$$pe_{\mathrm{eq}} > ({pe_{\mathrm{eq}}}^0)_S + 3\frac{1}{y} \tag{62}$$

$$({pe_{\mathrm{eq}}}^0)_S = \frac{1}{y}\log K_S \tag{63}$$

From Eqs. (60) and (63) also we obtain

$$y\log K_T - z\log K_S > 3(z+y) \tag{64}$$

If both oxidation-reduction systems in the titration involve two electrons, the difference between the log K values must be greater than 6.

In addition, if the oxidized or reduced product present in the solution containing the redox system takes part in a side-reaction, and the equilibrium position of this reaction can be kept constant, by maintaining suitable experimental conditions, the conditional oxidation-reduction constant, *K*', can be deduced and used similarly to those used in complex chemistry

$$K' = \frac{[Red']}{[Ox'][e^-]^z} \tag{65}$$

[*Red*'] and [*Ox*'] are analytical concentrations without any respect to side-reactions. The connection between the conditional and real constants is the following one

$$K' = K\frac{\alpha_{\mathrm{Red(A)}}}{\alpha_{\mathrm{Ox(B)}}} \tag{66}$$

where $\alpha_{\mathrm{Red}(A)}$ and $\alpha_{\mathrm{Ox}(B)}$ are the side-reaction functions, and A and B denote the substances reacting with the reduced and oxidized substance, respectively.

$$\alpha_{\mathrm{Red}(A)} = 1 + [L]\beta_1 + [L]^2\beta_2 + \ldots \tag{67}$$

$$\alpha_{\mathrm{Ox}(B)} = 1 + [L]\beta_1^* + [L]\beta_2^* + \ldots \tag{68}$$

[*A*] and [*B*] are concentrations of the species reacting with the reduced and oxidized form, respectively, β's and β^*'s are complex products or protonization constants products.

In all calculations concentration constants can be used if they are corrected to the corresponding ionic strength.

In practice, the most important side-reactions are complex formation and protonation. The oxidized and reduced form of a metal ion may form complexes of different stabilities with the complexing ligand L.

So, if the criterion of the quantitative determination is not fulfilled by a suitable *pH* change or by the use of a complexing agent that shifts the values of the conditional constants, the titration may be realized. For example, according to Vydra and Pribil [101], cobalt (II) can be titrated with iron (III) ions if 1,10-phenanthroline is added to the solution, and the *pH* is adjusted to 3 even though $K_{\mathrm{Co}} >>> K_{\mathrm{Fe}}$.

On the other hand, if another component present in the solution has similar oxidizing or reducing properties, then interfering species can be masked, so that the conditional redox constant of the interfering system is changed to such an extent that it no longer interferes with the main reaction.

5.1. Practical examples

(1) Calculate the pH necessary for the accurate direct titration of potassium hexacyanoferrate (III) with ascorbic acid, given that log $K_{\mathrm{Fe(CN)6}} = 6.1$; the protonation constants of hexacyanoferrate(II) are log $K_1 = 4.17$, log $K_2 = 2.22$, log $K_3 < 1$, log $K_4 < 1$; the logarithms of all the protonation constants of hexacyanoferrate(III) are >1. The equilibrium constant of the dehydroascorbic acid-ascorbinate redox system is log $K_{\mathrm{A}} = -2.5$; the protonation constants of the ascorbinate ion are log $K_1 = 11.56$ and 4.17.

The criterion for the feasibility of the titration, according to Eq. (64) is:

$$2\log K'_{\mathrm{Fe(CN)}_6} - \log K'_A > 3(1+2) = 9 \tag{69}$$

From the protonation constants of hexacyanoferrate(II), if the $pH > 5.5$, then $\alpha_{\mathrm{Fe(CN)6(H)}} =$ 1 and

$$\log K'_{\mathrm{Fe(CN)}_6} = \log K_{\mathrm{Fe(CN)}_6} = 6.1. \tag{70}$$

Therefore 2 x 6.1 − $\log K'_A > 9$; or $\log K'_A < 3.2$. If the $pH = 6$, then $\alpha_{A(\mathrm{H})} = 1 + 10^{-6} \times 10^{11.57} + 10^{-12} \times 10^{15.73} = 10^{5.58}$

$$\log K'_A = 3.1 \tag{71}$$

Thus, if the $pH > 6$, the titration can be performed with adequate accuracy.

(2) Calculate the $[L]$ maximum for the accurate direct titration of Ce^{+4} with Fe^{2+}, being L the organic complexant (acetylacetone) present in the solution and given that $\log K_T = 24$ and $\log K_S = 11.3$ (see Eqs. (28)–(30)); the global constants of Fe-(L)(II) are $\log \beta_1 = 5.07$; $\log \beta_2 = 8.67$.

The reactions involved would be

$$\begin{aligned} &\mathrm{Ce}^{4+} + \mathrm{Fe}^{2+} \leftrightarrow \mathrm{Ce}^{3+} + \mathrm{Fe}^{3+} \\ &\mathrm{Fe}^{2+} + L \leftrightarrow \mathrm{Fe}L^{2+} \\ &\mathrm{Fe}L^{2+} + L \leftrightarrow \mathrm{Fe}L_2^{2+} \end{aligned} \tag{72}$$

The criterion for the feasibility of the titration, according to Eq. (64) is

$$\log K_T - \log K'_S > 3(1+1) = 6 \tag{73}$$

$$\begin{aligned} &\log K_T - 6 > \log K'_S, \quad \log K'_S < 18 \\ &K'_S = K_S \alpha_{[\mathrm{FeL_2}]^{+2}}, \quad \alpha_{[\mathrm{FeL_2}]^{+2}} < 10^{6.7} \end{aligned} \tag{74}$$

$$\alpha_{[\mathrm{Fe}L_2]^{+2}} = 1 + [L]\beta_1 + [L]^2\beta_2 \tag{75}$$

Thus, if the $[L] < 0.1$ M, the titration can be performed with adequate accuracy.

(3) Calculate the $[SO_4{}^{-2}]$ maximum present in the solution for the accurate direct titration of Ce^{4+} with Tl^{+}, given that $\log K_T = 24$ and $\log K_S = 41.6$ (see Eqs. (28), (42), (43)); the global constants of Ce-SO₄(IV) are $\log \beta_1 = 3.5$; $\log \beta_2 = 8.0$; $\log \beta_3 = 10.4$.

The criterion for the feasibility of the titration, according to (64) is:

$$2\log K'_T - \log K_S > 3(1+2) = 9 \tag{76}$$

$$\log K'_T > 25.3$$

$$K_T' = K_T \frac{1}{\alpha_{Ce-SO_4}}, \quad \alpha_{Ce-SO_4} < 10^{7.2} \tag{77}$$

$$\alpha_{[FeL_2]^{+2}} = 1 + [SO_4^{-2}]\beta_1 + [SO_4^{-2}]^2\beta_2 + [SO_4^{-2}]^3\beta_3 \tag{78}$$

Thus, $[SO_4^{-2}]$ should be < 0.39 M, to perform the titrimetry with accuracy.

6. Final comments

As a matter of fact redox titrations play a prominent role in volumetric analysis of redox actives species. A systematic study of the bibliography is undertaken in order to ascertain the state of the art concerning to redox titration curves. A method for the determination of titration error in donor/acceptor titrations of displacement and electronic transference reactions has been devised; a hyperbolic sine expression being derived for the titration error, applicable to symmetrical reactions (no polynuclear species being involved in one side of a half-reaction). The hyperbolic sine expression developed is compact and allows calculating the entire titration curve without piecemeal approximations, as usually occurs by dividing the titration curve in three parts: before, in, and beyond the equivalence point regions.

The method has been applied to some experimental systems characteristics of redox titration reactions. The method proposed is also applicable to mixtures of analytes, e.g., Fe(II) + Tl(I), as well as to multistep redox titrations, e.g., V(II)/V(III)/V(IV)/V(V) system. The forms of the redox titration curves are independent of the concentrations. However, when the concentrations involved are very low the responses of the electrodes are not appropriate. All calculations involved have been checked with the method proposed by "de Levie" [62] for the sake of comparison, and no differences were found in the numerical values obtained by both methods. A diagram for the titration error in function of the difference between the end and equivalence point (pX) is drawn in order to facilitate the graphical calculation of titration error.

Automatic titrators enable recording automatically the change with potential (*E*) or *pH* in titre during a given titration. The accuracy of the measurements can increase with the help of on-line microcomputer for the control and data acquisition, allowing among the possibility for curve-smoothing and differentiation.

The extension of the method to nonhomogeneous systems of the type $Cr_2O_7^{2-}/Cr^{3+}$, I_2/I^-, or $S_4O_6^{2-}/S_2O_3^{2-}$ remains a challenge, this being a complex problem involving a complete reformulation of the presented equations, which implies a higher level of difficulty.

At the end of the chapter an appendix including a detailed study of the propagation of systematic and random errors on redox titration error has been carried out and spite of the complex expression obtained first on differentiation, the final expressions formulated were very compact. This topic is still under study and it will be dealt in further calculus.

Appendix

Error analysis

The exact calculation of standard deviations of nonlinear function of variables that are subject to error is generally a problem of great mathematical complexity. A linearization based on a Taylor expansion of the nonlinear portion of the expansion allows to obtain approximate estimates of standard deviations [102]; this approximation is quite adequate for most practical applications.

ΔT is a function of several variables, i.e. pX, $\log K_S$, $\log K_T$ all independent of each other. The systematic error present in pX, $\log K_S$, and $\log K_T$, are propagated to give an overall systematic error in a calculate quantity

$$E_{\mathrm{sys}(\Delta T)} = \left(\frac{\partial \Delta T}{\partial pX}\right) E_{\mathrm{sys}(pX)} + \left(\frac{\partial \Delta T}{\partial \log K_S}\right) E_{\mathrm{sys}(\log K_S)} + \left(\frac{\partial \Delta T}{\partial \log K_T}\right) E_{\mathrm{sys}(\log K_T)} \tag{79}$$

provided the errors $E_{\mathrm{sys}}(pX)$, $E_{\mathrm{sys}}(\log K_S)$, $E_{\mathrm{sys}}(\log K_T)$ are small enough for higher order derivatives to be discarded.

For random error [103], the variance of ΔT can be calculated according to the propagation of variance

$$E^2_{\mathrm{ran}(\Delta T)} = \left(\frac{\partial \Delta T}{\partial pX}\right)^2 s^2_{pX} + \left(\frac{\partial \Delta T}{\partial \log K_S}\right) s^2_{\log K_S} + \left(\frac{\partial \Delta T}{\partial \log K_T}\right) s^2_{\log K_T} \tag{80}$$

where s_{pX}^2, $s_{\log Ks}^2$, and $s_{\log KT}^2$ are the variances of the components pX, $\log K_S$, and $\log K_T$, respectively. The partial derivatives are taken, as before, as values equal or closest to the measured values.

For any measurement the total absolute error $E_{\mathrm{abs}}(\Delta T)$ is related to the different types of error present by

$$E_{\mathrm{abs}(\Delta T)} = \Delta T - \Delta \tau = E_{\mathrm{ran}} + E_{\mathrm{sys}} + E_{\mathrm{bl}} \tag{81}$$

where ΔT is the value of the measurement, $\Delta \tau$ the true value, E_{rand} the random error, E_{sys} is the systematic error, and E_{bl} is the error due to blunders.

In any case, in order to know the proper error, it is necessary to know the standard deviation of the experimentally measured quantities, i.e., $E_{\mathrm{sys}}(\mathrm{i})$ and s_i.

For the sake of convenience Eq. (22) may be put in the form

$$\Delta T = \frac{A}{1 + [X]^a K_S} [X]^p \sinh(C \Delta pX) \tag{82}$$

where

$$A = 2\sqrt{\frac{K_S}{K_T}} \tag{83}$$

$$p = \frac{a-b}{2} \tag{84}$$

$$C = \ln 10 \frac{a+b}{2} = \ln 10q \tag{85}$$

Make now

$$u = \frac{A[X]^p}{1+[X]^a K_S} \tag{86}$$

and

$$v = \sinh\left(C\Delta pX\right) \tag{87}$$

in order to may differentiate easily ΔT against d[X]. Thus

$$\Delta T = uv \tag{88}$$

and we get for the derivative of a product

$$(\Delta T)' = u'v + v'u \tag{89}$$

The derivative of u will be given by

$$u' = A\left(\frac{p[X]^{p-1}(1+[X]^a K_S) - K_S a[X]^{a-1}[X]^p}{(1+[X]^a K_S)^2}\right) = A\left(\frac{p[X]^{p-1}(1+[X]^a K_S) - K_S a[X]^a[X]^{p-1}}{(1+[X]^a K_S)^2}\right)$$

$$= \frac{A[X]^{p-1}}{1+[X]^a K_S}\left(\frac{p - aK_S[X]^a}{1+[X]^a K_S}\right) \tag{90}$$

On the other hand

$$v' = \cosh(C\Delta pX)C\frac{\partial(\Delta pX)}{\partial[X]} = \cosh(C\Delta pX)C\frac{\partial pX}{\partial[X]} = -\cosh(C\Delta pX)C\frac{1}{\ln 10[X]} \tag{91}$$

Taking into account Eqs. (87)–(91), we get

$$\frac{\partial \Delta T}{\partial[X]} = \frac{A[X]^{p-1}}{1+[X]^a K_S}\left(\frac{p - aK_S[X]^a}{1+[X]^a K_S}\right)\sinh(C\Delta pX) - \frac{A[X]^p}{1+[X]^a K_S}\cosh(C\Delta pX)\frac{C}{\ln 10[X]} \tag{92}$$

and then

$$\frac{\partial \Delta T}{\partial [X]} = \frac{A[X]^{p-1}}{1+[X]^a K_S}\left(\left(\frac{p - aK_S[X]^a}{1+[X]^a K_S}\right)\sinh(C\Delta pX) - q\cosh(C\Delta pX)\right) \tag{93}$$

By multiplying through [X] and taking into account Eq. (31) we get

$$[X]\frac{\partial \Delta T}{\partial [X]} = \frac{\Delta T}{\sinh(C\Delta pX)}\left(\left(\frac{p - aK_S[X]^a}{1+[X]^a K_S}\right)\sinh(C\Delta pX) - q\cosh(C\Delta pX)\right) \tag{94}$$

and so

$$\frac{\partial \Delta T}{\partial pX} = -\ln 10[X]\frac{\partial \Delta T}{\partial [X]} = \ln 10\Delta T\left(q\coth(C\Delta pX) - \left(\frac{p - aK_S[X]^a}{1+[X]^a K_S}\right)\right) \tag{95}$$

Eq. (22) may be presented in the form

$$\Delta T = \frac{2[X]^p}{\sqrt{K_T}}\frac{K_S^{1/2}}{1+[X]^a K_S}\sinh(C\Delta pX) \tag{96}$$

Differentiation of Eq. (96) with respect to K_S gives

$$\frac{\partial \Delta T}{\partial K_S} = \frac{2[X]^p}{\sqrt{K_T}}\left(\left(\frac{\frac{1}{2}K_S^{\frac{1}{2}-1}(1+[X]^a K_S) - K_S^{\frac{1}{2}}[X]^a}{(1+[X]^a K_S)}\right)\sinh(C\Delta pX) + \frac{K_S^{\frac{1}{2}}}{(1+[X]^a K_S)}\cosh(C\Delta pX)C\frac{\partial \Delta pX}{\partial K_S}\right) \tag{97}$$

On the other hand

$$\frac{\partial \Delta pX}{\partial K_S} = \frac{\partial\left(pX_{\text{end}} - \frac{\log K_T + \log K_S}{a+b}\right)}{\partial K_S} = \frac{-1}{a+b}\frac{\partial \log K_S}{\partial K_S} = \frac{-1}{(a+b)\ln 10 K_S} \tag{98}$$

$$C\frac{\partial \Delta pX}{\partial K_S} = \ln 10\left(\frac{a+b}{2}\right)\frac{-1}{(a+b)\ln 10 K_S} = \frac{-1}{2K_S} \tag{99}$$

By combining Eqs. (97) and (99) we obtain

$$\frac{\partial \Delta T}{\partial K_S} = \frac{2[X]^p}{\sqrt{K_T}}\left(\left(\frac{\frac{1}{2}K_S^{-\frac{1}{2}} - \frac{1}{2}K_S^{\frac{1}{2}}[X]^a}{(1+[X]^a K_S)^2}\right)\sinh(C\Delta pX) - \frac{\frac{1}{2}K_S^{-\frac{1}{2}}}{(1+[X]^a K_S)}\cosh(C\Delta pX)\right) \tag{100}$$

which on rearranging gives

$$\frac{\partial \Delta T}{\partial K_S} = \frac{\frac{K_S^{-\frac{1}{2}}}{\sqrt{K_T}}}{1+[X]^a K_S}[X]^p\left(\left(\frac{1 - K_S[X]^a}{1+K_S[X]^a}\right)\sinh(C\Delta pX) - \cosh(C\Delta pX)\right) \tag{101}$$

Differentiating now ΔT against log K_S leads to

$$\frac{\partial \Delta T}{\partial \log K_S} = \ln 10 K_S \frac{\partial \Delta T}{\partial K_S} = \frac{\Delta T}{2\sinh(C\Delta pX)}\left(\left(\frac{1-K_S[X]^a}{1+K_S[X]^a}\right)\sinh(C\Delta pX) - \cosh(C\Delta pX)\right)$$

$$= \frac{\ln 10}{2}\Delta T\left(\left(\frac{1-K_S[X]^a}{1+K_S[X]^a}\right) - \coth(C\Delta pX)\right) \tag{102}$$

In order to differentiate ΔT against K_T we put (from Eq. (22))

$$\frac{\partial \Delta T}{\partial K_T} = \frac{2\sqrt{K_S}}{1+[X]^a K_S}[X]^p\left(\frac{\partial}{\partial K_T}\left(\frac{1}{\sqrt{K_T}}\right)\cdot \sinh(C\Delta pX) + \frac{1}{\sqrt{K_T}}\cosh(C\Delta pX) C\frac{\partial \Delta pX}{\partial K_T}\right) \tag{103}$$

From Eqs. (18), (20), and (34), we get

$$C\frac{\partial \Delta pX}{\partial K_T} = \frac{-1}{2K_T} \tag{104}$$

By differentiating $1/\sqrt{K_T}$ against K_T in Eq. (52) and combining the resulting expression with Eq. (53) we get

$$\frac{\partial \Delta T}{\partial K_T} = \frac{2\sqrt{K_S}}{1+[X]^a K_S}[X]^p\left(\frac{-1}{2}K_T^{-\frac{3}{2}}\sinh(C\Delta pX) - \frac{-1}{2}K_T^{-\frac{3}{2}}\cosh(C\Delta pX)\right)$$

$$= \frac{-\sqrt{\frac{K_S}{K_T}}}{K_T(1+[X]^a K_S)}[X]^p(\sinh(C\Delta pX) + \cosh(C\Delta pX)) \tag{105}$$

As before with the case of [X] and K_S, we may express the partial derivative of ΔT against log K_T as a function of the derivative against K_T and then

$$\frac{\partial \Delta T}{\partial \log K_T} = \ln 10 K_T \frac{\partial \Delta T}{K_T} = -\ln 10 \frac{\sqrt{\frac{K_S}{K_T}}}{1+[X]^a K_S}[X]^p(\sinh(C\Delta pX) + \cosh(C\Delta pX))$$

$$= -\frac{\ln 10}{2}\frac{\Delta T}{\sinh(C\Delta pX)}(\sinh(C\Delta pX) + \cosh(C\Delta pX)) = -\frac{\ln 10}{2}\Delta T(1+\coth(C\Delta pX)) \tag{106}$$

By combining Eqs. (28), (44), (51), and (55), we get for the standard deviation of systematic error

$$E_{sys(\Delta T)} = \ln 10 \Delta T\left(\left(\frac{p - aK_S[X]^a}{1+[X]^a K_S} - q\coth(C\Delta pX)\right)\right) E_{sys(pX)}$$

$$+\frac{1}{2}\left(\frac{1-K_S[X]^a}{1+K_S[X]^a} - \coth(C\Delta pX)\right) E_{sys(\log K_S)} - \frac{1}{2}(1+\coth(C\Delta pX)) E_{sys(\log K_T)}\Big) \tag{107}$$

In the same way, from Eqs. (29), (44), (51), and (55), we may obtain the variance of the random error

$$E^2_{\text{rand}(\Delta T)} = \ln^2 10 \Delta T^2 \left(\left(\frac{p - aK_S[X]^a}{1 + [X]^a K_S} - q\coth(C\Delta pX) \right)^2 s^2_{pX} + \frac{1}{4}\left(\frac{1 - K_S[X]^a}{1 + K_S[X]^a} - \coth(C\Delta pX) \right)^2 s^2_{\log K_S} + \frac{1}{4}(1 + \coth(C\Delta pX))^2 s^2_{\log K_T} \right) \quad (108)$$

Note that in spite that relative complex expressions are involved in the required differentiations carried out with the purpose to propagate the systematic and random errors implied in the donor/acceptor titration, the algebra involved is simple, and the final expressions obtained are compact.

Author details

Julia Martín, Laura Ortega Estévez and Agustín G. Asuero*

*Address all correspondence to: asuero@us.es

Department of Analytical Chemistry, Faculty of Pharmacy, The University of Seville, Seville, Spain

References

[1] Felber H, Rezzonico S, Máriássy M. Titrimetry at a metrological level. Metrologia. 2003;**40**(5):249–254

[2] Asuero AG, Michalowski T. Comprehensive formulation of titration curves for complex acid-base systems and its analytical implications. Critical Reviews in Analytical Chemistry. 2011;**41**(2):151–187

[3] Michałowski T, Asuero AG. New approaches in modeling carbonate alkalinity. Critical Reviews in Analytical Chemistry. 2012;**42**(3):220–244

[4] Michałowski T, Toporek M, Michałowska-Kaczmarczyk AM, Asuero AG. New trends in studies on electrolytic redox systems. Electrochimica Acta. 2013;**109**:519–531

[5] Winkler-Oswatitsch R, Mangen M. The art of titration: from classical end points to modern differential and dynamics analysis. Angewandte Chemie International Edition. 1979;**18**(1):20–49

[6] Zhang CL, Narusawa Y. A new titration method based on concentration-variable patterns-principles and applications to acid-base and redox titrations. Bulletin of the Chemical Society of Japan. 1997;**70**(3):593–600

[7] Terra J, Rossi AV. Sobre o desenvolvimento da análise volumétrica e algumas aplicações antais. Química Nova. 2005;**28**(1):166–171

[8] Zhan X, Li C, Li Z, Yang X, Zhong S, Yi T. Highly accurate nephelometric titrimetry. Journal of Pharmaceutical Sciences. 2004;**93**(2):441–448

[9] Van Hulle SWH, De Meyer S, Vermeiren TJL, Vergote A, Hogie J, Dejans P. Practical application and statistical analysis of titrimetric monitoring of water and sludge samples. Water SA. 2009;**35**(3):329–333

[10] King B. Review of the potential of titrimetry as a primary method. Metrologia. 1997;**34**(1):77–82

[11] Quinn TJ. Primary methods of measurement and primary standards. Metrologia. 1997;**34**(1):61–65

[12] Kolthoff IM, Furman NH. Potentiometric Titrations. 2nd ed. New York: Wiley; 1947. pp. 45–60. Chapter 3

[13] Goldman JA. Oxidation reduction equilibria and titration curves. In: Kolthoff IM, Elving PJ, editors. Treatise on Analytical Chemistry, Part 1, Theory and Practice. Volume 3, Section D, Solution Equilibria and Chemistry (Continued). New York: Wiley; 1983. pp. 1–79. Chapter 24

[14] Kolthoff IM. Die oxydopoteniiometrischc Bestimmung von Ferro. Chemisch weekblad. 1919;**16**:408

[15] Beck CM. Toward a revival of classical analysis. Metrologia. 1997;**34**(1):19–30

[16] Beck CM. Classical analysis. A look of the past, present and future. Analytical Chemistry. 1994;**66**(4):224A-230A, 233A, 239A

[17] Sillen LG. Graphic Presentation of Equilibrium Data. New York: Interscience; 1959

[18] Sillen LG. Redox diagrams. Journal of Chemical Education. 1952;**29**(12):600–608

[19] Budevski O. Foundations of Chemical Analysis. Chichester, UK: Ellis Horwood; 1979

[20] Budevski O. Graphical method for construction of titration curves. In: Ringbom A, Wänninen E, editors. Essays on Analytical Chemistry: In Memory of Professor Ringbom A, and Wänninen E. Oxford: Pergamon Press; 1977. pp. 169–174

[21] de Levie R. Redox buffer strength. Journal of Chemical Education. 1999;**76**(4):574–577

[22] Tabbut FD. Titration curves from logarithmic concentration diagrams. Journal of Chemical Education. 1966;**43**(5):245–249

[23] Inczédy J. Analytical Applications of Complex Equilibria. New York: Wiley; 1976

[24] Inczédy J. Representation of titrations errors in logarithmic diagrams. Journal of Chemical Education. 1970;**47**(11):769–772

[25] Bjerrum N. Die Theorie der alkalimetrischen und azidimetrischen Titrierungen; Samlung chemischer und chemisch-technischer Vorträge Band XXI, 1–128. Stuttgart: Verlag von Ferdinand Enke; 1914. pp. 91, 94, 97, 103

[26] de Levie R. Principles of Quantitative Chemical Analysis. New York: McGraw-Hill; 1997

[27] Kahlert H, Scholz F. Acid-Base Diagrams. Heidelberg: Springer; 2013

[28] Maccà C, Bombi GG. A graphical approach to redox titrations. Fresenius' Journal of Analytical Chemistry. 1986;**324**(1):52–57

[29] Vicente-Pérez S. Química de las Disoluciones. Diagramas y Cálculos Gráficos. UNED: Madrid; 1997. p. 424

[30] Vicente-Perez S, Durand JS, Montes F. Titration of weak reductants (or oxidants) with weak oxidants (or reductants)—The study with logarithmic diagrams. Anales de Quimica (Madrid). 1992;**88**(7–8):688–693

[31] Vicente-Perez S, Losada J, Espinosa A. E-pC diagrams for monoelectronic-redox equilibrium system—Applications. Anales de Quimica (Madrid). 1990;**86**(7):751–761

[32] de Levie R. Advanced Excel for Scientific Data Analysis. 3rd ed. New York: Oxford University Press; 2012

[33] de Levie R. How to Use Excel in Analytical Chemistry and in General Scientific Data Analysis. Cambridge: Cambridge University Press; 2001

[34] de Levie R. Explicit expressions of the general form of the titration curve in terms of concentration. Writing a single-closed form expression for the titration curve for a variety of titrations without using approximations or segmentation. The Journal of Chemical Education. 1993;**70**(3):209–217

[35] Anfalt T, Jagner D. The precision and accuracy of some current methods for potentiometric end-point determination with reference to a computer-calculated titration curve. Analytica Chimica Acta. 1971;**57**:165–176

[36] Yongnian N, Ling J. Application and advancement of titration assisted with mathematical methods. Chinese Journal of Analytical Chemistry. 1996;**24**(10):1219–1226

[37] Gran G. Determination of the equivalence point in potentiometric titration. Acta Chemica Scandinavica. 1950;**4**:559–577

[38] Gran G. Determination of the equivalence point in potentiometric titrations. Part II. Analyst. 1952;**67**(11):661–671

[39] Turyan Y, Ivanova GV, Pokhodzei VF. The gran method in redox potentiometric titrations. Journal of Analytical Chemistry. 1992;**47**(4):527–533

[40] Maccà C, Bombi GG. Linearity range of gran plots for the end-point in potentiometric titrations. Analyst. 1989;**114**:463–470

[41] Johansson A. Choice of chemical conditions in order to obtain linear titration curves in potentiometry. Talanta. 1975;**22**(12):945–954

[42] Dyrssen D, Jagner D, Wengelin F. Computer Calculation of Ionic Equilibria and Titration Procedures. Stockholm: Almqvist and Wiksell; 1968

[43] Michalowski T, Kupiec K, Rymanowski M. Numerical analysis of the gran methods. A comparative study. Analytica Chimica Acta. 2008;**606**(2):172–183

[44] Ponikvar M, Michalowski T, Kupiec K, Wybraniec S, Rymanowski M. Experimental verification of the modified gran methods applicable to redox systems. Analytica Chimica Acta. 2008;**628**(2):181–189

[45] Michalowski T, Baterowicz A, Madel A, Kochana J. Extended gran method and its applications for the simultaneous determination of Fe(II) and Fe(III). Analytica Chimica Acta. 2001;**442**(2):287–293

[46] Zhang C, Fu Y, Long C-Y. The discussion of the calculation about stoichiometric point potential for redox-titration. Journal of Dezhou University. 2003;**04**. http://en.cnki.com.cn/Article_en/CJFDTOTAL-DZHX200304015.htm

[47] Maryanov BM. Linear regression analysis of potentiometric titration data for asymmetric redox titrations. Journal of Analytical Chemistry. 1997;**52**(6):508–513

[48] Maryanov BM. Processing of potentiometric redox titration data by regression-analysis with reduction of the number of variables. Journal of Analytical Chemistry. 1992;**47**(3):396–398

[49] Krotopov VA. Approximation of redox titration curves by logarithmic functions. Journal of Analytical Chemistry. 1998;**53**(8):701–703

[50] Krotopov VA. Approximation of potentiometric titration curves by logarithmic functions: Prediction of random errors in titration parameters. Journal of Analytical Chemistry. 2000;**51**(5):449–453

[51] Krotopov VA. Approximation of potentiometric titration curves by logarithmic functions. Factors affecting the accuracy of redox titration. Journal of Analytical Chemistry. 2000;**55**(2):60–164

[52] Meites L, Fanelli N. Factors affecting the precision of a new method for determining the reduced and oxidized forms of a redox couple by a single potentiometric titration. Analytica Chimica Acta. 1987;**194**:151–162

[53] Federov AA, Shmata TS. Computer-assisted calculation and graphical presentation of titration curves. Journal of Analytical Chemistry. 2004;**59**(5):402–406

[54] Krotopov VA. Numerical simulation of redox titrations. Journal of Analytical Chemistry. 1993;**48**(2):167

[55] Huang X-Z, Chen H-L. The computer simulation of redox titration. Journal of Nanyang Institute of Technology. 2011;**2**. http://en.cnki.com.cn/Article_en/CJFDTOTAL-NYLG201102028.htm

[56] Allnutt MI. The use of conjugate charts in transfer reactions: a unified approach. The Journal of Chemical Education. 2007;**84**(10):1659–1662

[57] Desbarres J, Bauer D. Simulation des courbes de dosage potentiomètriques par emploi d'une equation universelle. Dosage par echange d'une seule particule. Talanta. 1975;**22**(10–11):877–879

[58] Maccà C. The formulation of the electron and proton balance equations for solving complicated equilibrium problems in redox titrations. Fresenius' Journal of Analytical Chemistry. 1997;**357**(2):229–232

[59] Pereira CF, Alcalde M, Villegas R, Vale J. Predominance diagrams, a useful tool for the correlation of the precipitation-solubility equilibria with other ionic equilibria. Journal of Chemical Education. 2007;**84**(3):520–525

[60] Charlot M. Les Réactions Chimiques en Solution Aqueuse. Paris: Elsevier-Masson; 1997

[61] Charlot G. Les Réactions Chimiques en Solution Aqueuse et Caractérisation des Ions. 7th ed. Paris: Masson; 1983

[62] de Levie R. A simple expression for the redox titration curve. Journal of Chemical Education. 1992;**323**(1–2):347–355

[63] Elenkova NG. General treatment of conjugated acid-base, redox and complexation equilibria. Talanta. 1980;**27**(9):699–704

[64] Chaston SHH. Double scales for equilibria. Journal of Chemical Education. 1979;**56**(1):24–26

[65] Pacer RA. Conjugate acid-base and redox theory. Journal of Chemical Education. 1973;**50**(3):178–180

[66] Hazlehurst TH. Acid-base reactions. Their analogy to oxidation-reduction reactions in solution. Journal of Chemical Education. 1940;**17**(10):466–468

[67] Rosset R, Bauer D, Desbarres J. Chimie Analytique des Solutions et Informatique. 2nd ed. Paris: Masson; 1991.

[68] Bishop E. Some theoretical considerations in analytical chemistry. V. A simple method of calculating acid-base titration errors. Analytica Chimica Acta. 1960;**22**:205–213

[69] Butler JN. Calculating titration errors. Journal of Chemical Education. 1963;**40**(2):66–69

[70] Butcher J and Fernando QJ. Theoretical error in acid-base titrations. Journal of Chemical Education. 1966;**43**(10):546–550

[71] Gonzalez GG, Jimenez AM, Asuero AG. Titration errors in acid-base titrations. Microchemical Journal. 1990;**41**(1):113–120

[72] Hulanicki A. Reactions of Acids and Bases in Analytical Chemistry. Chichester: Ellis Horwood; 1987.

[73] Carr PW. Intrinsic end-points errors in precipitation titrations with ion selective electrodes. Analytical Chemistry. 1971;**43**(3):525–430

[74] Christopherson HL. Theoretical titration error in potentiometric asymmetrical precipitation titrations. Journal of Chemical Education. 1963;**40**(2):63–65

[75] Fernando Q, Butcher J. Calculation of titration error in precipitation titrations. A graphical method. Journal of Chemical Education. 1967;**44**(3):166–168

[76] Goldman JA. Potentiometric titration curves for single ion combination reactions. Journal of Chemical Education. 1963;**40**(10):519–522

[77] Schultz FA. Titration errors and curve shapes in potentiometric titrations employing ion-selective indicator electrodes. Analytical Chemistry. 1971;**43**(4):502–508

[78] Kosonem PO and Hakoila EJ. The titration error in potentiometric precipittion titration. Talanta. 1975;**22**(12):1045–1047

[79] Ringbom A. Complexation in Analytical Chemistry. New York: Interscience; 1963

[80] Johansson E, Olin A. A complexometric method for the determination of small amounts of an alkaline earth metal in mixture with other metals. Talanta. 1980;**27**(2):165–168

[81] Hulanicki A and Głab S. Total systematic error in redox titrations with visual indicators. I: Basic principles. Talanta. 1975;**22**(4–5):363–370

[82] Perez-Ruiz T, Hernández Córdoba M, Martínez-Lozano C, Sánchez-Pedreño C. Nuevas consideraciones teóricas sobre las valoraciones de oxido-reducción. Química Analítica. 1984;**3**:138–146

[83] Edens GJ. Redox titration of antioxidant mixtures with n-bromosuccinimide as titrant: Analysis by non-linear least-squares with novel weighting function. Analytical Chemistry. 2005;**21**(11):1349–1354

[84] Umetsu K, Itabashi H, Saloh K, Kawashima T. Effect of ligands on the redox reactions of metal ions and the use of a ligand buffer for improving end-point detection in the potentiometric titration of vanadium(V) with iron(II). Analytical Chemistry. 1991;**7**(1):115–118

[85] Butler JN. Ionic Equilibrium. Solubility and pH Calculations. New York: Wiley; 1998

[86] Charlot G, Gauguin R. Les Méthodes d'Analyse des Réactions en Solution. Paris: Masson; 1951

[87] Garric M. Chimie Générale. Vol. 1, Paris: Dunod; 1970. pp. 439–442

[88] King DW. A general approach for calculating speciation and posing capacity of redox systems with multiple oxidation states: Application to redox titrations and the generation of pε-pH diagrams. The Journal of Chemical Education. 2002;**79**(9):1135–1140

[89] Monnier D, Haerdi W, Buffle Y, Ruscony Y. Chimie Analytique. Application aux Méthodes Instrumentales, Radiochimiques et a la Chimie de l'Environnent.Genève: Georg; 1979

[90] Monnier D, Haerdi W, Ruscony Y. Chimie Analytique. Analyse Qualitative Minérale. Eléments de Radiochimie.Genève: Georg Genève; 1968. pp. 118–132. Chapter IX

[91] Stumm W, Morgan JJ. Oxidation and Reduction. Equilibria and Microbial Mediation. In Aquatic Chemistry. An introduction emphasizing chemical equilibria in natural waters. Edited by Stumm W and Morgan JJ. In 3rd ed. New York: Wiley; 1996. pp. 425–515. Chapter 7

[92] Tremillon B. Reactions in Solution: An Applied Analytical Approach. New York: Wiley; 1997

[93] Asuero AG. Buffer capacity of a polyprotic acid: first derivative of the buffer capacity and pKa values of single and overlapping equilibria. Critical Reviews in Analytical Chemistry. 2007;**37**(4):269–301

[94] Michalowski T, Lesiak A. Formulation of generalized equations for redox titration curves. Chemia Analityczna (Warsaw). 1994;**39**:623–637

[95] Michalowsi T, Wajda N, Janecki D. A unified approach to electrolytic systems. Chemia Analityczna (Warsaw). 1996;**41**:667–685

[96] Michałowski T. The generalized approach to electrolytic systems: I. Physicochemical and analytical implications. Critical Reviews in Analytical Chemistry. 2010;**40**(1):2–16

[97] Michałowski T, Michałowska-Kaczmarczyk AM, Toporek M. Formulation of general criterion distinguishing between non-redox and redox systems. Electrochimica Acta. 2013;**112**:199–211

[98] Michalowka-Kaczmarcyk AM, Asuero AG, Toporek M, Michalowski T. "Why not stoichiometry" versus "stoichiometry—why not?" Part II: Gates in context with redox systems. Critical Reviews in Analytical Chemistry. 2015;**45**(3):241–269

[99] Toporek M, Michalowka-Kaczmarcyk AM, Michalowski T. Symproportionation versus disproportionation in bromine redox systems. Electrochimica Acta. 2015;**171**:176–186

[100] Michalowka-Kaczmarcyk AM, Toporek M, Michalowski T. Speciation diagrmas in dynamic iodide + dichromate system. Electrochimica Acta. 2015;**155**:217–227

[101] Vydra F, Přibil R. New redox systems-IV1 1 Part III-see reference 4. Oxidation of cobaltIII with ironII chloride in 2,20-bipyridyl. Talanta. 1961;**8**:124

[102] Still ER. Statistical adjustment of parameters for potentiometric titration data. Talanta. 1980;**27**(7):573–582

[103] Asuero AG, Gonzalez G, de Pablos F, Gomez-Ariza JL. Determination of the optimum working range in spectrophotometric procedures. Talanta. 1988;**35**(7):531–537

Section 2

Biological Applications

Chapter 7

Redox Reactions in the Physiopathology of the Liver

Isabel Méndez, Francisco Vázquez-Cuevas,
Rolando Hernández-Muñoz,
Héctor Valente-Godínez,
Olivia Vázquez-Martínez and Mauricio Díaz-Muñoz

Additional information is available at the end of the chapter

http://dx.doi.org/10.5772/intechopen.68841

Abstract

Electron fluxes are constant within cellular metabolism. Donating or accepting electrons, either naked or as hydrogen atoms, is one of the most important properties of bioenergetic networks. These redox reactions fulfill key physiological phenomena such as cellular growing, phenotypic differentiation, nutritional adaptations and redox-dependent cellular signaling, but when they became unregulated, serious pathologies such as degenerative diseases and metabolic disorders arise. The liver being an important metabolic organ, redox reactions play a strategic role in its main functions: processing of nutrients, fasting response, xenobiotic managing and circadian activity. However, liver is also very sensitive to compounds that disturb redox state such as ethanol, CCl_4, aflatoxins, among others, as well as to stressors such as hypercaloric diets, endocrine disruptors and stressful life situations. This chapter reviews concepts related to redox reactions in the liver, including metabolic aspects of reactive oxygen species (ROS), prooxidant and antioxidant subcellular systems, alterations produced by hepatotoxins, adaptations to experimental surgical protocols such as portacaval anastomosis, and participation in cancer. It is out of question that for a better comprehension of the physiopathological events in the liver and other metabolic organs, the more complete understanding of the roles played by redox reactions will be a necessity.

Keywords: metabolism, ROS, antioxidant, prooxidant, hepatocyte

1. Introduction

Living organisms are dynamic and complex systems with the notable capacity to continuously preserve their structural identity, but at the same time, to display functional and morphological adaptations in daily terms (circadian rhythms), during ontogeny, as well as in the context of evolutionary progression. Since living beings are thermodynamic open systems, they are empowered to exchange matter and energy. This competence sustains the intricate intertwine that make up the metabolic networks present in every cell [1].

For the last 2 billion years (after the transition of oxygenic photosynthesis), in our planet the bioenergetic cycle has been defined by two principal and complementary processes: photosynthesis and respiration [2]. Indeed, every other chemical and energetic transformation that takes place in the cellular milieu is included within this biochemical cycle (**Figure 1**). Inspection of the formula shown in **Figure 1** rapidly indicates that carbon-containing molecules oscillate from a reduced (as glucose) to an oxidized state (as carbon dioxide). The corollary is that in a general sense, metabolism can be visualized as a collection of reduction-oxidation (redox) reactions [3].

Electrons flux is in the pith of the formal definition of redox reactions: it establishes that molecules accepting electrons are reduced whereas those that give up electrons are oxidized. Therefore, redox reactions always occur between redox pairs. As Albert Szent-Gyorgyi, the prestigious Nobel Prize winner biochemist in 1937, quoted: (1) understanding metabolism is to figure out the direction taken by the electrons in transit among molecules and (2) the secret of life is to take advantage of the correct flow of the electrons [4]. Electrons in transit can be mobilized alone (as in the iron-sulfur complexes and cytochromes within mitochondrial and microsomal electron transport chains) or joined with protons as hydrogen atoms (as in biochemical transformation among acids, aldehydes and alcohols).

Redox potential (ΔE, quantified in volts) is a physicochemical parameter that measures the capacity to either release or accept electrons within a chemical reaction. It characterizes the extent of free Gibbs energy (ΔG) and the direction of the electron flow in each redox reaction ($\Delta G = nF\Delta E$, n is the electrons transferred and F is Faraday's constant). Molecular entities with higher (more positive) redox potential have the facility to oxidize molecules with a lower (more negative) redox potential [5]. Spontaneous biochemical transformations involve the release of metabolic energy as the electrons move from reactions with negative redox potential (oxidation of glucose into 2 pyruvates + $4e^-$, −720 mV) to reactions with positive redox potential (reduction of O_2 with $2e^-$ into water, +820 mV).

1.1. Liver as a paradigmatic metabolic organ

Certainly, every cell and tissue in the organism shows metabolic activity. However, the liver has a special consideration since it is the principal organ in the biochemical processing of nutrients and xenobiotics. Distinctive metabolic pathways, such as gluconeogenesis, ureagenesis, assembly of lipoproteins, synthesis of ketone bodies, metabolism of foreign chemical substances, lipogenesis, cholesterol formation, glutamine synthesis, and others, take place in the different population of hepatocytes: periportal (with high $[O_2]$ and oxidative metabolism) and pericentral (with low $[O_2]$ and less oxidative metabolism) [6].

Hepatic metabolism comprises synthetic (anabolic) and degradative (catabolic) pathways, each one being regulated by particular factors, including: (1) cellular compartments (oxidative

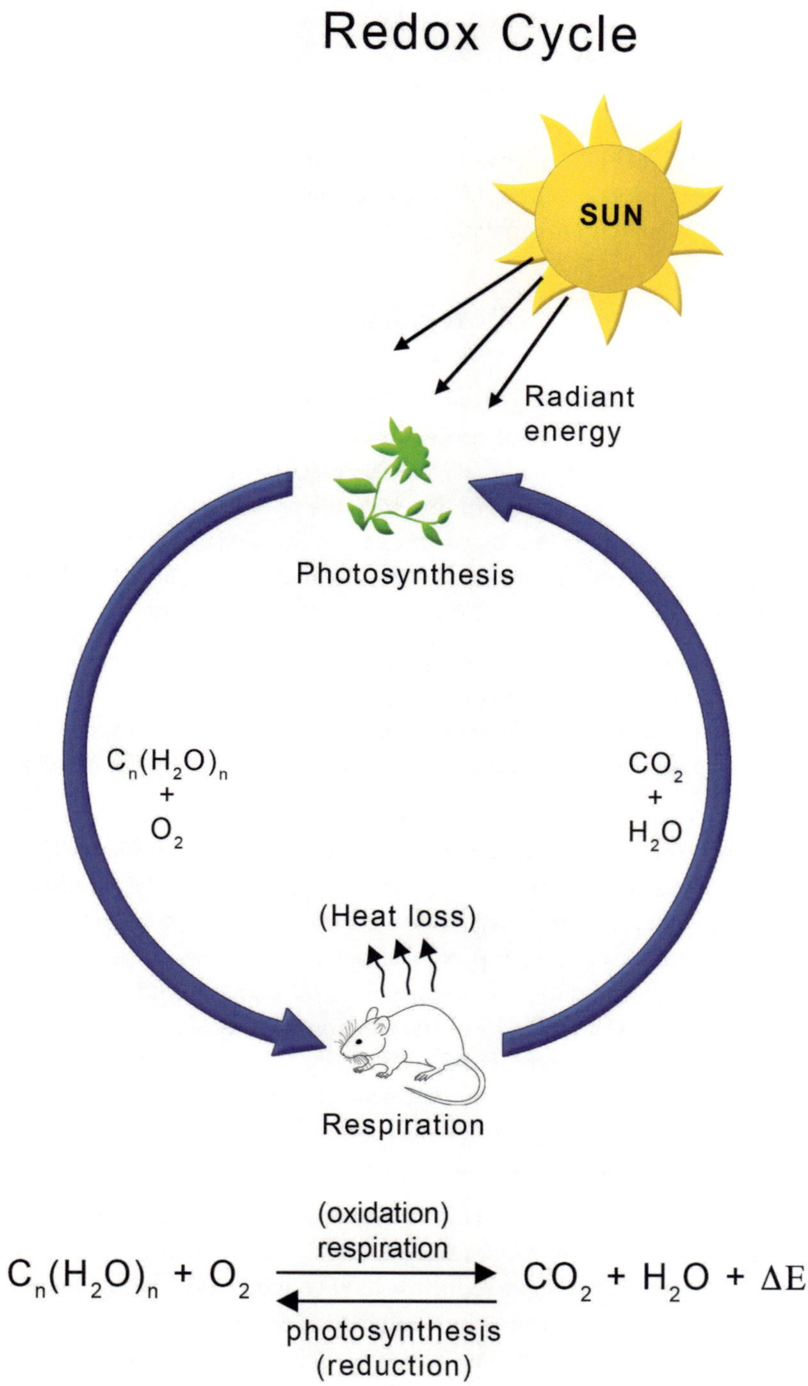

Figure 1. Global energy flux underlying redox cycle between photosynthesis and respiration. Driving by solar energy, autotrophic organisms synthetize biomolecules (carbohydrates for example) and oxygen by photosynthesis. Heterotrophic organisms oxidize carbohydrates to yield carbon dioxide and water by respiration. The cycle is complete when carbon dioxide and water are used again by autotrophic organisms.

reactions are preponderant in mitochondria and peroxisomes whereas reductive reactions are more common in cytosol), (2) available coenzymes (oxidized nicotinamide adenine dinucleotide (NAD^+), reduced nicotinamide adenine dinucleotide (NADH) for catabolism and oxidized nicotinamide adenine dinucleotide phosphate ($NADP^+$), reduced nicotinamide adenine dinucleotide phosphate (NADPH) for anabolism), (3) adenine nucleotides pool (low energy charge, AMP-activated protein kinase (AMPK) activation for catabolism and high energy charge, reduced AMPK activity for anabolism) [7].

2. Redox molecules as metabolic regulators

It was thought that reactive oxygen species (ROS) were damaging molecules that were associated with the main pathological consequences of oxidative stress. However, recently, numerous reports have demonstrated that ROS and reactive nitrogen species (RNS) also play important roles in signaling pathways. In this way, the nowadays perspective of prooxidant reactions in the cellular milieu is that they form part of the physiological response to internal and environmental regulatory factors [8]. Indeed, the liver being one of the major metabolic organs, the signaling consequences of ROS and RNS are very relevant in the hepatic tissue.

It is well known that mitochondrial activity is the major source for ROS production, with consequences in the oxidative phosphorylation coupling affecting the $\Delta\Psi$ (mitochondrial membrane potential) as well as several mitochondrial metabolic networks. Eventually, important processes such as cell proliferation and apoptosis can be triggered by the mitochondrial prooxidant condition [9].

During the electron transfer through the mitochondrial respiratory complexes, O_2 is eventually reduced to water by receiving two electrons, along the creation of a [H^+] gradient that makes possible the formation of ATP. However, some electrons "leak" without completing the pathway until the cytochrome oxidase complex is achieved. In this case, one electron is received by an oxygen molecule forming the anion superoxide (O_2^-). O_2^- is produced in the interphase between sites I and II, as well in site III of the electron transfer chain. Within the mitochondria, O_2^- is transformed to hydrogen peroxide (H_2O_2) by mitochondrial superoxide dismutase (SOD2); when it reaches the cytosolic compartment, the O_2^- is turned into H_2O_2 by SOD1. In cytosol, the mitochondrial O_2^- and H_2O_2 are incorporated to the ROS pool generated mainly by the activity of the family of NADPH oxidases to potentially act as the signaling molecules [8].

H_2O_2 is itself a metabolic regulator, but at high levels can act as a deleterious factor. H_2O_2 can be converted into H_2O by several antioxidant enzymes: H_2O_2 can be reduced by peroxiredoxins or glutathione peroxidases, which couple reduction of H_2O_2 with oxidation of glutathione (GSH). Oxidized peroxiredoxins can be reduced by thioredoxins. Subsequently, oxidized thioredoxins become reduced by thioredoxin reductase in a NADPH-dependent manner. Oxidized glutathione disulfide (GSSG) is reduced by glutathione reductase again in the presence of NADPH.

H_2O_2 acts as a signaling factor by oxidizing thiol groups (-SH HS-) into disulfide bridges (-S-S-) in cysteine residues of key regulatory proteins (enzymes and receptors). Several years ago, some reports postulated that H_2O_2 showed insulin-like properties in hepatocytes. This

http://dx.doi.org/10.5772/intechopen.68841

observation was eventually confirmed in the yeast *Saccharomyses cerevisiae* and supported the notion, fully accepted nowadays, that H_2O_2 could be acting as a signaling molecule in many cellular systems [8]. **Figure 2** shows the important role that mitochondria play in the prooxidant reactions of many cellular functions.

2.1. Cellular proliferation

Another system of ROS modulation well characterized in the liver is the control of growth factors by the redox regulation of cysteine residues in tyrosine phosphatases. H_2O_2 downregulates cyclin D1 and cyclin E to inhibit proliferation and upregulates Bcl-2-associated X protein (BAX) to induce apoptosis in hepatocytes and MCF-7 cells (cell line from cancerous mammary gland) [10].

Another example is the modulation of redox-sensitive cysteine residues in the epidermal growth factor (EGF) receptor, which is activated by the action of H_2O_2.

Other protein factors such as erythropoietin (EPO) and vascular endothelial growth factor (VEGF) have been identified as redox regulators of this process.

In response to hypoxic conditions, the hypoxia-inducible transcription factors (HIFs) are upregulated by ROS, especially the ones from mitochondrial source. Paradoxically, in liver

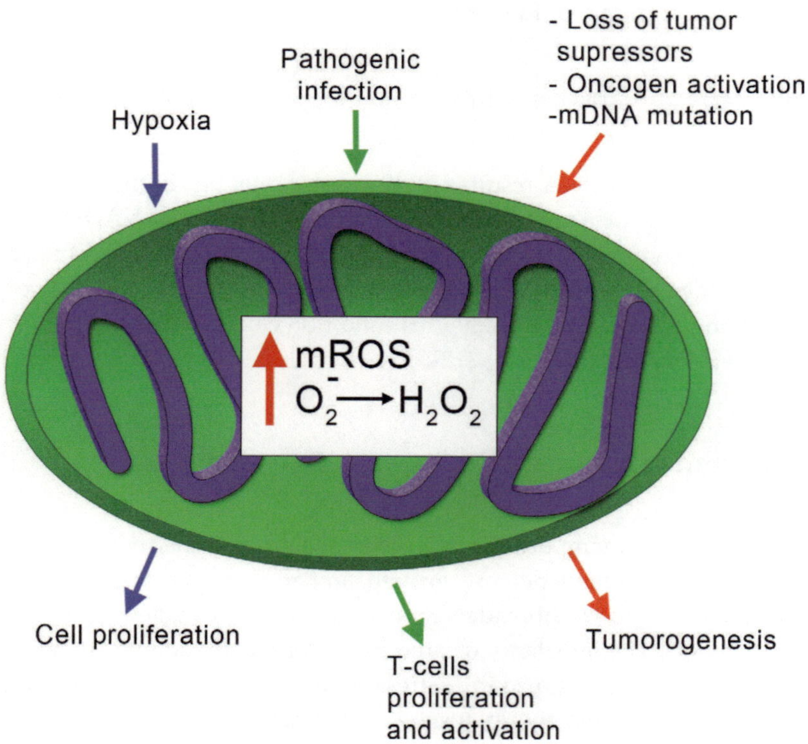

Figure 2. Mitochondrial handling of reactive oxygen species (ROS). Pro-oxidant reactions (synthesis of O_2^- and H_2O_2) are upregulated by a variety of physiological or pathological events. Mitochondrial activation promotes responses such as cell proliferation and activation, as well as cancerous processes.

and vascular smooth muscle cells, in hypoxic condition there is an increase in mitochondrial ROS production from complex III; however, the exact mechanism is not well understood.

In liver and muscle cells, angiotensin II signaling also promotes higher production of mitochondrial ROS, which are necessary to activate downstream responses such as mitogen-activated protein kinases (MAPK) [11].

2.2. Immune system

When an organism is infected primordial T-cells rapidly proliferate, and differentiate into effector T-cells, and mitochondria contribute to this activation through the production of ROS. This has been analyzed adding antioxidants to mice after a viral infection and showing that they exhibited a depressed immune system. Data suggest that ROS play an important role in T-cell activation, proliferation and adaptive immune function [12].

2.3. Cancer

A fundamental characteristic of cancerous cells is their uncontrolled proliferation. It has been observed that they generate high levels of ROS, especially H_2O_2; in addition, elevated levels of antioxidants have been detected, maybe to implement a protective cellular response. The final equilibrium allows a high rate of ROS synthesis concomitant with a transformed functional phenotype.

ROS are responsible of oncogenes activation and/or loss of tumor suppressors enhancing mitogenic signaling.

Mutations in mitochondrial DNA result in ROS increase that has been associated to a great variety of human cancers. One example is the mutation in the gene of NADH dehydrogenase in the mitochondrial complex I, which promotes an elevation of ROS production. The pro-oxidant stimulus leads to the proliferation of several human and mouse cell lines, as well as tumor formation in rodents. Interestingly, this condition could be rescued by the reconstitution of the wild-type enzymatic activity [13].

3. Anabolism and catabolism. Redox-sensitive enzymes

Prooxidant reactions play an important role in cellular signaling and redox regulation of metabolic processes such as immune defense, growth, and apoptosis among others. The increase in the ROS production requires antioxidant strategies to prevent a potential oxidative damage to cellular components. The imbalance of prooxidants and antioxidants leading to cell damage and tissue injury may cause oxidative stress. Oxidative stress is common in organs and tissues with high metabolic and energy turnover, including skeletal and heart muscle, blood cells, and liver [14].

It has been recognized as the interplay between energy metabolism and ROS to make possible the homeostasis in the liver physiology. As it was mentioned above, redox couples as

NADH/NAD$^+$, FADH$_2$/FAD$^+$, and GSH/GSSG are involved in the donation and acceptance of electrons in a variety of reactions. NADPH is a key cofactor for many enzymatic reactions in the metabolism, and it is considered as one of the main regulator of the redox potential (**Figure 3**). Its production is required for the regeneration of GSH in mitochondria for scavenging mitochondrial ROS through glutathione reductase and peroxidase systems [15]. NADP$^+$ is synthesized from NAD$^+$ by NAD$^+$ kinase, whereas NADPH is derived from NADH by three major enzymes in the mitochondrial matrix: NAD(P)$^+$ transhydrogenase, NADP$^+$-dependent isocitrate dehydrogenase (IDH) and malic enzyme and by other three cytosolic enzymes in cytosol: glucose-6-phosphate dehydrogenase (G6PD), 6-phosphogluconate dehydrogenase in the pentose phosphate pathway (PPP), and the malic enzyme.

One of the mechanisms for generating cytosolic NADPH from mitochondrial oxidations implies substrates shuttles. This is the case of the shuttle mechanism that involves NADP$^+$-dependent IDHs present in both, mitochondrial and extramitochondrial spaces, whereas NAD$^+$-dependent IDH is solely mitochondrial. IDHs catalyze oxidative decarboxylation of isocitrate to α-ketoglutarate in the tricarboxylic acid cycle (TCA), in a NAD$^+$ or NADP$^+$-dependent manner producing NADH or NADPH, respectively. NADPH may be transported by the isocitrate-2-ketoglutarate shuttle by the cytosolic IDH in which liver has a high enzymatic activity. NADP$^+$-dependent IDH is induced by ROS and controls the mitochondrial redox balance. A decreased expression of NADP$^+$-dependent IDH importantly elevates ROS generation, lipid peroxidation and DNA fragmentation; consequently, a significant reduced ATP level is associated to the mitochondrial damage, whereas overexpression of NADP$^+$-dependent

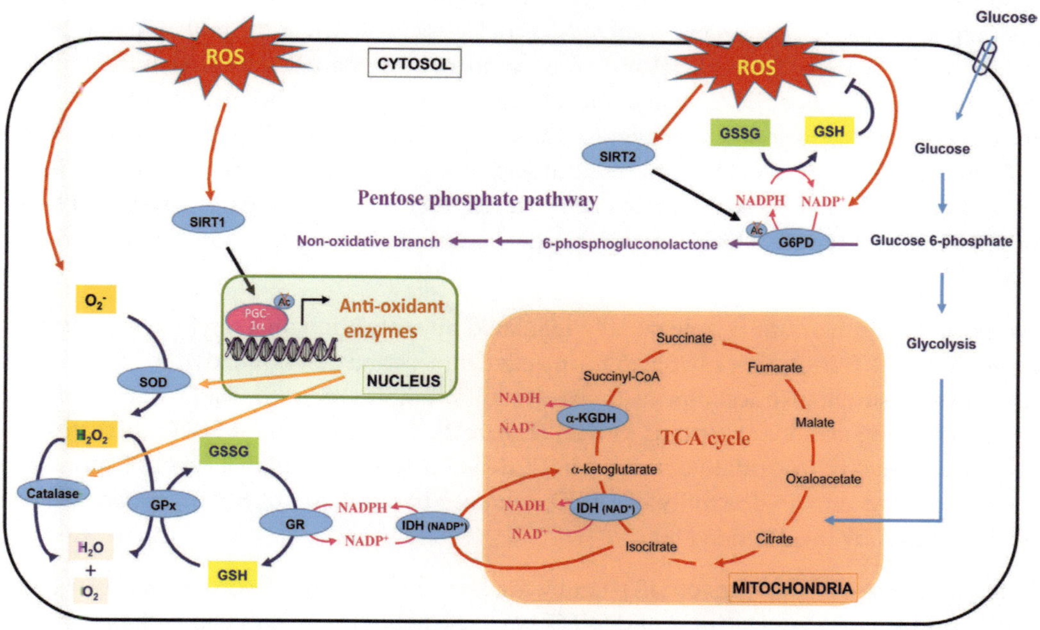

Figure 3. Redox cycles linking metabolic and antioxidant pathways. Glutathione peroxidase (GPx), glutathione reductase (GR), superoxide dismutase (SOD), isocitrate dehydrogenase (IDH), glucose-6-phosphate dehydrogenase (G6PD), α-ketoglutarate dehydrogenase (α-KGDH), sirtuin (SIRT), reduced glutathione (GSH), oxidized glutathione (GSSG).

IDH protects from ROS-induced damage [16]. Thus, both cytosolic and mitochondrial IDHs play important roles in cellular defense against oxidative damage by providing NADPH needed for the generation of GSH [17, 18] (**Figure 2**). Likewise, $NADP^+$-dependent IDH is inactivated by GSSG-dependent glutathionylation leading to enzyme inactivation, followed by GSH-dependent reactivation, suggesting an alternative modification to the redox regulation of IDHs [19].

Among enzymes sensitive to redox state is α-ketoglutarate dehydrogenase (α-KGDH). α-KGDH is a mitochondrial enzyme of the TCA that catalyzes the conversion of α-ketoglutarate to succinyl-CoA, producing NADH that supplies electrons for the respiratory chain (**Figure 2**). α-KGDH is considered a component of the mitochondrial antioxidant system and a key sensor of redox status [20]. It is sensitive to ROS and its inhibition impact significantly in the energetic deficit induced by oxidative stress; α-KGDH can also generate ROS by its catalytic action regulated by the NADH/NAD^+ ratio [21]. A reversible inhibition of α-KGDH is obtained by glutathionylation. Since α-KGDH controls supply of reducing equivalents generated by the TCA, the redox regulation of α-KGDH would control energy production in response to oxidative stress.

Another source of NADPH required for detoxification of free radicals and peroxides is the activity of G6PD, the rate-limiting enzyme of the PPP (**Figure 2**). It has an important role for cell growth by providing NADPH for redox balance [22] and its expression is induced due to oxidative stress [23], whereas the reduction of G6PD sensitizes cells to oxidative stress [24]. In humans, mutants of G6PD alleles are associated with hemolytic anemia, while mutants in mouse embryonic stem cells by targeted homologous recombination have shown that G6PD is essential to protect cells against even mild oxidative stress whereas the null mutant is lethal. A high-carbohydrate, fat-free diet promotes an increase in hepatic G6PD activity [25], but polyunsaturated fatty acids content of the diet decreases its activity. Besides the effect at activity level, gene expression is also altered. Inhibition of G6PD gene expression is caused by polyunsaturated fatty acids but not by saturated or monounsaturated fatty acids [26]. In the liver of young Zucker obese fa/fa rats, G6PD expression and activity are increased prior to the onset of diabetes type 2, which seems to be contributing factors for the induction of oxidative stress [27].

Furthermore, human G6PD is negatively regulated by acetylation on a phylogenetically conserved lysine 403 becoming unable to form active dimers and consequently the loss of activity. The exposure to extracellular oxidative stimuli promotes reduced G6PD acetylation due to deacetylating activity of sirtuins (SIRTs) (**Figure 2**). The inhibition of SIRT2 increases cellular susceptibility to oxidative stress. SIRT2 deletion leads to a higher level of lysine 403 acetylation and impaired activity of G6PD, whereas the addition of SIRT2 rescues the cell death induced by the deletion [28].

In mammals, seven members of SIRT family are known; they have diverse subcellular localization and activity. SIRT1, SIRT6 and SIRT7 are nuclear, SIRT6 is associated with heterochromatic regions and SIRT7 with nucleoli, SIRT2 is in cytosol, and SIRT3–5 are mitochondrial. SIRT1 shows a potent NAD^+-dependent deacetylase activity on lysine 16 of histone H4 that could promote the formation of heterochromatin, as well as on lysine 382 of p53, while SIRT6 and SIRT7 lack deacetylase activity [29]. Several proteins of this family regulate lifespan in diverse organisms and in human cells.

SIRT1 is the homolog of Sir2, a protein with an important role in longevity in yeast, and with an important role in mammalian development, metabolic regulation and modulation of cellular stress response and survival by acting on p53, NF-κB signaling and FoxO transcription factors [29]. SIRT1 is an important metabolic regulator that orchestrates hepatic gluconeogenesis and lipid metabolism, suggesting that it can play an important role in the developing of metabolic and age-related diseases. The effects of SIRT1 are mediated through the induction of antioxidant proteins, SOD2, nuclear respiratory factor 1 (NRF1) [30], catalase, peroxiredoxins 3 and 5, thioredoxin 2, thioredoxin reductase 2, and uncoupling protein 2 (UCP-2) via formation of forkhead box/peroxisome proliferator-activated receptor gamma coactivator 1-α complex (FoxO3a/PGC-1α) [31] (**Figure 3**).

Increasing ROS level modulates activity of SIRT1. Likewise, SIRT1 activity is regulated by AMPK, which is a redox-sensing enzyme, and a metabolic gauge by increasing cellular NAD^+ synthesis [32]. AMPK is activated by high AMP/ATP ratio and is sensitive to glutathionylation by action of H_2O_2 and NO that oxide reactive thiol residues [33]. Oxidative stress also induces S-glutathionylation of SIRT1 and reduces NAD^+ level, thus inhibiting SIRT1 activity [34]. SIRT1 has an important role in the liver glucose metabolism. Ablation of SIRT1 in liver induces lipid accumulation by upregulating lipogenic genes expression and reducing β-oxidation, whereas overexpression of SIRT1 protects against hepatic steatosis induced by high-fat diet [35]. SIRT1 regulates positively β-oxidation through the activation of nuclear receptors peroxisome proliferator-activated receptor α (PPARα) that regulates gene expression of lipid catabolic genes.

Peroxisomes are oxidative subcellular organelles for H_2O_2, fatty acids, and cholesterol. PPARs are lipophilic ligand-activated transcription factors that belong to nuclear hormone receptor superfamily with three subtypes: PPARα, PPARγ, and PPARβ/δ [36]. PPARα is mainly present in the liver, playing a relevant role in the fasting response. Additionally, PPARα contributes to protection from oxidative stress by upregulating expression of genes of the chaperone and proteasome families, with consequences in protein folding and degradation of damaged proteins [37]. In cancerous cells, PPARα leads to increased peroxisome proliferation and production of ROS contributing to DNA damage [38]. Oxidized lipids are produced during oxidative stress and are natural endogenous PPARγ ligands suggesting a role in oxidative stress response. PPARγ upregulates the expression of antioxidant and prooxidant genes such as catalase, SOD2, GPx3, eNOS and mitochondrial uncoupling protein 2 (UCP-2), whereas downregulates cyclooxygenase-2 (COX-2) and iNOS [39].

Handling of cellular oxidative stress involves several metabolic enzymes that are originally described as part of other metabolic pathways. Most of these enzymes reduce free-radical production and protect cell from injury. The loss of control in the events that coordinates redox homeostasis contributes to oxidative damage, metabolic diseases and progression of degenerative diseases associated to aging. Understanding the mechanisms involved in these events will contribute to improve the quality of life during physiological or pathological conditions.

4. Redox alterations in hepatotoxicity

The role played by the cellular redox state in the hepatic pathophysiological mechanisms is not well defined. Indeed, hepatotoxicity has been much linked to the oxidative status and deficiencies in the liver antioxidant system. For instance, in the case of the alcoholic liver disease (ALD) mainly represented by a chronic stage of alcoholic steatohepatitis (ASH), alterations in the cell redox state have been implicated in injured hepatocytes [40]. Ethanol is metabolized via alcohol dehydrogenase (ADH), the microsomal ethanol oxidizing system, and by catalase in the liver peroxisomes, although the activity of ADH is responsible for most of the ethanol catabolism. In this pathway, NAD^+ is reduced by a transfer of hydrogen (and one electron, e^-) to NADH, with the concomitant production of acetaldehyde. The $NADP^+$ can be also reduced, and hydrogen equivalents from ethanol, but not NADH, are transferred from the cytosol to the mitochondria via a shuttle mechanism such as the malate shuttle, the fatty acid elongation cycle, and/or the α-glycerophosphate cycle; therefore, mitochondria become more reduced [41, 42].

Lately, the nonalcoholic steatohepatitis (NASH) is becoming a pathological entity gaining a significant public health concern [43]. NASH is a progressive form of nonalcoholic fatty liver disease (NAFLD) and features of NASH include steatosis, inflammation and varying degrees of fibrosis, and seems to follow a 2-hit model, where the "1st Hit" involves excess of lipid accumulation in the liver, which sensitizes the liver to the "2nd Hit". This "2nd Hit" involves inflammation, oxidative stress, liver damage and fibrosis [44]. Here, multiple cellular processes play important roles in maintaining the NAD^+/NADH ratio. For example, during glycolysis, β-oxidation, and the TCA activity, NAD^+ is reduced to NADH. Within the mitochondria, NADH is oxidized by the electron transport chain enzymes during oxidative phosphorylation.

Laboratory animals exposed to high fat diet exhibit impaired oxidative metabolism through reduced electron transport chain activity [45]. Then, it is possible that those animals exposed to high fat during late gestation and early postnatal life present a state of redox imbalance, and although their livers are able to readily reduce NADH during catabolic redox reactions (in response to increased fat intake), the ability to replenish NAD^+ reserves is reduced due to impaired oxidative capacity [46]. High fat feeding is associated with depleted NAD^+ reserves and reduced SIRTs abundance, both established hallmarks of metabolic aging, and supplementation with factors reversing the effects of depleted NAD^+ reserves, and/or SIRT1 and SIRT3 abundance rescue the increased susceptibility to develop severe fatty liver disease in the adult life [47].

Liver steatosis induces a reduced state in cytosol and mitochondria of hepatocytes, as demonstrated by alterations in the NADH/NAD^+ ratio calculated from the β-hydroxybutyrate dehydrogenase and lactate dehydrogenase reactions [48]. The increased formation of reducing equivalents could impair fatty acids oxidation and the TCA cycle [49], but it could also enhance the formation of glycerol-3-phosphate and thus lipogenesis [50]. Saturation of lipids may also modify cellular redox status. Among free fatty acids, monounsaturated fatty acids, such as oleic acids, are less toxic than palmitate, a saturated acid, because the latter increases the NADH/NAD⁺ ratio and promotes uncoupling between glycolysis and TCA cycle fluxes, leading to increased ROS production [51]. The hepatic accumulation of saturated fatty acids can promote redox imbalance and the formation of reactive oxygen intermediates, mainly inducing endoplasmic reticulum (ER) stress and apoptosis [52].

Moreover, the pathogenesis of early-stage NASH is characterized by hyperinsulinemia and *de novo* synthesis of fatty acids and nascent triacylglycerides, which are deposited as lipid droplets within the hepatocytes. Hyperinsulinemia shifts the energy supply from glucose to ketone bodies, and the high ketone body concentration induces the overexpression of cytochrome P450 2E1 (CYP2E1), resulting in unsaturated fatty acids peroxidation and aldehydes production [53]. The NADPH oxidase-derived ROS from arachidonic acid peroxidation can induce the nuclear translocation of *EGR1*, which in turn can stimulate the expression of the downstream genes *ATF3* and *GADD45G*. ATF3 is a transcription factor involved in cell proliferation, apoptosis, and invasion [54]. The increased redox signaling plays a central role in promoting insulin resistance in the liver in early NASH, followed by fibrogenesis through activation of protein kinase R (PKR), protein kinase R-like endoplasmic reticulum kinase (PERK) key stress kinases [55]. There is evidence that activation of the purinergic receptor P2X7 can give rise to NADPH oxidase activation, leading to Kupffer cell activation, a key event in NASH progression [56].

Peroxisomal oxidation of fatty acids is the normal route of metabolism of very long chain fatty acids and dicarboxylic acids, where electrons from $FADH_2$ and NADH are transferred directly to O_2 [57]. Moreover, fatty acids not oxidized by mitochondria are mainly oxidized by CYP2E1; a process that further increases ROS production [58]. Therefore, many cellular systems are important sources of ROS, including the mitochondrial respiratory chain [59], the cytochrome P450s [60], oxidative enzymes (xanthine oxidase, aldehyde oxidase, cyclooxygenase, monoamine oxidase (MAO), and the NADPH oxidase complex) [61]. The uncoupling protein 2 (UCP-2) may also enhance the reoxidation of NADH into NAD^+, which is required for both β-oxidation and the TCA [62]; UCP-2 oxidation might be considered an attempt to prevent steatosis by increasing hepatic fatty acid oxidation [63].

Moreover, it is accepted that in the NAFLD, depletion of hepatic antioxidants may contribute to the progression of steatosis to NASH by increasing oxidative stress that produces lipid peroxidation, inflammation, and fibrosis. Indeed, metabolic adaptations resulting from severe GSH deficiency seem to protect against the development of steatohepatitis [64]. In summary, all the redox alterations seem to be deeply implicated in the onset of NASH and in its progression to liver fibrosis and a putative installation of a cirrhotic process.

However, a possible role of a disturbed cell redox state is much less known in cirrhosis. Nonetheless, it has been suggested that collagen metabolism could be influenced by changes in redox state. It has been postulated that conversion of glutamic acid to proline [65], a decreased NAD^+/NADH ratio [66], as well as impediment of proline transport and oxidation, could increase the liver proline pool, as a fundamental collagen component in the onset of liver fibrosis [67].

In experimental models of rat liver fibrosis/cirrhosis, mitochondrial function and structure show a variety of alterations. ATP synthesis is reduced in rats treated with CCl_4 or thioacetamide, as well as in rats with secondary biliary cirrhosis. These alterations are compensated by increasing mitochondrial volume per hepatocyte and possible augmentation of extrahepatic ATP production, as an effort for maintaining mitochondrial function in the cirrhotic liver [68]. These reports agree with the statement that in perfused cirrhotic livers, a reduced cytoplasmic and mitochondrial redox states occur accompanied by a diminished activity in the mitochondrial electron-transport chain [69].

5. Redox alterations in experimental portacaval anastomosis

Portacaval anastomosis (PCA) is a pathological condition that usually accompanies the portal hypertension associated to cirrhosis; however, the shunt can occur among a variety of portal and systemic veins [70]. Experimentally, PCA is a surgical procedure that communicates a sectioned porta vein to an oval incision in the inferior cava vein. It results in a straight flow of the full-of-nutrients portal vein from the small intestine directly to the systemic circulation. It has been used for a long time to implement experimental models of hyperammonemia and the consequent hepatic encephalopathy (HE) [71]. However, only few reports exist regarding the metabolic and physiological consequences of the PCA in the hepatic tissue.

Ammonium (NH_4^+)-metabolic handling by the liver involves equilibrium between anabolic (synthesis of proteins, nucleic acids and amination reactions) and catabolic (urea cycle and glutamine synthesis) pathways. Intracellular glutamate plays a key role since high glutamate serves as substrate for the synthesis of N-acetylglutamate, an essential allosteric activator of carbamyl phosphate synthetase I, a key regulatory enzyme in the urea cycle in the periportal hepatocytes. Nitrogen disposal is complemented by glutamine synthesis (glutamate + NH_4^+) in the pericentral hepatocytes [72].

In human adults, approximately 1 mol (about 17 g) of NH_4^+ is produced daily in the liver. Part is reutilized in biosynthesis, while the rest is a metabolic disposal with the potential of being neurotoxic. Its normal concentration in the portal blood varies from 300 to 600 μM, but in the blood leaving the liver the concentration is clearly reduced to 20–60 μM [73]. Other organs such as the brain, muscle and kidney play a role in regulating the NH_4^+ levels. Insult to the liver, whether acute or chronic in nature, reduces its capacity to metabolize NH_4^+ with the consequence to promote in hyperammonemic state, with up to five times elevation of circulating NH_4^+ [72]. Although the brain is partially protected by the blood-brain barrier from toxic agents such as ammonia, excessive amounts of NH_4^+ can pass into the brain, constituting the principal factor in the onset of HE.

PCA in the rat results in liver atrophy, sustained hyperammonemia, and subtle neurological symptoms of HE including abnormal locomotor activity, altered sleep patterns, and modifications of neuromuscular coordination. Feeding NH_4^+ salts or resins to the shunted rats leads to more severe signs, eventually progressing to coma. Neuropathological examination of these rats reveals Alzheimer type II astrocytosis, the histological characteristic of chronic hyperammonemic syndromes [74].

Oxidative stress is believed to play a role in the pathogenesis of HE because acute doses of NH_4^+ are prooxidant [75]. ROS include molecules, such as hydrogen peroxide (H_2O_2), superoxide (O_2^-) and the hydroxyl radical (OH·). Indeed, there is a physiological role for ROS including cellular proliferation, differentiation and signaling. However, a nonphysiological increase in ROS or a decrease in the antioxidant capacity of the organism can lead to an oxidative stress condition [76]. NH_4^+ promotes oxidative stress by increasing ROS [77]. In this context, the brain is susceptible to oxidative stress due to high content of unsaturated fatty acids prone to peroxidation, high O_2 consumption, elevated Fe^{2+}/Fe^{3+}, and low antioxidant systems [78]. However, a polemic issue has arisen since recent report in a model of hyperammonemia using a four-week PCA rat model did not express any signs of oxidative stress in the frontal cortex and in arterial plasma by 4-hydroxy-nonenal (4-HNE)-linked proteins and detection of carbonyl moieties [70].

Liver is by excellence the main metabolic organ, and shows an extensive handling of prooxidant reactions; especially during the biochemical transformation of nutrients and the processing of xenobiotics [79]. However, no information has been reported characterizing putative prooxidant reactions during the experimental PCA.

5.1. Lipid peroxidation

Oxygen is needed for proper energetic metabolism and correct mitochondrial function, but at the same time, it promotes the formation of ROS and, in consequence, oxidation of biomolecules. Lipid peroxidation is a suitable assay to estimate prooxidant reactions. By measuring the presence of conjugated dienes, it is possible to infer the rate of peroxidative activity under "*in vivo*" conditions, and it is also feasible to deduce the balance between prooxidant reactions and antioxidant defenses using the thiobarbituric acid reactive substances (TBARS) assay. When the TBARS test is done with Fe^{2+} supplementation, it offers another set of information: Because it enhances the breakdown of hydroperoxides, the Fe^{2+}-induced lipid peroxidation is maximum and gives an idea about the total antioxidant mechanisms and the presence of global unsaturated fatty acids present in the studied membrane [80].

Initial observations in hepatic subcellular fractions, liver homogenate and serum from sham ($n = 10$) and PCA ($n = 23$) operated rats after 8 to 13 weeks of surgery were used to test lipid peroxidative activity (data not published). Conjugated dienes and TBARS were quantified by standard techniques [81]. Strikingly, rats with PCA showed reduced TBARS levels in the liver homogenate and most of the subcellular fractions (**Figure 4**), whereas conjugated dienes showed no changes with lower levels in the mitochondrial fractions (**Figure 5**). The reduction of TBARS was also observed when the assay was supplemented with Fe^{2+} (**Figure 6**).

PCA is an experimental protocol to generate a hypofunctional liver condition. The abovementioned information strongly suggests that redox equilibrium within the liver under PCA surgery shows a reduction in the prooxidant reactions and/or increase in antioxidant defense. More focused experiments are needed to elucidate the underlying mechanism(s), but it is interesting to consider the biochemical consequences that the alteration in the portal blood flow can promote within the hepatocytes' redox equilibrium.

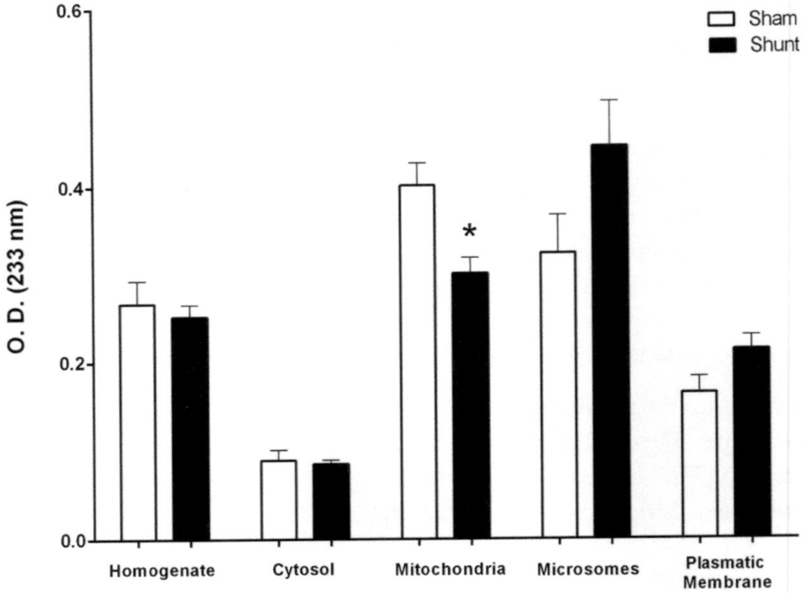

Figure 4. Effect of portacaval anastomosis on thiobarbituric acid reactive substances (TBARS) assay of liver homogenate, subcellular fractions and serum. Sham, false-operated rats (white bars); shunt, rats with portacaval anastomosis for 6-8 weeks (black bars). Data are average ± SEM from at least eight independent observations. *Significant statistical difference by *t*-student test, $p < 0.05$.

6. Redox alterations in liver cancer

6.1. Redox and carcinogenesis

Hepatocellular carcinoma (HCC) is a relevant problem of public health since it is in the sixth place in incidence, and the second in mortality at worldwide level [82]. The main risk factors conducing to HCC are viral hepatitis, steatosis and cirrhosis [83]. HCC could be considered the final stage of chronic liver disease (CLD), characterized by a persistent presence of inflammation and oxidative stress [84].

One of the best-characterized effects of redox changes is related to hepatic carcinogenesis [85]. The accumulation of ROS in early stages of hepatic damage produces lipid peroxidation of the cellular membranes; the reaction of ROS with the double bounds of polyunsaturated fatty acids results in the generation of aldehydes as 4-hydroxynonenal (4-HNE), acrolein, crotonaldehyde, and malondialdehyde, all with the capability of forming DNA adducts and genetic instability [86, 87]. 4-HNE forms exocyclic guanine adducts (4-HNE-dG) and consequently induces carcinogenic mutations [85]. These mutations affect loci of oncogenes or tumor suppressors conduce to malignant transformation. An example is the demonstration that 4-HNE induce G-C to A-T changes in the codon 243 of the p53 locus, an extensively studied tumor suppressor [88], promoting phenotypic loss [89, 90].

In this context, mechanisms activated by oxidative stress have the capability to influence a variety of proteins whose function impact in cancer, by the ability to oxidize thiol residues in

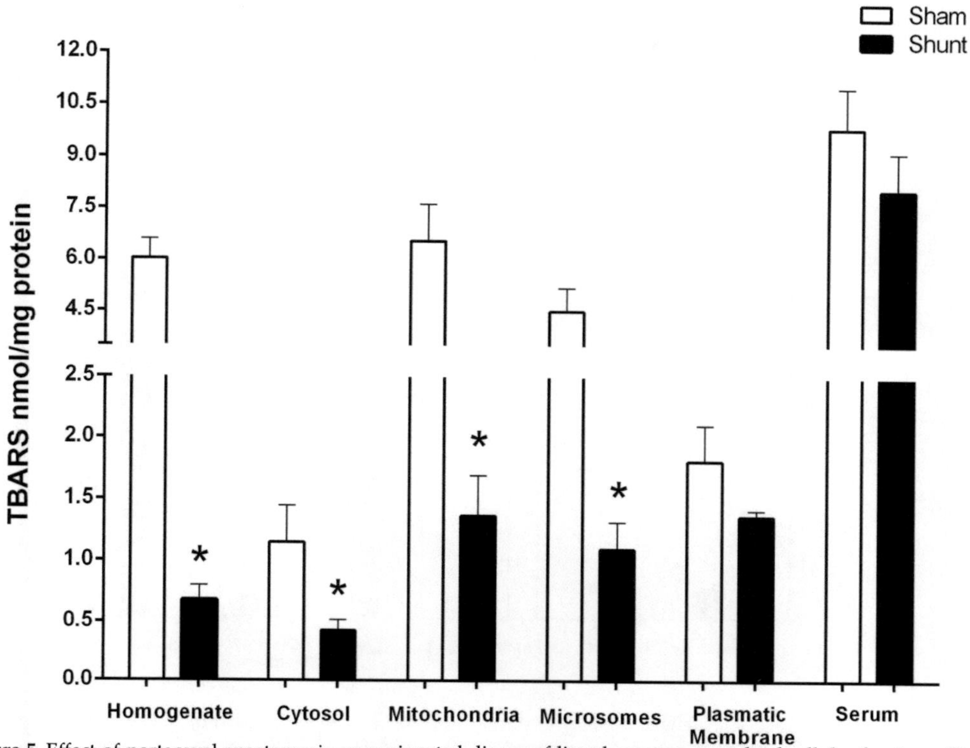

Figure 5. Effect of portacaval anastomosis on conjugated dienes of liver homogenate and subcellular fractions. Sham, false-operated rats (white bars); shunt, rats with portacaval anastomosis for 6-8 weeks (black bars). Data are average ± SEM from at least eight independent observations. *Significant statistical difference by *t*-student test, $p < 0.05$.

a process known as "redox priming" [91]. Examples of these redox sensors are the factor nf-κB [92], oncogenes or tumor suppressors as p53 [93] and Src tyrosine kinase [94].

Interestingly, accumulated evidence suggests that antioxidant systems activated in response to oxidative stress, improve the proliferation rate and protect cancer cells from a hostile environment [95, 96]. For example, it was demonstrated that prostaglandin 1 reductase-1 (Ptgr1), an oxidoreductase involved in the catabolism of eicosanoids and LPO-derived compound as 4-HNE, whose expression is regulated by the transcription factor (erythroid-derived-2)-like-2 (NRF2) [97], is overexpressed in human HCC biopsies and in samples from experimental animals. It was documented that Ptgr-1 regulates positively the proliferation rate of cells and improves their survival in two models of experimental HCC [97], suggesting that this antioxidant response plays a protumoral role in HCC.

6.2. Redox and hepatocellular carcinoma

A role for redox reactions has been detected during the HCC, from the proliferation of initial cancer cells to the dissemination process [96]. At the beginning of CLD, cells have a high proliferation rate, since the tumor growth demands nutrient supply and metabolic conditioning; in consequence, important architectural changes occur including the formation of new

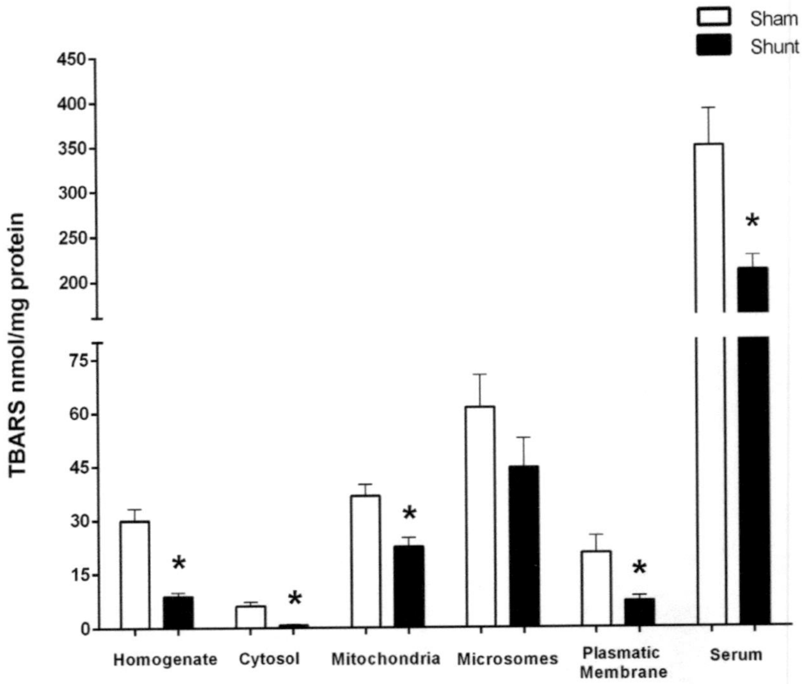

Figure 6. Effect of portacaval anastomosis on thiobarbituric acid reactive substances (TBARS) assay supplemented with FeSO4 50 (μM) of liver homogenate, subcellular fractions and serum. Sham, false-operated rats (white bars); shunt, rats with portacaval anastomosis for 6-8 weeks (black bars). Data are average ± SEM from at least eight independent observations. *Significant statistical difference by *t*-student test, $p < 0.05$.

blood vessels and the establishment of specific microenvironment within the tumor. These adaptations make the tumor cells adapt to highly variable O_2 concentration environment with successive lapses of hypoxia-reperfusion [98].

From the metabolic point of view, lack of O_2 favors the onset of aerobic glycolysis, by the Warburg's effect [99]; hypoxia also induces adaptive responses as the expression of specialized proteins named hypoxia-inducible factors (HIFs) [100]. HIF-1 is a transcription factor formed by two subunits (HIF-1α and HIF-1β). Although HIF-1 is constitutively expressed in normoxic conditions, HIF-α is constantly degraded by prolyl-hydroxylases; hydroxylated residues serve as docking site for von Hippel-Lindau tumor suppressor protein that is a constituent of an E3 ubiquitin ligase complex. The ubiquitinated HIF-1α suffers proteasomal degradation [101]. Low oxygen tension and some proinflammatory cytokines stabilize HIF-1α and allow its nuclear translocation to regulate key genes for the hypoxic response [100]. HIF-1α mediates the expression of genes that supports tumor growth such as NADH dehydrogenase (ubiquinone) 1α subcomplex, 4-like 2 (NDUFA4L2), a protein that attenuates the activity of the mitochondrial complex I, reducing the ROS production in low O_2 conditions [102].

It was shown that the kinase inhibitor sorafenib, an antineoplasic extensively used in oncology with potent antiangiogenic effects, induces intratumor oxidative stress that favor drug resistance in HCC; the insensibility to the drug requires the action of HIF-1α regulating the expression of specific genes. For example, Nf-κB induces the expression of the antiapoptotic protein Bcl-2 and vascular endothelium growth factor (VEGF) is a potent inductor of angiogenesis [103]. The resultant evasion of apoptosis and genesis of new blood vessels makes evident that protective cellular mechanisms are exploited during the progression of cancerous cells.

A role for ROS has also been described in the invasive process particularly in the epithelium to mesenchymal transition (EMT). EMT is a differentiation process consisting of the change from epithelial to mesenchymal characteristics. Epithelial cells are coupled through specialized structures as adherens junctions and exhibit apicobasal polarity, whereas in the mesenchymal phenotype the cells lose the coupling with neighbors and acquire migratory abilities [104]. EMT is considered the fundamental process in metastasis, being the cytokine TGF-β an effective EMT inductor as well as the growth factors EGF and PDGF [105]. These factors regulate the activity of the transcription factor SNAIL [106]. It has been described that in renal tubular epithelial cells, TGF-β induces synthesis of H_2O_2 and the onset to the EMT [107]. In hepatocarcinoma-derived cells HepG2, the stimulation with the phorbol-ester TPA, favors the activation of protein kinase C (PKC), phosphorylation of ERK and accumulation of ROS, resulting in the induction of EMT and cellular migration. These effects were prevented by ROS scavengers, suggesting a key role for these molecules in the cell migration promoted by TPA [108].

Recently, it was shown that ROS regulate EMT and cell migration induction through the activity of eukaryotic translation initiation factor *eIF5A2*; the expression pattern in the HCC-derived cell line SUN449 correlated with those modified by the knock down of *eIF5A2*, strongly suggesting that eIF5A2 is an effector of ROS signaling [109].

Hepatocellular tumoral cells show a variety of adaptive mechanism in extreme environmental conditions (hypoxia, low nutrients) to continue their growth and progression. Thus, understanding carcinoma cells biology requires yet exhaustive research and integrative efforts of the available data and an intimate knowledge of redox regulation.

Acknowledgements

M.D-M. research investigation is supported by DGAPA, PAPIIT grant no. IN200815. We thank LN Fernando López-Barrera for his artistic assistance and Jéessica González-Norris for critically editing the text.

Author details

Isabel Méndez[1], Francisco Vázquez-Cuevas[1], Rolando Hernández-Muñoz[2], Héctor Valente-Godínez[1], Olivia Vázquez-Martínez[1] and Mauricio Díaz-Muñoz[1]*

*Address all correspondence to: mdiaz@comunidad.unam.mx

1 Neurobiology Institute, Campus UNAM-Juriquilla, Querétaro, Mexico

2 Cellular Physiology Institute, University City, Mexico City, Mexico

References

[1] Pascal R, Pross A, Sutherland JD. Towards an evolutionary theory of origen of life based on kinetics and thermodynamics. Open Biology. 2013;**3**(11):130156

[2] Falkowski PG, Godfrey LV. Electrons, life and the evolution of Earth's oxygen cycle. Philosophical Transactions of the Royal Society B Biological Sciences. 2008;**363**(1504): 2705-2016

[3] Zera AJ. Microevolution of intermediary metabolism: Evolutionary genetics meets metabolic biochemistry. Journal of Experimental Biology. 2011;**214**(Pt 2):179-190

[4] Manchester KL. Albert Szent-Gyorgyi and the unravelling of biological oxidation. Trends in Biological Sciences. 1998;**23**(1):37-40

[5] Moore GR, Pettigrew GW, Rogers NK. Factors influencing redox potentials of electron transfer proteins. Proceedings of the National Academy of Sciences USA. 1998; **83**(14):4998-4999

[6] Hijmans BS, Grefhorst A, Oosterveer MH, Groen AK. Zonation of glucose and fatty acids metabolism in the liver: Mechanism and metabolic consequences. Biochimie. 2014;**96**:121-129

[7] Brown GC. Control of respiration and ATP synthesis in mammalian mitochondria and cells. Biochemical Journal. 1992;**284**(Pt 1):1-13

[8] Rhee SG. Cell signaling. H_2O_2, a necessary evil for cell signaling. Science. 2006;**312**(5782): 1882-1883

[9] Diebold L, Chandel NS. Mitochondrial ROS regulation of proliferating cells. Free Radical Biology and Medicine. 2016;**100**:86-93

[10] Brandes N, Jakob U. Thiol-based redox switches in eukaryotic proteins. Antioxidants and Redox Signaling. 2009;**11**(5):997-1014

[11] Kimura S. Mitochondria-derived reactive oxygen species and vascular MAP kinases: Comparison of angiotensin II and diazoxide. Hypertension. 2005;**45**(3):438-444

[12] Gill T, Levine AD. Mitochondria-derived H2O2 selectively enhances T cell receptor-initiated signal transduction. Journal of Biological Chemistry. 2013;**288**(36):26246-26255

[13] Hanahan D, Weinberg RA. Hallmarks of cancer: The next generation. Cell. 2011;**144**(5): 646-674

[14] Puppel K, Kapusta A, Kuczynska B. The etiology of oxidative stress in the various species of animals, a review. Journal of the Science of Food and Agriculture. 2015;**95**:2179-2184

[15] Méndez I, Vazquez-Martinez O, Hernandez-Munoz R, Valente-Godínez H, Díaz-Muñoz M. Redox regulation and pro-oxidant reactions in the physiology of circadian systems. Biochimie. 2016;**124**:178-186

[16] Jo SH, Son MK, Koh HJ, Lee SM, Song IH, Kim YO, et al. Control of mitochondrial redox balance and cellular defense against oxidative damage by mitochondrial NADP+-dependent isocitrate dehydrogenase. The Journal of Biological Chemistry. 2001;**276**:16168-16176

[17] Koh HJ, Lee SM, Son BG, Lee SH, Ryoo ZY, Chang KT, et al. Cytosolic NADP+-dependent isocitrate dehydrogenase plays a key role in lipid metabolism. The Journal of Biological Chemistry. 2004;**279**:39968-39974

[18] Lee SM, Koh HJ, Park DC, Song BJ, Huh TL, Park JW. Cytosolic NADP(+)-dependent isocitrate dehydrogenase status modulates oxidative damage to cells. Free Radical Biology & Medicine. 2002;**32**:1185-1196

[19] Kil IS, Park JW. Regulation of mitochondrial NADP+-dependent isocitrate dehydrogenase activity by glutathionylation. The Journal of Biological Chemistry. 2005;**280**:10846-10854

[20] McLain AL, Szweda PA, Szweda LI. Alpha-Ketoglutarate dehydrogenase: A mitochondrial redox sensor. Free Radical Research. 2011;**45**:29-36

[21] Tretter L, Adam-Vizi V. Alpha-ketoglutarate dehydrogenase: A target and generator of oxidative stress. Philosophical transactions of the Royal Society of London Series B, Biological Sciences. 2005;**360**:2335-2345

[22] Tian WN, Braunstein LD, Pang J, Stuhlmeier KM, Xi QC, Tian X, et al. Importance of glucose-6-phosphate dehydrogenase activity for cell growth. The Journal of Biological Chemistry. 1998;**273**:10609-10617

[23] Stapleton SR, Stevens GJ, Teel JF, Rank KB, Berg EA, Wu JY, et al. Effects of acetaldehyde on glucose-6-phosphate dehydrogenase activity and mRNA levels in primary rat hepatocytes in culture. Biochimie. 1993;**75**:971-976

[24] Pandolfi PP, Sonati F, Rivi R, Mason P, Grosveld F, Luzzatto L. Targeted disruption of the housekeeping gene encoding glucose 6-phosphate dehydrogenase (G6PD): G6PD is dispensable for pentose synthesis but essential for defense against oxidative stress. The EMBO Journal. 1995;**14**:5209-5215

[25] Salati LM, Amir-Ahmady B. Dietary regulation of expression of glucose-6-phosphate dehydrogenase. Annual Review of Nutrition. 2001;**21**:121-140

[26] Stabile LP, Klautky SA, Minor SM, Salati LM. Polyunsaturated fatty acids inhibit the expression of the glucose-6-phosphate dehydrogenase gene in primary rat hepatocytes by a nuclear posttranscriptional mechanism. Journal of Lipid Research. 1998;**39**:1951-1963

[27] Gupte RS, Floyd BC, Kozicky M, George S, Ungvari ZI, Neito V, et al. Synergistic activation of glucose-6-phosphate dehydrogenase and NAD(P)H oxidase by Src kinase elevates superoxide in type 2 diabetic, Zucker fa/fa, rat liver. Free Radical Biology & Medicine. 2009;**47**:219-228

[28] Wang YP, Zhou LS, Zhao YZ, Wang SW, Chen LL, Liu LX, et al. Regulation of G6PD acetylation by SIRT2 and KAT9 modulates NADPH homeostasis and cell survival during oxidative stress. The EMBO Journal. 2014;**33**:1304-1320

[29] Michishita E, Park JY, Burneskis JM, Barrett JC, Horikawa I. Evolutionarily conserved and nonconserved cellular localizations and functions of human SIRT proteins. Molecular Biology of the Cell. 2005;**16**:4623-4635

[30] Pfluger PT, Herranz D, Velasco-Miguel S, Serrano M, Tschop MH. Sirt1 protects against high-fat diet-induced metabolic damage. Proceedings of the National Academy of Sciences of the United States of America. 2008;**105**:9793-9798

[31] Olmos Y, Sanchez-Gomez FJ, Wild B, Garcia-Quintans N, Cabezudo S, Lamas S, et al. SirT1 regulation of antioxidant genes is dependent on the formation of a FoxO3a/PGC-1alpha complex. Antioxidants & Redox Signaling. 2013;**19**:1507-1521

[32] Canto C, Gerhart-Hines Z, Feige JN, Lagouge M, Noriega L, Milne JC, et al. AMPK regulates energy expenditure by modulating NAD^+ metabolism and SIRT1 activity. Nature. 2009;**458**:1056-1060

[33] Cardaci S, Filomeni G, Ciriolo MR. Redox implications of AMPK-mediated signal transduction beyond energetic clues. Journal of Cell Science. 2012;**125**:2115-2125

[34] Salminen A, Kaarniranta K, Kauppinen A. Crosstalk between oxidative stress and SIRT1: Impact on the aging process. International Journal of Molecular Sciences. 2013;**14**: 3834-3859

[35] Boutant M, Canto C. SIRT1 metabolic actions: Integrating recent advances from mouse models. Molecular Metabolism. 2014;**3**:5-18

[36] Tyagi S, Gupta P, Saini AS, Kaushal C, Sharma S. The peroxisome proliferator-activated receptor: A family of nuclear receptors role in various diseases. Journal of Advanced Pharmaceutical Technology & Research. 2011;**2**:236-240

[37] Anderson SP, Howroyd P, Liu J, Qian X, Bahnemann R, Swanson C, et al. The transcriptional response to a peroxisome proliferator-activated receptor alpha agonist includes increased expression of proteome maintenance genes. The Journal of Biological Chemistry. 2004;**279**:52390-52398

[38] Burri L, Thoresen GH, Berge RK. The role of PPARalpha activation in liver and muscle. PPAR Research. 2010;**2010**:457831

[39] Polvani S, Tarocchi M, Galli A. PPARgamma and oxidative stress: Con(beta) catenating NRF2 and FOXO. PPAR Research. 2012;**2012**:641087

[40] Diesen DL, Kuo PC. Nitric oxide and redox regulation in the liver: Part II. Redox biology in pathologic hepatocytes and implications for intervention. Journal of Surgical Research. 2011;**167**(1):96-112

[41] Lieber CS, DeCarli LM. Hepatotoxicity of ethanol. Journal of Hepatology. 1991;**12**(3): 394-401

[42] Lieber CS. Alcohol and the liver: 1994 update. Gastroenterology. 1994;**106**(4):1085-1105

[43] Charlton MR, Burns JM, Pedersen RA, Watt KD, Heimbach JK, Dierkhising RA. Frequency and outcomes of liver transplantation for nonalcoholic steatohepatitis in the United States. Gastroenterology. 2009;**141**(4):1249-1253

[44] Day CP, James OF. Steatohepatitis: a tale of two "hits"? Gastroenterology. 1998;**114**(4): 842-845

[45] Bruce KD, Cagampang FR, Argenton M, Zhang J, Ethirajan PL, Burdge GC, Bateman AC, Clough GF, Poston L, Hanson MA, McConnell JM, Byrne CD. Maternal high-fat feeding primes steatohepatitis in adult mice offspring, involving mitochondrial dysfunction and altered lipogenesis gene expression. Hepatology. 2009;**50**(6):1796-1808

[46] McCurdy CE, Bishop JM, Williams SM, Grayson BE, Smith MS, Friedman JE, Grove KL. Maternal high-fat diet triggers lipotoxicity in the fetal livers of nonhuman primates. Journal of Clinical Investigation. 2009;**119**(2):323-335

[47] Bruce KD, Szczepankiewicz D, Sihota KK, Ravindraanandan M, Thomas H, Lillycrop KA, Burdge GC, Hanson MA, Byrne CD, Cagampang FR. Altered cellular redox status, sirtuin abundance and clock gene expression in a mouse model of developmentally primed NASH. Biochimica et Biophysica Acta. 2016;**1861**(7):584-593

[48] Zmijewski JW, Banerjee S, Bae H, Friggeri A, Lazarowski ER, Abraham E. Exposure to hydrogen peroxide induces oxidation and activation of AMP-activated protein kinase. Journal of Biological Chemistry. 2010;**285**(43):33154-33164

[49] Begriche K, Igoudjil A, Pessayre D, Fromenty B. Mitochondrial dysfunction in NASH: Causes, consequences and possible means to prevent it. Mitochondrion. 2006;**6**(1):1-28

[50] Lieber CS. New concepts of the pathogenesis of alcoholic liver disease lead to novel treatments. Current Gastroenterology Reports. 2004;**6**(1):60-65

[51] Noguchi Y, Young JD, Aleman JO, Hansen ME, Kelleher JK, Stephanopoulos G. Effect of anaplerotic fluxes and amino acid availability on hepatic lipoapoptosis. Journal of Biological Chemistry2009;**284**(48):33425-36.

[52] Wei Y, Wang D, Topczewski F, Pagliassotti MJ. Saturated fatty acids induce endoplasmic reticulum stress and apoptosis independently of ceramide in liver cells. American Journal of Physiology—Endocrinology and Metabolism. 2006;**291**(2):E275–E281

[53] Yang SL, Xia JH, Zhang YY, Fan JG, Wang H, Yuan J, Zhao ZZ, Pan Q, Mu YL, Xin LL, Chen YX, Li K. Hyperinsulinemia shifted energy supply from glucose to ketone bodies in early nonalcoholic steatohepatitis from high-fat high-sucrose diet induced Bama minipigs. Scientific Reports. 2015;**5**:13980

[54] Fan F, Jin S, Amundson SA, Tong T, Fan W, Zhao H, Zhu X, Mazzacurati L, Li X, Petrik KL, Fornace Jr AJ. Rajasekaran B, Zhan Q. ATF3 induction following DNA damage is regulated by distinct signaling pathways and over-expression of ATF3 protein suppresses cells growth. Oncogene. 2002;**21**(49):7488-7496

[55] Bettaieb A, Jiang JX, Sasaki Y, Chao TI, Kiss Z, Chen X, Tian J, Katsuyama M, Yabe-Nishimura C, Xi Y, Szyndralewiez C, Schröder K, Shah A, Brandes RP, Haj FG, Török NJ. Hepatocyte nicotinamide adenine dinucleotide phosphate reduced oxidase 4 regulates stress signaling, fibrosis, and insulin sensitivity during development of steatohepatitis in mice. Gastroenterology. 2015;**149**(2):468-480.e10

[56] Chatterjee S, Ganini D, Tokar EJ, Kumar A, Das S, Corbett J, Kadiiska MB, Waalkes MP, Diehl AM, Mason RP. Leptin is key to peroxynitrite-mediated oxidative stress and Kupffer cell activation in experimental non-alcoholic steatohepatitis. Journal of Hepatology. 2013;**58**(4):778-784

[57] Bradbury MW, Berk PD. Lipid metabolism in hepatic steatosis. Clinical Liver Disease. 2004;**8**(3):639-671, xi

[58] Lieber CS. CYP2E1: From ASH to NASH. Hepatology Research. 2004;**28**(1):1-11

[59] Chance B, Sies H, Boveris A. Hydroperoxide metabolism in mammalian organs. Physiological Reviews. 1979;**59**(3):527-605

[60] Orellana M, Rodrigo R, Varela N, Araya J, Poniachik J, Csendes A, Smok G, Videla LA. Relationship between in vivo chlorzoxazone hydroxylation, hepatic cytochrome P450 2E1 content and liver injury in obese non-alcoholic fatty liver disease patients. Hepatology Research. 2006;**34**(1):57-63

[61] Yin M, Gäbele E, Wheeler MD, Connor H, Bradford BU, Dikalova A, Rusyn I, Mason R, Thurman RG. Alcohol-induced free radicals in mice: Direct toxicants or signaling molecules? Hepatology. 2001;**34**(5):935-942

[62] Fleury C, Sanchis D. The mitochondrial uncoupling protein-2: Current status. The International Journal of Biochemistry & Cell Biology. 1999;**31**(11):1261-1278

[63] Himms-Hagen J, Harper ME. Physiological role of UCP3 may be export of fatty acids from mitochondria when fatty acid oxidation predominates: An hypothesis. Experimental Biology and Medicine (Maywood). 2001;**226**(2):78-84

[64] Haque JA, McMahan RS, Campbell JS, Shimizu-Albergine M, Wilson AM, Botta D, Bammler TK, Beyer RP, Montine TJ, Yeh MM, Kavanagh TJ, Fausto N. Attenuated progression of diet-induced steatohepatitis in glutathione-deficient mice. Laboratory Investigation. 2010;**90**(12):1704-1717

[65] Ross G, Dunn D, Jones ME. Ornithine synthesis from glutamate in rat intestinal mucosa homogenates: Evidence for the reduction of glutamate to gamma-glutamyl semialdehyde. Biochemical and Biophysical Research Communications. 1978;**85**(1):140-147

[66] Hansford RG, Sacktor B. Regulation of glycogen metabolism in insect flight muscle. Activation of phosphorylase b kinase by calcium and inorganic phosphate. FEBS Letters. 1970;**7**(2):183-187

[67] Meyer J. Proline transport in rat liver mitochondria. Archives of Biochemistry And Biophysics. 1977;**178**(2):387-395

[68] Krähenbühl S, Reichen J. Adaptation of mitochondrial metabolism in liver cirrhosis. Different strategies to maintain a vital function. Scandinavian Journal of Gastroenterology. 1992;**193**:90-96

[69] Krähenbühl S, Krähenbühl-Glauser S, Stucki J, Gehr P, Reichen J. Stereological and functional analysis of liver mitochondria from rats with secondary biliary cirrhosis: impaired mitochondrial metabolism and increased mitochondrial content per hepatocyte. Hepatology. 1992;**15**(6):1167-1172

[70] Yang X, Bosoi CR, Jiang W, Tremblay M, Rose CF. Portacaval anastomosis-induced hyperammonemia does not lead to oxidative stress. Metabolic Brain Disease. 2010;**25**(1):11-15

[71] Butterworth RF, Norenberg MD, Felipo V, Ferenci P, Albrecht J, Blei AT. Experimental models of hepatic encephalopathy: ISHEN guidelines. Liver International. 2009;**29**(6):783-788

[72] Aldridge DR, Tranah EJ, Shawcross DL. Pathogenesis of hepatic encephalopathy: Role of ammonia and systemic inflammation. Journal of Clinical and Experimental Hepatology. 2015;**5**(S1):S7–S20

[73] Oja SS, Saransaari P, Korpi ER. Neurotoxicity of ammonia. Neurochemical Research. 2016;**3**:1-8

[74] Butterworth RF, Girard G, Giguère J-F. Regional differences in the capacity for ammonia removal by brain following portocaval anastomosis. Journal of Neurochemistry. 1988;**51**(2):486-490

[75] Bosoi CR, Yang X, Huynh J, Parent-Robitaille C, Jiang W, Tremblay M, et al. Systemic oxidative stress is implicated in the pathogenesis of brain edema in rats with chronic liver failure. Free Radical Biology & Medicine. 2012;**52**(7):12281235

[76] Rupérez AI, Gil A, Aguilera CM. Genetics of oxidative stress in obesity. International Journal of Molecular Sciences. 2014;**15**(2):3118-3144

[77] Liang Z, Liu R, Zhao D, Wang L, Sun M, Wang M, et al. Ammonia exposure induces oxidative stress, endoplasmic reticulum stress and apoptosis in hepatopancreas of pacific white shrimp (*Litopenaeus vannamei*). Fish and Shellfish Immunology. 2016;**54**:523-528

[78] Poon HF, Calabrese V, Scapagnini G, Butterfield DA. Free radicals and brain aging. Clinics in Geriatric Medicine. 2004;**20**(2):329-359

[79] Luna-Moreno D, Vázquez-Martínez O, Báez-Ruiz A, Ramírez J, Díaz-Muñoz M. Food restricted schedules promote differential lipoperoxidative activity in rat hepatic subcellular fractions. Comparative Biochemistry and Physiology Part A: Molecular & Integrative Physiology 2007;**146**(4):632-643

[80] Vázquez-Martínez O, Pérez-Mendoza M, Valente-Godínez H, Revueltas-Guillén F, Carmona-Castro A, Díaz-Muñoz M, et al. Day-night variations in pro-oxidant reactions of hypothalamic, hepatic and pancreatic tissue in mice with spontaneous obesity (Neotomodon alstoni). Biological Rhythm Research. 2015;**1016**:1-10

[81] World Health Organization. International Agency for Research in Cancer, Cancer Today [Internet]. 2017. Available from: http://gco.iarc.fr/today/online-analysis-multi-bars?mode=cancer&mode_population=continents&population=900&sex=0&cancer=29&type=0&statistic=0&prevalence=0&color_palette=default; and http://gco.iarc.fr/today/online-analysis-multi-bars?mode=cancer&mode_population=continents&population=900&sex=0&cancer=29&type=1&statistic=0&prevalence=0&color_palette=default [Accessed: March 21, 2017]

[82] Klaassen CD, Plaa GL. Comparison of the biochemical alterations elicited in livers from rats treated with carbon tetrachloride, chloroform, 1,1,2-trichloroethane and 1,1,1-trichloroethane. Biochemical Pharmacology. 1969;**18**:2019-2027

[83] GLOBOCAN. International Agency for Research of Cancer. World Health Organization [Internet]. 2012. Available from: http://gco.iarc.fr/today/home

[84] Novo E, Cannito S, Paternostro C, Bocca C, Miglietta A, Parola M. Cellular and molecular mechanisms in liver fibrogenesis. Archieves of Biochemistry and Biophysics. 2014;**548**:20-37

[85] Blachier M, Leleu H, Peck-Radosavljevic M, Valla DC, Roudot-Thoraval F. The burden of liver disease in Europe: A review of available epidemiological data. Journal of Hepatology. 2013;**58**(3):593-608

[86] Ames BN, Gold LS. Endogenous mutagens and the causes of aging and cancer. Mutation Research. 1991;**250**:3-16

[87] Esterbauer H, Schaur RJ, Zollner H. Chemistry and biochemistry of 4-hydroxynonenal, malonaldehyde and related aldehydes. Free Radicals Biology and Medicine. 1991;**11**:81-128

[88] Zarkovic N. 4-hydroxynonenal as a bioactive marker of pathophysiological processes. Molecular Aspects of Medicine. 2003;**24**(4-5):281-291

[89] Mantovani F, Walerych D, Sal GD. Targeting mutant p53 in cancer: A long road to precision therapy. FEBS J. 2017;**284**(6):837-850.

[90] Hu W, Feng Z, Eveleigh J, Iyer G, Pan J, Amin S, Chung FL, Tang MS. The major lipid peroxidation product, trans-4-hydroxy-2-nonenal, preferentially forms DNA adducts at codon 249 of human p53 gene, a unique mutational hotspot in hepatocellular carcinoma. Carcinogenesis. 2002;**23**(11):1781-1789

[91] Feng Z, Hu W, Amin S, Tang MS. Mutational spectrum and genotoxicity of the major lipid peroxidation product, trans-4-hydroxy-2-nonenal, induced DNA adducts in nucleotide excision repair-proficient and -deficient human cells. Biochemistry. 2003;**42**(25):7848-7854

[92] Parri M, Chiarugi P. Redox molecular machines involved in tumor progression. Antioxidant & Redox Signaling. 2013;**19**(15):1828-1845

[93] Schreck R, Rieber P, Baeuerle PA. Reactive oxygen intermediates as apparently widely used messengers in the activation of the NF-kappa B transcription factor and HIV-1. EMBO Journal. 1991;**10**:2247-2258

[94] Rainwater R, Parks D, Anderson ME, Tegtmeyer P, Mann K. Role of cysteine residues in regulation of p53 function. Molecular and Cellular Biology. 1995;**15**:3892-3903

[95] Akhand AA, Pu M, Senga T, Kato M, Suzuki H, Miyata T, Hamaguchi M, Nakashima I. Nitric oxide controls src kinase activity through a sulfhydryl group modification-mediated Tyr 527-independent and Tyr-416-linked mechanism. Journal of Biological Chemistry. 1999;**274**(36):25821-25826

[96] Trachootham D, Alexandre J, Huang P. Targeting cancer cells by ROS-mediated mechanisms: A radical therapeutic approach? Nature Reviews Drug Discovery. 2009;**8**(7):579-951

[97] Pani G, Giannoni E, Galeotti T, Chiarugi P. Redox-based escape mechanism from death: The cancer lesson. Antioxidants and Redox Signaling. 2009;**11**(11):2791-2806

[98] Sánchez-Rodríguez R, Torres-Mena JE, Quintanar-Jurado V, Chagoya-Hazas V, Rojas Del Castillo E, Del Pozo Yauner L, Villa-Treviño S, Pérez-Carreón JI. Ptgr1 expression is regulated by NRF2 in rat hepatocarcinogenesis and promotes cell proliferation and resistance to oxidative stress. Free Radical Biology and Medicine. 2017;**102**:87-99

[99] Sánchez-Rodríguez R, Torres-Mena JE, De-la-Luz-Cruz M, Bernal-Ramos GA, Villa-Treviño S, Chagoya-Hazas V, Landero-López L, García-Román R, Rouimi P, Del-Pozo-Yauner L, Meléndez-Zajgla J, Pérez-Carreón JI. Increased expression of prostaglandin reductase 1 in hepatocellular carcinomas from clinical cases and experimental tumors in rats. The International Journal of Biochemistry & Cell Biology. 2014;**53**:186-194

[100] Dewhirst MW, Cao Y, Moeller B. Cycling hypoxia and free radicals regulate angiogenesis and radiotherapy response. Nature Reviews Cancer. 2008;**8**:425-437

[101] Nakajima EC, Van Houten B. Metabolic symbiosis in cancer: Refocusing the Warburg lens. Molecular Carcinogenesis. 2012;**52**:329-337

[102] Semenza GL. HIF-1: Mediator of physiological and pathophysiological responses to hypoxia. Journal of Applied Physiology. 2000;**88**:1474-1480

[103] Berra E, Benizri E, Ginouvès A, Volmat V, Roux D, Pouysségur J. HIF prolyl-hydroxylase 2 is the key oxygen sensor setting low steady-state levels of HIF-1α in normoxia. The EMBO Journal; 2003;**22**(16):4082-4090

[104] Tello D, Balsa E, Acosta-Iborra B, Fuertes-Yebra E, Elorza A, Ordóñez á, Corral-Escariz M, Soro I, López-Bernardo E, Perales-Clemente E, Martínez-Ruiz A, Enríquez JA, Aragonés J, Cadenas S, Landázuri MO. Induction of the mitochondrial NDUFA4L2 protein by HIF-1α decreases oxygen consumption by inhibiting Complex I activity. Cell Metabolism. 2011;**14**(6):768-779

[105] Liang Y, Zheng T, Song R, Wang J, Yin D, Wang L, Liu H, Tian L, Fang X, Meng X, Jiang H, Liu J, Liu L. Hypoxia-mediated sorafenib resistance can be overcome by EF24 through Von Hippel-Lindau tumor suppressor-dependent HIF-1α inhibition in hepatocellular carcinoma. Hepatology. 2013;**57**(5):1847-1857

[106] Kalluri R, Weinberg RA. The basics of epithelial-mesenchymal transition. Journal of Clinical Investigation. 2009;**119**(6):1420-1428

[107] Thiery JP. Epithelial-mesenchymal transitions in tumour progression. Nature Reviews Cancer. 2002;**2**:442-454. DOI: 10.1038/ncr822

[108] Cano A, Pérez-Moreno MA, Rodrigo I, Locascio A, Blanco MJ, del Barrio MG, Portillo F, Nieto MA. The transcription factor snail controls epithelial-mesenchymal transitions by repressing E-cadherin expression. Nature Cell Biology. 2000;**2**:76-83

[109] Rhyu DY, Yang Y, Ha H, Lee GT, Song JS, Uh ST, Lee HB. Role of reactive oxygen species in TGF-beta1-induced mitogen-activated protein kinase activation and epithelial-mesenchymal transition in renal tubular epithelial cells. Journal of American Society of Nephrology. 2005;**16**(3):667-675

Chapter 8

Redox Modulation of Adipogenesis

Xin Wang

Additional information is available at the end of the chapter

http://dx.doi.org/10.5772/intechopen.68727

Abstract

Obesity and obesity-associated disorders have become globally prevalent. Knowing how adipose tissue forms is important for the intervention of obesity. In the last decades, numerous evidence have shown that redox status changes under the condition of obesity and in the process of adipogenesis. Alterations of both oxidant and antioxidant levels may influence the transformation from stem cell or preadipocytes to mature adipocytes. Redox system exerts "tridimensional" mechanism in the regulation of adipocyte differentiation, including transcriptional, epigenetic, and posttranslational modulations. However, the roles of redox system in the regulation of adipocyte differentiation are paradoxical. Therefore, we propose that restoration and maintenance of redox balance rather than simple prooxidant or antioxidant interventions are critical for the prevention and therapy of obesity and obesity-associated disorders.

Keywords: redox, adipogenesis, obesity, ROS, antioxidant

1. Introduction

To date, it is estimated that more than a billion adults are overweight and over 300 million adults are classified as obese worldwide [1]. Owing to the dramatic prevalence of obesity, there is a drastic increase in obesity-associated diseases, including hypertension, type 2 diabetes, dyslipidemia, kidney disease, heart disease, cancer, obstructive sleep apnea, and osteoarthritis. The adipose tissue, also called fat, includes white adipose tissue (WAT) and brown adipose tissue. Obesity is mainly caused by an excessive accumulation of fat mass in WAT to the extent that health may be adversely affected. WAT functions as both a metabolic and an endocrine organ that lies at the heart of a complex network participating in the regulation of various biological responses, including inflammation, immunity, endocrine, and metabolism [2].

Adipose tissue is a dynamic tissue, and the mass of adipose is generally determined by a balance of adipocyte proliferation, differentiation, enlargement, and lipolysis. Increased adipose mass may result from the enhancement of adipocyte formation (adipocyte differentiation, also called adipogenesis), enlargement of adipocyte (lipogenesis), and/or reduction of lipolysis that contribute to fat accumulation in WAT. Adipocyte differentiation is defined as the formation of new adipocytes from multipotent stem cells or preadipocyte precursors [3]. In brief, adipogenesis is divided into four steps, including initial growth arrest, mitotic clonal expansion (MCE), early differentiation, and terminal differentiation-development of mature adipocyte phenotype [4]. White and brown adipocyte differentiation share common pathways and possess specific characteristics. Unless specified, "adipocyte differentiation or adipogenesis" indicates white adipocyte differentiation here.

2. Redox state changes under obese condition and in the process of adipogenesis

It is viewed that obesity is a state of chronic oxidative stress. In obese patients, increased oxidative metabolic products of DNA, protein and lipid, and/or decreased antioxidant activity are closely associated with excessive fat accumulation. In high fat (HF) diet-fed animals, oxidative stress precedes the onset of insulin resistance and obesity. In obese animals, levels of reactive oxygen species (ROS) in mature adipocytes are higher than that in other tissues, including liver, skeletal muscle, and aorta. In the course of adipogenesis, intracellular redox potentials tend to a significant oxidizing state. It is suggested that increased ROS level is associated with the development of obesity and may be required for the formation of new adipocytes.

2.1. ROS and adipogenesis

ROS plays a controversial role in adipogenesis. On the one hand, ROS is a required signal molecule in the activation of key transcription factors responsible for adipogenesis. On the other hand, ROS may also act as a negative regulator of the formation of adipocytes.

2.2. ROS is a signal required for adipogenesis

It is originally identified that ROS is produced physiologically during adipogenesis. In the last decades, accumulating data prove that endogenously generated ROS acts as an essential mediator for adipogenic differentiation. Compared with that of preadipocytes, adipocytes have a relatively higher rate of spontaneous intra- and extracellular ROS production. Adipogenic hormonal cocktails could increase ROS production in preaipocytes and enhance the process of differentiation. C/EBPβ and PPARγ are the main factors that mediate the proadipogenic effect of ROS.

2.3. ROS also functions as a detrimental signal of adipogenesis

It is shown that oxidative stress inhibits fat cell formation by reducing the C/EBP DNA-binding activity [5]. Mitochondrial ROS could inhibit preadipocyte proliferation, indicating that it can influence adipocyte differentiation through inhibiting MCE. In 3T3-F442A

http://dx.doi.org/10.5772/intechopen.68727

preadipocytes treated with various pharmacological inhibitors, Carriere et al. [6] demonstrated that mitochondrial ROS was negatively correlated with adipocyte differentiation through increasing adipogenic repressor CHOP-10/GADD153.

2.4. ROS-producing systems and adipocyte differentiation

In a cell, ROS are mainly generated by mitochondrial respiration system, nicotinamide adenine dinucleotide phosphate (NADPH) oxidases, endoplasmic reticulum (ER), and certain metabolic and detoxifying enzymes. Effect of ROS-producing systems on adipocyte differentiation is shown in **Figure 3**.

3. Mitochondria and adipocyte differentiation

Approximate 1–3% of O_2 consumed in a cell is incompletely metabolized and diverted into superoxide in complex I and complex III in the respiratory chain. ROS production in mitochondria is obligatory during mitochondrial oxidative metabolism. In addition, ROS can also be produced by several other enzymes that locate in mitochondria. It is demonstrated that most of mitochondrial alterations is associated with redox oscillation.

Mitochondria are highly dynamic organelles that can be rapidly restructured to meet the metabolic demands timely. Notable cytoplasmic remodeling occurs during the process of adipocyte differentiation, and reorganization of mitochondrial network is one of the main events. In preadipocytes, mitochondria are filamentous and organized as continuous reticulum. In contrast, in differentiated adipocytes, they are fragmented, punctual, and redistributed around lipid droplets. Mitochondrial fusion and fission have direct influence on lipid accumulation in adipocytes. Induction of mitochondrial fusion by silencing of mitochondrial fission proteins, including dynamic-related protein (Drp1) and fission 1 homologue protein (Fis1), decreases cellular TG content. In contrast, silencing mitochondrial fusion proteins, including mitofusin 2 (Mfn2) and optic atrophy-1 (OPA1), increased cellular TG content followed by fragmentation of mitochondria.

The subunits of mitochondrial respiratory complexes I–IV increase as the differentiation progressed. There is a significant "burst" of expression level of proteins involved in the tricarboxylic acid (TCA) cycle, especially pyruvate carboxylase, at the very beginning of differentiation, which was followed by a more gradual and less intensive increase in the expression of genes associated with the electron transport chain, fatty acid metabolism, and mitochondrial transport. Above results indicate that during the transition from preadipocytes to adipocytes, cells enter a state of metabolic-overdrive characterized by simultaneous increase of flux through the TCA cycle and fatty acid oxidation.

Besides the depictive studies, several genetic and pharmacological treatments have been introduced to evaluate the role of mitochondria in adipogenesis. Depletion of mtDNA using ethidium bromide inhibits lipid accumulation in 3T3-L1 adipocytes. Carbonyl-cyanide-trifluoromethoxy-phenyl-hydrazone, an uncoupler of oxidative phosphorylation (OXPHOS), suppresses adipogenesis. Inhibition of electron transmission chain by several agents, such as rotenone, suppresses adipogenesis by modulating the expression of the key transcription factors C/EBPα, PPARγ, and sterol regulatory element binding protein (SREBP)-1c. Several factors critical for mitochondrial function have been reported to be required for adipocyte differentiation, such as CR6-interacting factor 1 and prohibitin.

Although specific changes of mitochondria during adipocyte differentiation are variant in different laboratories, we still could depict the common pattern: mitochondrial fission, replication, and metabolism are transiently enhanced by adipogenic stimuli, and then the drastic transformation gradually "calms down" when the adipocytes are terminally differentiated. As an essential signal or by-products, enhancement of mitochondrial activity and boost of OXPHOS may result in more ROS production. In addition, direct evidence support that ROS-generating enzymes in mitochondria were found to be activated during adipocyte differentiation, such as monoamine oxidases and dihydrolipoamide dehydrogenase. Moreover, $p66^{Shc}$, a main isoform of SHC-transforming protein 1 (SHC1) that locates in mitochondria and functions as a potent redox regulator through production of ROS, could be activated by insulin, a key component of DMI cocktail. After the MCE phase, $p66^{Shc}$ phosphorylation is reduced in differentiated adipocytes.$p66^{Shc}$and consequent mitochondrial ROS generation is also implicated in the adipogenic conversion of muscle satellite cells (SCs), forming intermuscular adipose tissue. Despite the complexity of the involvement of mitochondrial alterations in adipocyte differentiation, there is a common view that mitochondria may function to influence the process of differentiation, at least, through the following ways. First, increased TCA and fatty acid oxidation meet the energy need. Second, mitochondria can provide key substrates necessary to support the massive lipogenesis during adipogenesis.

3.1. Endoplasmic reticulum and adipocyte differentiation

Endoplasmic reticulum (ER) plays an important role in regulating redox homeostasis. An exclusive oxidizing-folding environment exists in ER, which facilitates disulfide bond formation and generally produces approximately 25% of ROS in a cell. ROS could be generated in ER by uncoupling protein reaction (UPR), NADPH oxidase (NOX) 4, and microsomal monooxygenase (MMO) system.

The changes of the ER environment for protein-folding can cause the accumulation of misfolded or unfolded proteins and activate the uncoupling protein reaction (UPR), a condition also called ER stress. ER stress is a pathophysiological characteristic of obesity and exhibits notable impact on adipogenesis. On the one hand, activation of ER stress could repress adipocyte differentiation. On the other hand, UPR pathways are found to be required for adipogenic differentiation. Eukaryotic initiation factor 2 α (eIF2α)-CHOP pathway may explain the inhibitory effect of ER stress on adipogenesis. In contrast, the inositol-requiring enzyme 1 alpha/X-box binding protein 1 (IRE1alpha/XBP1) pathway of the UPR is required for adipogenesis. In addition, glucose-regulated protein 78 (GRP78), an important molecular chaperone in ER, is reported to play an essential role in adipogenesis. Protein kinase RNA-like ER kinase (PERK) utilizes its intrinsic lipid kinase activity to generate phosphatidic acid, mediates Akt activation, and promotes adipocyte differentiation. PERK can promote adipocyte differentiation through activation of Akt. Activating transcription factor 6 (ATF6) α pathway is also involved in adipogenesis. Thus the three arms of UPR-mediated pathways, PERK, XBP1, and ATF6, are involved in adipogenesis, and blocking of ER stress prevents adipocyte differentiation and weight gain in mice. Regarding the inevitable ROS production in ER stress, the dual roles of ER stress in adipogenic differentiation could be attributed to different signal pathways, such as eIF2α-CHOP versus PERK, XBP1, and ATF6, which may be differentially regulated by ROS. Further studies are needed to test this hypothesis.

3.2. NADPH oxidases and adipocyte differentiation

The NADPH oxidases (NOXs) are important enzymatic sources of ROS in a cell. NOXs family members include catalytic subunits, known as NOX1 to NOX5, Duox1, and Duox2, and interacting partners, such as $p22^{phox}$, $p47^{phox}$, and $p67^{phox}$. Among those NOXs, only NOX4 and $p47^{phox}$ subunits have been observed to be associated with adipocyte differentiation. Role of NOX4 in adipogenesis is controversial. Compared with preadipocytes, both white and brown adipocytes possess lower expression of NOX4. In adipose tissue in vivo, NOX4 is mainly expressed in cells localized within SVF rather than mature adipocytes accumulated area. In apolipoprotein E knockout ($apoE^{-/-}$) mice, AT2 receptor deficiency-attenuated adipocyte differentiation is accompanied by an increase in NOX activity. In this trend, NOX4-deficient mice display latent adipose tissue accumulation and are susceptible to diet-induced obesity and early onset of insulin resistance, which could be attributed to accelerated adipocyte differentiation and hypertrophy. Above evidence suggests that NOX4 acts as a negative regulator of adipocyte differentiation.

However, opposite data have been reported. In hypertrophied adipose tissue, ROS level is increased accompanied by an increase of NOX. In 3T3-L1 adipocytes, NOX4 is required for proper insulin signaling transduction through ROS generation. Thus, pharmacological inhibition of NOX4 or KD of NOX4 by RNA interference inhibits ROS production and adipocyte differentiation. As a component of the NOX, $p47^{phox}$ is shown to be positively related with adipocyte differentiation. Mice lacking $p47^{phox}$ are protected against HF diet-induced increase of adiposity, adipocyte size, and hepatic and alcohol-induced hepatic steatosis. Whether other NOX isoforms are involved in adipogenesis is currently unclear.

3.3. Xanthine oxidoreductase and adipocyte differentiation

In some cases, activation of xanthine oxidoreductase (XOR) contributes to ROS generation through the enzymatic form of xanthine dehydrogenase (XDH) or xanthine oxidase (XO). It is reported that XOR colocalizes with the lipid-binding protein adipophilinon the milk-fat globule, functioning as a major component of the milk-fat globule membrane, and is required for fat-droplet secretion in mice. Cheung et al. first show that XOR is highly expressed in adipose tissue compared to other tissues and is transiently induced during 3T3-L1 adipocyte differentiation in vitro. Under obese state, XOR expression is increased in the adipose tissue. In vivo, adipose mass is reduced by 50% in $XOR^{-/-}$ mice versus wild-type littermates. KD of XOR inhibits PPARγ activity and adipogenesis in vitro. The deficiency of adipogenesis in cells with XOR down-regulation is fully restored by an addition of rosiglitazone, implicating that

XOR is at the upstream of PPARγ. However, constitutive overexpression of XOR increases PPARγ activity but inhibits adipogenesis. The authors explain that XOR may be transiently required for adipogenic differentiation, and thus prolonged expression of XOR may exert unexpected results. In addition, C/EBPβ is shown to transcriptionally regulate XOR level in several cell types. These results support the notion that XOR lies downstream of C/EBPβ and upstream of PPARγ in the cascade of factors that control adipogenesis and is required for adipocyte differentiation. Moreover, although lacking direct evidence, the finding that inhibition of XOR's dehydrogenase activity suppresses PPARγ activation and inhibits adipogenesis, indicates that NADH-oxidizing activity of XOR and the concomitant generation of ROS might be involved in adipogenic regulatory role of XOR.

3.4. Nitric oxide synthases and adipocyte differentiation

NO, a ubiquitous signaling molecule belonging to reactive nitrogen species (RNS), is synthesized by nitric oxide synthases (NOS). There are three isoforms of NOS, including neuronal NOS (nNOS), inducible NOS (iNOS), and endothelial NOS (eNOS). NO is found to promote differentiation of rat brown and white preadipocytes. The mechanism of NO-promoted adipogenesis involves the activation of guanylyl cyclase and increase of cGMP, which in turn up-regulates cAMP and leads to enhancement of adipogenesis. In addition, NO is shown to induce mitochondrial biogenesis in brown adipocytes, which is dependent on induction of PGC-1α expression.

Both eNOS and iNOS exist in adipose tissue. iNOS expression is increased in adipogenic differentiation. Clinical studies have suggested that increase of NOS and NO level strongly correlate with body fat in obese humans, and eNOS expression is increased in omental versus subcutaneous adipose tissue in obese human subjects. In contrast, it is also found that eNOS levels are reduced in the adipose tissue of obese humans, db/db mice, and HF diet-fed mice. Moreover, overexpression of eNOS prevents HF diet-induced obesity and attenuates hypertrophy of WAT. It is more likely that iNOS and eNOS play differential roles in adipocyte differentiation through regulating NO generation under different conditions. Inducible NO generation by iNOS may be a required event for adipogenic differentiation.

3.5. Cytochrome P450 and adipocyte differentiation

The cytochrome P450 enzymes (CYPs) is a major source of ROS in various tissues and cells. AA is metabolized by several CYP isoforms to produce 20-HETE and epoxyeicosatrienoic acids (EETs). It has been shown that 20-HETE promotes ROS production, whereas EETs is associated with inhibition of ROS production. Under obese condition, CYPs family could be altered in a complicated and isozyme-specific way. CYP3A4 is decreased and CYP2E1 is increased in obesity. The changing patterns of CYP1A2, 2C9, 2C19, and 2D6 isozymes under obese condition are inconclusive. In addition, CYP2E1 expression and activity are increased in female ob/ob mice, while that of CYP1A2 are decreased in male ob/ob mice regardless of gender. Inhibition of CYP27A1 activity, or KD and/or deletion of the Cyp27a1 gene induce adipocyte differentiation through regulating oxysterol 27-hydroxycholesterol. During the differentiation of MSCs to adipocytes, CYP4A11 and CYP4F2 expression are decreased to nearly undetectable level. CYP2J5 is decreased in visceral adipose tissue isolated from HO-2 null mice, in which adipocyte differentiation is enhanced.

3.6. Antioxidant enzymes and adipocyte differentiation

The enzymatic antioxidants are capable of counteracting ROS/RNS and oxidant insults through catalyzing a variety of redox reactions.

3.7. Superoxide dismutase and adipocyte differentiation

Regarding the dual roles of superoxide dismutase (SOD) in eliminating superoxide and generating hydrogen peroxide, SOD may play opposite roles in adipocyte differentiation. It is reported that SOD1 expression is decreased in adipose tissue of obese mice. However, in $Sod2^{-/-}$ BMSCs, adipogenesis is found to be increased. Amifostine, an antioxidant, was shown to decrease the numbers of adipocytes in $Sod2^{-/-}$ BMSCs in both basal and adipogenic media and reduce the expression of PPARγ and LPL. In contrast, in $Sod2^{+/+}$ BMSCs, adipogenesis was decreased in adipogenic medium. During 3T3-L1 adipocyte differentiation, SOD expression and activity were also found to be increased. In mice fed with HF diet, MnTBAP, a pharmacological analogue of SOD, attenuates weight gain and adiposity through a reduction in adipocyte hypertrophy, adipogenesis, and fatty acid uptake in epididymal (eWAT) but not in inguinal (iWAT) adipose tissue.

3.8. Catalase and adipocyte differentiation

Catalase (CAT) catalyzes the decomposition of hydrogen peroxide to water and oxygen. mRNA expression of CAT is decreased in the adipose tissue in obese mice. The decreased CAT activity resulting from its promoter mutation is positively associated with childhood obesity and obesity biomarkers such as weight, BMI, and aP2, along with a tendency toward significance with insulin resistance biomarkers. However, in inguinal WAT in obese Zucker rats, CAT activities were found to be increased significantly. Similarly, CAT expression and activity are greatly elevated along with the progression of 3T3-L1 adipocyte differentiation. Furthermore, a remote enhancer region containing two functional PPARγ binding sites in mouse CAT gene was found, suggesting that increased CAT expression during adipocyte differentiation may be a subsequent consequence of PPARγ activation. Thus, whether there is a causal role of CAT alteration in adipocyte differentiation is still unelucidated.

3.9. Thiol-dependent antioxidant system and adipocyte differentiation

Thiol-dependent antioxidant system mainly consists of glutathione (GSH), thioredoxins (Trxs), and peroxiredoxins (Prxs). The interactions between these antioxidant proteins form a potent antioxidant defense system against potential oxidant threat.

3.10. Glutathione metabolic system and adipocyte differentiation

GSH, also named γ-L-glutamyl-L-cysteinyl-glycine, is the most important hydrophilic antioxidant. Glutamate-cysteine ligase (GCL) is the rate-limiting enzyme for de novo synthesis of GSH from the precursor amino acids cysteine, glutamate, and glycine in cooperation with glutathione synthase (GS). GCL consists of a modulatory or light subunit (GCLm) and a catalytic or heavy subunit (GCLc). GSH could eliminate oxidant radicals through nonenzymatic reaction. GSH-mediated reduction of hydroperoxides needs the enzymatic catalysis by GSH peroxidase (GPx). Glutathione-S-transferases (GSTs) detoxify xenobiotic compounds through catalyzing the conjugation of GSH to nonpolar substrates. After the reduction of oxidants or the detoxification of toxicants, GSH form GSSG (oxidized form of GSH) or GSSR (glutathionylated-cysteine derivative), which can be reduced back to GSH catalyzed by NADPH-dependent GSH reductase (GR) and Trxsystem.

3.11. General GSH status and adipocyte differentiation

Compared with WT littermates, GSH level in adipose tissue is higher in ob/ob mice. Sulfhydryl depletor DEM significantly inhibits adipogenesis, suggesting GSH is involved in the adipogenic process. Moreover, in vivo evidence shows that pharmacological depletion of the GSH in mice prevents diet-induced obesity and increases energy expenditure.

However, controversial data have been reported that up-regulation of GSH was involved in carnosic acid (CA) and carnosol (CS)-inhibited adipocyte differentiation. As a precursor of cysteine, intracellular NAC can be converted into GSH. NAC can be used for the treatment of acetaminophen-induced hepatic injury and for the prevention of radio contrast-induced nephropathy due to its potent antioxidant property. NAC dose-dependently inhibits ROS level and attenuates DMI-induced adipogenesis in 3T3-L1 cells, accompanied by inhibited expression of C/EBPβ and PPARγ. NAC also blocks adipocyte differentiation and the transcriptional activation of CREB in MSC. During ST2 cell adipogenesis, the intracellular GSH redox potential becomes more oxidized. Furthermore, intraperitoneal administration of NAC to rats and mice resulted in reduction of bodyweight. These data suggest that dysregulation of GSH may disturb the physiological role of GSH redox in adipogenic differentiation.

3.12. GCL and adipocyte differentiation

GCL expression in adipose tissue is higher in ob/ob mice in comparison to WT littermates. BSO, a specific and irreversible inhibitor of GCL, deletes GSH and inhibits the differentiation of 3T3-L1preadipocytes and preadipocytes derived from the SVF of inguinal as well as epididymal fat pads. BSO inhibits adipocyte differentiation through inhibition of MCE via down-regulation of E2F-dependent transactivation of the MCM7 target promoter. However, up-regulation of GCLc was proposed to be involved in CA and CS-inhibited adipocyte differentiation.

3.13. GPx and adipocyte differentiation

GPx1, 3, 4, and 7 are identified to be abundantly expressed in mature adipocytes and WAT. In parallel with lipid accumulation during adipocyte differentiation, GPx activity is increased. However, controversial data are also reported.

Compared with WT counterparts, mice over-expressing GPx1 develop hyperglycemia, hyperinsulinemia, and obesity. In patients with nonalcoholic fatty liver disease (NAFLD), the increase in GPx1 expression level is an independent variable associated with NAFLD progression. Chung et al. suggest that GPx3 is required for the regulation of PPARγ-mediated antioxidant effects. Although GPx3 is abundantly expressed in adipose tissues, its expression is reduced selectively in the adipose tissue of several obese animal models. Antioxidant NAC and the anti-diabetic drug rosiglitazone increase adipose GPx3 expression in obese and diabetic db/db mice.

GPx4 is stimulated in the course of 3T3-L1 adipocyte differentiation. GPx7 is highly enriched in WAT and is mainly expressed in preadipocytes but not mature adipocytes of WAT. In response to induction medium, GPx7 level in 3T3-L1 preadipocytes declines rapidly within the first 24 h. GPx7 deficiency in 3T3-L1 preadipocytes increases the expression of key transcription factors involved in adipogenesis, including C/EBPβ, C/EBPα, and PPARγ. GPx7 deficiency increases C/EBPβ expression through PKA signaling pathway in a ROS-dependent manner. In addition, deficiency of GPx7 promoters C/EBPβ dimerization increases its binding to the promoter of C/EBPα, leading to an increase of cell proliferation and expression of downstream targets C/EBPα and PPARγ. Furthermore, human genetic variant association results suggest that the subjects carrying risky GPx7 alleles may have reduced GPx7 expression in adipose tissue and higher BMI. However, GPx7 level was increased in the late period of adipocyte differentiation in human liposarcoma cell line SW872.

3.14. GST and adipocyte differentiation

GSTA3 is over-expressed at the late stage of adipogenesis in human liposarcoma cell line SW872. Jowsey et al. identified GSTA3 as a novel adipocyte differentiation-associated protein whose expression was enhanced manyfolds during the conversion of mouse preadipocytes into adipocytes. During the terminal phase of adipogenesis in 3T3-L1 preadipocytes, GSTζ is found to be increased in C/EBPα- and PPARγ-dependent manner. However, in diet-induced obese rats, GSTYc2 subunit, GST8, and GSTP subunits were found to be down-regulated. Up-regulation of GSTA2 may participate in CA and CS-inhibited adipocyte differentiation.

3.15. The thioredoxins system and adipocyte differentiation

The Trxs system is composed of NADPH, Trx, thioredoxin reductase (TrxR), thioredoxin interacting protein (Txnip), and novel nucleoredoxin (Nrx). TrxR reduces oxidized Trx. TXNIP is a negative regulator of Trx, resulting in oxidative stress. There are significant alterations of Trx system during adipogenesis. Rajalin et al. showed that the protein level of TrxR1, TrxR2, and Trx2 was elevated in the course of 3T3-L1 adipocyte differentiation. Chutkow et al. found that Txnip protein dramatically disappeared within minutes after DMI stimulation, indicating that Txnip degradation is required for the onset of adipocyte differentiation. Over-expression of Txnip in preadipocytes prevents adipocyte differentiation, whereas its down-regulation enhances adipogenesis. Compared with WT littermates, Txnip KO mice gain significantly more adipose mass due to calorie intake and adipogenesis. In Txnip-silenced preadipocytes and Txnip$^{-/-}$ mouse embryonic fibroblasts (MEFs), adipogenesis is markedly increased, whereas Txnip over-expression impaired adipocyte differentiation. Txnip deletion could also augment PPARγ-stimulated adipogenesis. Txnip negatively regulates the expression and activation of PPARγ, which in turn, suppresses Txnip expression, reflecting reciprocal feedback inhibition between them. In 3T3-L1 preadipocytes, Trx over-expression stabilized Txnip protein levels and thus promoted inhibition of adipogenesis. As an important component of adipogenic stimulants, insulin promoted the dissociation of Txnip-Trx into a more labile state.

Nrx level is higher in WAT of ob/ob mice and is increased in the early adipogenic stage of 3T3-L1 preadipocyte differentiation. KD of Nrx decreases and its over-expression increases adipogenic differentiation of 3T3-L1 cells. In transgenic mice with adipose tissue-specific Nrx over-expression, adipocyte size as well as number is increased compared with WT mice. Moreover, inhibition of Wnt/β-catenin pathway is involved in Nrx-exerted regulation of adipogenic differentiation.

3.16. Peroxiredoxins and adipocyte differentiation

As one of the most abundant antioxidant proteins, Prxs are able to reduce hydrogen peroxide and organic hydroperoxides to water and alcohol. Genetic variations of Prx3 are associated with the risk of obesity. In WAT of db/db mice and in perirenal WAT of human subjects with BMI > 25, Prx3 expression is significantly decreased. Huh et al. first studied the role of Prxs in adipocyte differentiation in detail and showed a gradual increase inPrx1, Prx3, and Prx5 expression in the course of 3T3-L1 adipocyte differentiation. Compared with Prx1 and Prx5, Prx3 was increased by a larger extent. At the age of 20 months, Prx3 KO mice display larger fat pads in epididymal WAT with hypertrophied adipocytes. In differentiated ADSCs, expression of aP2, C/EBPα, and PPARγ, adipophilin, and FAS are increased in Prx3 KO mice. In 3T3-L1 preadipocytes, siRNA-mediated KD of Prx3 increased aP2, C/EBPα, and PPARγ expression, whereas Prx3 over-expression inhibited their expression. Moreover, ROS level in mitochondria was increased in WAT of Prx3 KO mice. In MSCs, over-expression of Prx2 also shows inhibitory effect on the lipid accumulation during adipocyte differentiation.

3.17. NADPH: quinone oxidoreductase 1 and adipocyte differentiation

NADPH:quinone oxidoreductase 1 (NQO1) may generate quinones, quinoneimines, and andazodyes accompanied by inhibition of ROS generation. NQO1 protein is present in human adipocytes, and NQO1 mRNA level is higher in adipocytes than in adipose-derived stromal vascular cells. NQO1 expression is positively correlated with BMI and weight loss results in a decrease in NQO1 mRNA content in adipose tissue. In comparison to WT mice, NQO1$^{-/-}$ mice exhibit lower abdominal adipose tissue. Using 3T3-L1 preadipocytes, Vomhof-DeKrey et al. investigated NQO1 expression in the course of adipocyte differentiation. They found an increase of NQO1 protein at limited MCE and postmitotic growth arrest (Days 1–3) stages and a decrease in terminally differentiated (Day 8) adipocytes that lasted for several days afterward. In contrast to the mapping of protein, NQO1 mRNA expression was increased in differentiated adipocytes (Days 11–14), indicating a discrepancy between steady-state of mRNA and resulting protein levels. Sulforaphane (SFN) enhanced NQO1 protein level and blunted TG accumulation in a NQO1-dependent manner.

3.18. Heme oxygenase and adipocyte differentiation

Heme oxygenase or haem oxygenase (HO) is an enzyme that catalyzes the degradation of heme and produces free iron, biliverdin, and carbon monoxide. Three isozymes of HO have been defined, including inducible heme oxygenase-1 (HO-1), constitutive heme oxygenase-2(HO-2) and not fully defined HO-3. HO-1 exhibits antioxidant capacity due to its products bilirubin/biliverdin and carbon monoxide. The HO system acts as a key cellular antioxidant defense system in obesity and diabetes. Decreased HO-1 expression levels were observed in type 2 diabetic patients. In addition, products of HO-1 catalysis, CO, and bilirubin are decreased in humans and animal models of type 2 diabetes. Zucker fat (ZF) rats display a decrease in both HO activity and HO-1 and HO-2 protein expression.

A number of studies have demonstrated that genetic inhibition of HO expression and activity increase adipogenesis. In vitro induction of HO-1 decreases adipogenesis. siRNA-mediated inhibition of either HO-2 or HO-1 leads to an increase in adipogenesis. Up-regulation of HO-1 induces a decrease of adipocyte differentiation in MSCs. BMSCs from HO-$2^{-/-}$ mice displayed an increase in adipogenesis and accumulation of lipid droplets, resulting in adipocyte hypertrophy. Using a lentivirus construct of the human HO-1 under the control of theaP2 promoter, Cao et al. revealed that adipocyte-specific over-expression of HO-1 attenuated adiposity and adipogenesis.

In addition, pharmacological studies support the notion that HO-1 negatively regulates adipocyte differentiation. Cobalt protoporphyrin (CoPP) or L-4F, a systemic apomimetic peptide inducer of HO-1, promotes chronic body weight loss in various species, decreases visceral and subcutaneous fat content, and reduces the number of enlarged adipocytes, and increases adiponectin levels and small adipocytes, without any effect on food intake, or other metabolic changes such as energy expenditure and O_2 consumption. In addition, other HO-1 inducers, such as D-4F, an apolipoprotein A1 mimetic peptide, replicate the body weight-lowering effect of HO-1 induction. In ZF animals and human bone marrow-derived adipocytes, HO-1 expression and activity was found to be increased by CoPP treatment. In vivo, HO-1 up-regulation was associated with decrease in superoxide levels, visceral and subcutaneous fat content and weight gain, and increase in plasma adiponectin. In vitro, up-regulation of HO-1 decreased superoxide level, adipose remodeling, smaller adipocytes, and increased adiponectin secretion in the culture medium. Burgess et al. found that induction of HO-1 by CoPP slowed the rate of weight gain in male obese mice and produced a significant decrease of IL-6, TNFα, and IL-1β. In B6v-Lep obese/J mice, HO-1 induction increases the number, whereas decreases the size of adipocytes. Moreover, the chronic treatment of carbon monoxide, the product of HO metabolism, prevents the development of HF diet-induced obesity via stimulating metabolism and remodeling adipocytes. In addition, over-expression of HO-1 increases pAMPK and eNOS levels and promotes human osteoblastic differentiation, another direction of cell differentiation. It was found that induction of HO-1 was associated with reduction of C/EBPα, PPARγ, Peg-1/Mest, aP2, CD36, Wnt5b expression and the increase of β-catenin, pGSK3β, Wnt10b, Pref-1, shh, and adiponectin, indicating that those key factors were responsible for HO-exerted effect on adipogenesis.

However, Huang et al. reported that over-expression of HO-1 did not protect against HF diet-induced body weight gain and insulin resistance in mice. Vanella et al. reported that L-4F increased early adipocyte differentiation markers and decreased Peg-1/Mest through activation of HO-1. In human MSCs, HO-2 depletion results in increase of adipogenesis and inflammatory cytokines, with lower expression of HO-1. Sofalcone, a chalcone derivative, inhibits the differentiation of 3T3-F442A preadipocytes into adipocytes, which is restored by SnPP treatment, indicating the involvement of HO-1.

3.19. Glucose-6-phosphate dehydrogenase and adipocyte differentiation

Glucose-6-phosphate dehydrogenase (G6PD) generates NADPH, providing the substrate for NOX and reducing equivalent for GSH and Trx system. Thus, G6PD could affect ROS level in a bidirectional manner. NADPH is also required for the biosynthesis of fatty acids and cholesterol. The expression and enzymatic activity of G6PD are significantly elevated in WAT of obese models, including db/db, ob/ob, and diet-induced obese mice. In 3T3-L1 cells, G6PD overexpression stimulates the expression of most adipocyte markers and elevates the levels of cellular FFA, triglyceride, and FFA release, indicating an important role of G6PD in adipogenesis.

3.20. Selenoproteins and adipocyte differentiation

Selenoprotein P (SeP), a circulating selenium carrier, functions as an antioxidant enzyme. Although it is expressed most abundantly in liver, SeP also plays a role in the regulation of lipid metabolism in adipocyte through redox modulation. Zhang et al. reported that SeP1 gene expression was significantly reduced in adipose tissue of ob/ob and HF diet-induced obese mice as well as in primary adipocytes isolated from ZF rats. SeP expression is induced in the course of 3T3-L1 adipocyte differentiation. In adipose tissue of obese mice, rosiglitazone treatment increased SeP protein expression. In contrast, exposure to either TNFα or high level of H_2O_2 reduces SeP1 expression in a time- and dose-dependent manner in differentiated 3T3-L1 adipocytes. Moreover, KD of SeP1 reduces GPx activity, whereas increases the expression inflammatory cytokines, such as MCP-1 and IL-6 in preadipocytes. These characteristics contribute to the inductive role of SeP in 3T3-L1 adipocyte differentiation.

3.21. Metallothioneins and adipocyte differentiation

Metallothioneins (MTs) are a family of proteins with abundant cysteine residues. MTs are capable of regulating oxygen respiration, possessing potential antioxidant activities. MTs function as regulators of various cellular processes including gene expression, apoptosis, proliferation, and differentiation. Evidences suggest that MTs are altered during adipocyte differentiation and play important roles in regulation of adipocyte differentiation. In human, metallothionein-2A (MT-2A) mRNA level in fat tissues is significantly elevated in obese subjects. MT-2 expression is higher in omental than in subcutaneous adipose tissue. Moreover, MT could be secreted by fat cells in WAT. However, Sato et al. discovered that $MT^{-/-}$ mice fed with a HF diet exhibited more fat mass and a larger adipocyte volume, indicating that MT may have a preventive role against HF diet-induced obesity.

3.22. Non-enzymatic antioxidants and adipocyte differentiation

In addition to those enzymatic antioxidants, endogenous and exogenous nonenzymatic antioxidants also play important roles in regulating lipid homeostasis. They are usually classified into water-soluble antioxidants, including GSH, vitamin C (V_C), lipoic acid (LA), and uric acid (UA), or lipid-soluble antioxidants, such as β-carotene, vitamin E (V_E), and coenzyme Q (CoQ). The adipose tissue of ob/ob mice exhibits higher content of hydrophilic molecules (GSH, V_C) in a lower redox state, which is associated with lower content of lipophilic molecule (V_E, CoQ) and lipid peroxidation. In particular, CoQ is deficient in WAT of ob/ob mice. In rodents as well as in humans, CoQ content is strongly and negatively correlated with subcutaneous adipose tissue and obesity indexes. In 3T3-F442A cell line, pharmacological inhibition of CoQ synthesis by chlorobenzoic acid strongly triggers adipose differentiation. In contrast, over-expression of 4-hydroxybenzoate acid polyprenyltransferase increases CoQ level and inhibits adipogenesis of 3T3-F442A cell. These data suggest that CoQ is an anti-adipogenic factor. NADPH is an essential reducing equivalent for numerous enzymes. Cytosolic $NADP^+$-dependent isocitrate dehydrogenase (IDPc) is a key enzyme for providing cytosolic NADPH. In 3T3-L1 preadipocytes and mice, KD of IDPc reduces adipogenesis and lipid accumulation.

3.23. Antioxidant extracts/natural products and adipocyte differentiation

In the last decades, dietary supplementation of natural antioxidants is prevalent worldwide. The dramatic discrepancies about the biological outcomes of antioxidant consumption have been reported.

A large number of experimental and clinical investigations have shown that many antioxidants possess potent anti-obesity effects. α-lipoic acid, a famous ROS scavenger for hydroxyl radicals and singlet oxygen, exerts reproducible anti-obesity effects in several independent clinical trials. The inhibition of adipocyte differentiation by α-lipoic acid involves the regulation of pro-adipogenic transcription factors via MAPK pathways. Dietary orlistat supplementation induces a reduction of body weight in rats fed with HF diet. Bardoxolone methyl, a novel antioxidant capable of activating the Keap1-Nrf2 pathway, shows unexpected weight-reducing effects in a recent clinical trial. Brazilein, a natural biologically active compound derived from *Caesalpinia sappan* L., exhibits antioxidant properties and inhibits adipocyte differentiation. Combined treatment of mulberry leaf and fruit extract ameliorates HF diet-induced obesity in mice. Diet supplemented with 5% ginger powder reduces body weight of rats fed with HF diet. *Grateloupia lanceolata* (Okamura) Kawaguchi, the edible red seaweed, was found to inhibit lipid accumulation and ROS production during 3T3-L1 cell differentiation. Tempol could attenuate adipogenesis in both hMSCs and 3T3-L1 cells. Ethyl acetate extract of ginger, the rhizome of *Zingiber officinale* Roscoe, possesses antioxidant capacity and inhibits adipocyte differentiation. Dietary ascorbic acid was able to protect against HF diet-induced increase of body weight and total body fat, and enlargement of different adipose depots, without affecting food intake. Resveratrol and its analogues, (E)-1,2-di (3,5-dimethoxyphenyl)ethane and 4-[2-(3,5-dimethoxyphenyl)vinyl] pyridine, inhibit adipocyte differentiation in 3T3-L1 cells via activation of AMPK. Oleanolic acid is a triterpenoid compound that has potent antioxidative and anti-inflammatory properties. In differentiated 3T3-L1 adipocytes, oleanolic acid was shown to decrease lipid accumulation and the expression level of differentiation markers, including PPARγ and C/EBPα. Pycnogenol® is a group of flavonoids with antioxidative effects. In 3T3-L1 adipocytes, Pycnogenol® inhibits mRNA expression of NOX4, G6PD, PPARγ, and C/EBPα, whereas increases the level of SOD1, SOD2, GPx, and GR. *Polygonum aviculare* L. ethanol extract exhibits anti-diabetic effect in HF diet-induced obese mice. Buckwheat sprouts exhibits anti-adipogenic and anti-oxidative properties.

Phenolic compounds, which are secondary plant products, are consumed regularly as part of the human diet and are helpful for disease prevention. Oleuropein is the principal phenolic compound in olive tree products, which is the main source of healthy Mediterranean diet ingredients. Hydroxytyrosol is a bioactive substance after the hydrolysis of oleuropein. It is reported that oleuropein and hydroxytyrosol exhibit antioxidant activities and inhibit adipogenesis in stem cells derived from human bone marrow and in 3T3-L1 preadipocytes. In 3T3-L1 preadipocytes, oleuropein and hydroxytyrosol dose-dependently suppressed fat accumulation and differentiation-related markers, including PPARγ, C/EBPα, SREBP-1c, GLUT4, CD36, and FASN through suppression of MCE. Cigarette smoking aggravates Graves' orbitopathy through oxidative stress-mediated adipogenesis. Quercetin, as an antioxidant, inhibits adipogenesis by reducing ROS in cultured orbital fibroblasts from Graves' orbitopathy patients treated by cigarette smoke extract.

However, there is also solid evidence supporting the conception that consumption of antioxidants promotes adipocyte differentiation and obesity. Abe et al. prepared a 100% methanol fraction of methanolic extract from unripe kiwi fruit (*Actinidia deliciosa*) and found that the extracts decreased the production of ROS and promoted adipocyte differentiation. Patients treated with clozapine often suffer from massive weight gain. Clozapine-induced increase of fat mass is the result of enhanced fat cell formation from adipogenic precursor cells, which may be attributed to antioxidant effects. Puerarin is a major isoflavone glycoside derived from Kudzu root (*Pueraria lobata*) that possesses antioxidant effects. Puerarin could enhance 3T3-L1 adipocyte differentiation, accompanied by increased expression of G6PD, and PPARγ and its target genes, and several antioxidant enzymes, including GR and CAT. Butylhydroxyanisole inhibits lipid accumulation and adipocyte differentiation in 3T3-F442A in a concentration-dependent manner.

3.24. Redox-sensitive transcriptional factors and adipocyte differentiation

Beyond the direct regulation of ROS by ROS-generating enzymes and enzymatic and non-enzymatic antioxidants, redox balance is controlled by a wide class of transcription factors, including nuclear factor erythroid 2-related factor 2 (Nrf2), peroxisome proliferator-activated receptor-gamma coactivator-1α (PGC-1α), p53, nuclear factor kappa-light-chain-enhancer of activated B cells (NF-κB), and forkhead box O (FoxO). These transcription factors could respond to subtle redox changes, determining the reaction toward oxidation or reduction through transcriptionally modulating ROS-generating enzymes and antioxidant enzymes. The large amount of transcription factors influencing adipocyte differentiation play particularly important roles in a cooperative and interactive way.

3.25. Nrf2 and adipocyte differentiation

Nrf2 is the most thoroughly studied redox-sensitive transcription factor. Under normal conditions, Nrf2 is inactivated by kelch-like ECH-associated protein 1 (Keap1), which targets Nrf2 for proteasomal degradation. Once activated, Nrf2 binds to antioxidant response elements (AREs) or electrophile response elements (EpREs), inducing the expression of more than 100 genes involved in the response to cellular stress. Functioning as a "supreme director" in redox regulation, Nrf2 can regulate both ROS/RNS-generating enzymes (such as SOD and iNOS) and antioxidant enzymes. However, Nrf2 more likely acts as an antioxidant regulator in most cases. Redox proteins regulated by Nrf2 are listed as follows: NQO1, GCLc and GCLm, GPxs, GR, GSTs, HO-1, SOD, CAT, Trxs, TrxRs, Prxs, UCPs, NOS, and G6PD [7, 8].

A large number of papers have revealed contradictory roles of Nrf2 in the regulation of metabolism. On the one hand, Nrf2 is critical for antioxidant defense system, innate immunity, protection against inflammation, insulin resistance and diabetes, and cancer. On the other hand, Nrf2 deficiency is beneficial for the attenuation of atherosclerosis, HF diet-induced obesity and insulin resistance, and cancer.

The Nrf2/Keap1 pathway is linked to the development of obesity and hepatic lipid accumulation in animal models. Various natural and synthetic substances with the ability to stimulate Nrf2 have been verified to exert anti-obesity activity, including 1-[2-cyano-3,12-dioxooleana-1,9(11)-dien-28-oyl]imidazole (CDDO-Im), oleanolic acid, oltipraz, ellagic acid, quercetin, curcumin, resveratrol, and chromium histidinate. Keap1 KD mice with a constitutive increase in Nrf2 activity have lower body weight and epididymal fat masse than WT mice. Enhanced expression of Nrf2 in mice attenuates the fatty liver induced by special diet. However, using Nrf2 KO mice, Nrf2 deficiency has been found to protect against HF diet-induced obesity. However, no significant relationship between Nrf2 and obesity has also been reported. Tanaka et al. found that in both WT and Nrf2-null mice fed with a HF diet for up to 4 weeks, weight gain shows no difference. Consistent with its role in obesity, Nrf2 functions as a double-edged sword in the pathogenesis of insulin resistance. Studies using genetic manipulation have shown that Nrf2 deficiency is beneficial for glucose homeostasis and insulin sensitivity.

Nrf2 is highly expressed in the adipose tissue and animals fed with HF diet exhibit higher Nrf2 expression. As an important constituent of adipogenic stimuli, insulin could significantly activate Nrf2 pathway [9]. These findings suggest a potential role of Nrf2 in the modulation of adipogenesis. Indeed, genetic manipulation and pharmacological activator treatment have been introduced to investigate the role of Nrf2 in adipogenesis and lipid metabolism, which is contradictory. Role of the Nrf2/ARE pathway in energy metabolism and storage has been reviewed by Vomhof-DeKrey et al. In this review, we focus on the involvement of Nrf2/Keap1 pathway in the regulation of adipocyte differentiation.

3.26. Activation of Nrf2 promotes adipogenesis

Pi and his colleagues have made great contributions to the identification of the proadipogenic effect of Nrf2. In their studies, Nrf2 was found to be activated in the early stage (<12 h) of adipogenesis [10]. Consistent with the notion that Nrf2 is a key regulator for GSH synthesis, intracellular GSH level follows a similar pattern as Nrf2-ARE. However, in another study, Nrf2 and Keap1 mRNA levels were found to be increased in differentiated adipocytes (Days 11–14), indicating a programmed control of Nrf2 during adipogenic differentiation. Pi et al. showed that KD of Nrf2 could totally block the enhancement of adipogenesis in Keap1-KD cells, confirming the contribution of Nrf2 activation to above process. To directly determine the regulatory role of Nrf2 in adipogenesis, preadipocytes derived from WAT of Nrf2-KO ($Nrf2^{-/-}$) and WT ($Nrf2^{+/+}$) mice were used. Hormonal cocktail-induced adipogenesis in $Nrf2^{-/-}$

preadipocytes shows substantially reduced lipid accumulation. Moreover, $Nrf2^{-/-}$ preadipocytes exhibit significantly decreased expression of adipogenic genes, including PPARγ1, PPAR2, adipsin, and aP2. In 3T3-L1 preadipocytes and human subcutaneous preadipocytes, shRNA-mediated KD of Nrf2 expression inhibits adipogenesis. Pi et al. also found that mice Nrf2 deficiency displayed reduced fat mass in association with small adipocytes and increased adipocyte numbers, and were resistant to diet-induced obesity.

3.27. Activation of Nrf2 inhibits adipogenesis

In contrast to the discovery by Hou et al. [10], Shin and colleagues [11] demonstrated an inhibitory role of Nrf2 in adipogenesis using $Nrf2^{-/-}$-immortalized MEFs. In $Keap1^{-/-}$-primary MEFs, adipogenesis is inhibited by enhanced Nrf2 signaling, compared to WT MEFs. In $Nrf2^{-/-}$-immortalized MEFs, ectopic expression of dominant-positive Nrf2 delays differentiation. HF diet feeding decreased the expression of Nrf2 and its target genes, such as NQO1 and GSTm6, and Nrf2 inhibited lipid accumulation in mouse liver after feeding a HF diet. Chartoumpekis et al. found a decrease in nuclear abundance and DNA binding activity of Nrf2 during adipogenesis in ST2 cells. Vomhof-Dekrey et al. speculate that these controversial data may be mainly due to the differential use of primary versus immortalized MEFs and the different experimental periods. It is highly recommended to construct adipose tissue-specific Nrf2 knockout mice to investigate its role in adipogenesis.

3.28. Obesity and ROS, friends or foes?

From an evolutionary standpoint, obesity protects humans from potential threat of food shortage through energy storage. Adipocyte differentiation is a protective mechanism essential for survival, in which energy is stored in the form of TG in the adipose tissue. In this tendency, adipocyte differentiation is a protective mechanism essential for survival. However, this protective mechanism would lead to obesity without control, causing a series of unhealthy problems. In obese individuals, WAT release proinflammatory factors, resulting in metabolic dysfunctions [461]. Moreover, ectopic accumulation of fat in other tissues is detrimental for normal physiological functions, leading to several serious diseases.

In aerobic organisms, the capability of oxygen consumption precedes the onset of advanced life and promotes the biological evolution. Three decades ago, ROS was considered to be generated as by-products of oxygen consumption. Current concepts support that ROS may be evolutionally utilized as essential factors for the regulation of various biological processes that reflect the essence of "concise" in life activity. In the last decades, more and more evidences have shown the active role of ROS in the regulation of life processes rather than the passive role in inducing oxidative injury.

3.29. Novel hypothesis of redox system: Redox Chain

Based on the current results and the complexity of redox system, we propose that there is a complex and interactive "Redox Chain," consisting of "ROS-generating Enzyme Chain," "Combined Antioxidant Chain," and "Transcription Factor Chain" (**Figure 1**). On the oxidant

http://dx.doi.org/10.5772/intechopen.68727

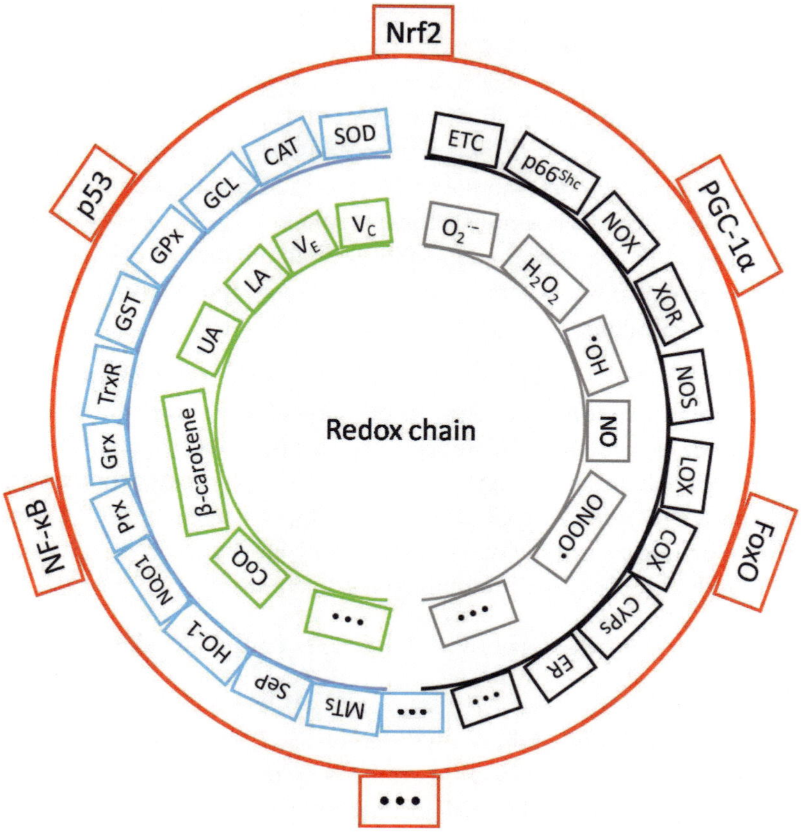

Figure 1. A hypothesis of "Redox Chain." The innermost line on the left indicates "Nonenzymatic Antioxidant Chain". The middle line on the left indicates "Antioxidant Enzyme Chain". The innermost line on the right indicates "Oxidant Chain". The middle line on the right indicates "ROS-generating Enzyme Chain". The outermost line indicates "Transcription Factor Chain".

side, "ROS-generating Enzyme Chain" consisted of enzymes that are responsible for intracellular ROS generation, including ROS-generating enzymes in mitochondria and ER, NOXs, XOR, NOS, LOX, COX, CYPs, and other undefined enzymes that may be involved in ROS production. In response to physiological and/or pathophysiological stimuli, these enzymes could be activated in a temporal and spatial sequence and ROS could be generated in different time periods and at different positions from various sources in a cell. ROS is probably produced by the activation of certain enzymes that could quickly respond to intra- or extracellular stimulation, such as NOX. Under continued pathological conditions, it is likely that ROS would mainly come from metabolic burden, such as OXPHOS and ER stress. On the antioxidant side, "Non-enzymatic Antioxidant Chain" comprises SOD, CAT, GCL, GPx, GSTs, Trxs, Prxs, NQO1, HO-1, SeP, MTs, and other undefined enzymes that may exhibit antioxidant activities, in combination with nonenzymatic antioxidants, constituting the first in vivo defense line against potential ROS insult. This defense chain could be reinforced by the consumption of naturally extracted antioxidants that are commercially available. On a higher level, the antioxidant enzymes constitute the "Antioxidant Enzyme Chain," a more efficient defense line against potential ROS damage. These enzymatic and nonenzymatic antioxidants comprise the "Combined Antioxidant Chain," which sequentially and cooperatively regulate redox state. As the super directors, the redox-regulating transcription factors, such as Nrf2, PGC-1α, p53, NF-κB, and FoxO, form the "Transcription Factor Chain," which direct the battle between oxidants and antioxidants through transcriptionally regulating both ROS-generating enzymes and antioxidant enzymes. Moreover, there are additional mechanisms of redox regulation, including epigenetic and posttranslational regulation. The regulation of "Redox Chain" could be implemented in a temporal and spatial way. In all, more work is needed to clarify the role of each member in redox family and its regulatory mechanisms under normal and pathophysiological conditions.

Disturbance of "Redox Chain" results in either oxidative stress or blockage of ROS signaling transduction, leading to various disorders. In the context of obesity or enhanced adipocyte differentiation, redox state is altered as reflected by disorganized "Redox Chain" at different levels. However, we only observed changes or roles of single or several members of "Redox Chain" in adipocyte differentiation. Clarification of temporal and spatial changing pattern of "Redox Chain" in an integrated way will undoubtedly contribute to the understanding of the mechanisms underlying adipocyte differentiation and obesity.

3.30. ROS acts as a double-edged sword in the course of adipocyte differentiation

Based on the large amount of paradoxical results, it is difficult to draw a "clear" conclusion of the role of redox regulation in adipocyte differentiation. What we know is that adipocyte differentiation is definitely determined either directly by redox system or indirectly by redox-sensitive transcriptional, posttranslational, or epigenetic regulation. Compared with preadipocytes, mature adipocytes are under relatively more oxidizing state. In the physiological process of adipocyte differentiation, ROS is a definitely required and essential signal for the initiation and maintenance of adipogenic events. However, ROS-generating signals may be limited to a certain concentration and sources in different stages of adipogenic differentiation. We propose that in addition to the differential experimental backgrounds, including differences in experimental subjects and interventions, the paradoxical results from different laboratories may be attributed to the "narrow safety window" of concentration, time and compartment for the redox-regulated adipogenic differentiation. The difference of genetic or dietary background of the subjects, doses or periods of interventions or "time window" selected for detection may severely affect the outcomes. Concentration is a key determinant of the effect of ROS on adipocyte differentiation. Turker et al. reported that 1 and 10 μM H_2O_2 resulted in a marked decrease in adipocyte differentiation, while higher doses of H_2O_2 markedly increased differentiation. Both deficiency and over-expression of XOR have been shown to inhibit adipocyte differentiation. In the course of adipocyte differentiation, cells undertake dramatic transformation of cell fate, involving drastic changes of organelles and compartments. In differential stages of adipogenesis, redox state in different organelles and compartments may be differential. These data demonstrated that redox state required for adipocyte differentiation must be controlled in a fine-tuned way, including appropriate concentration, time, and space. The paradoxical results indicate the importance of delicate regulation of redox system in adipocyte differentiation. Due to the quick redox reaction and extremely

http://dx.doi.org/10.5772/intechopen.68727

short half-life period, to date, the absolute amount of ROS could not be determined accurately in biological samples, and thus the ROS level we evaluated is a relative index. In different experiments, ROS level could not be compared. Thereof, over-expression or inhibition of a certain molecule (transcription factor, antioxidant enzyme, or ROS-generating enzyme) could not reflect the accurate level of ROS or redox state. Moreover, the adipogenic differentiation requires physiologically and endogenously generated ROS to create an oxidizing environment and to transduce molecular signals. The differentiation of adipocytes needs the accurate generation of ROS at the appropriate "time window" and "place." However, under pathological and stress conditions, ectopic and excessive generation of ROS may influence the redox environment and disturb adipogenesis. It is proposed that ROS acts as a double-edged sword in the course of adipocyte differentiation and thus in the pathogenesis of obesity and related metabolic disorders (**Figure 2**) [12].

3.31. Implications for intervention of obesity and associated redox-regulated disorders

During adipocyte differentiation, cells undertake dramatic and extensive morphological, physiological, and biochemical changes. These great changes render redox alteration the most

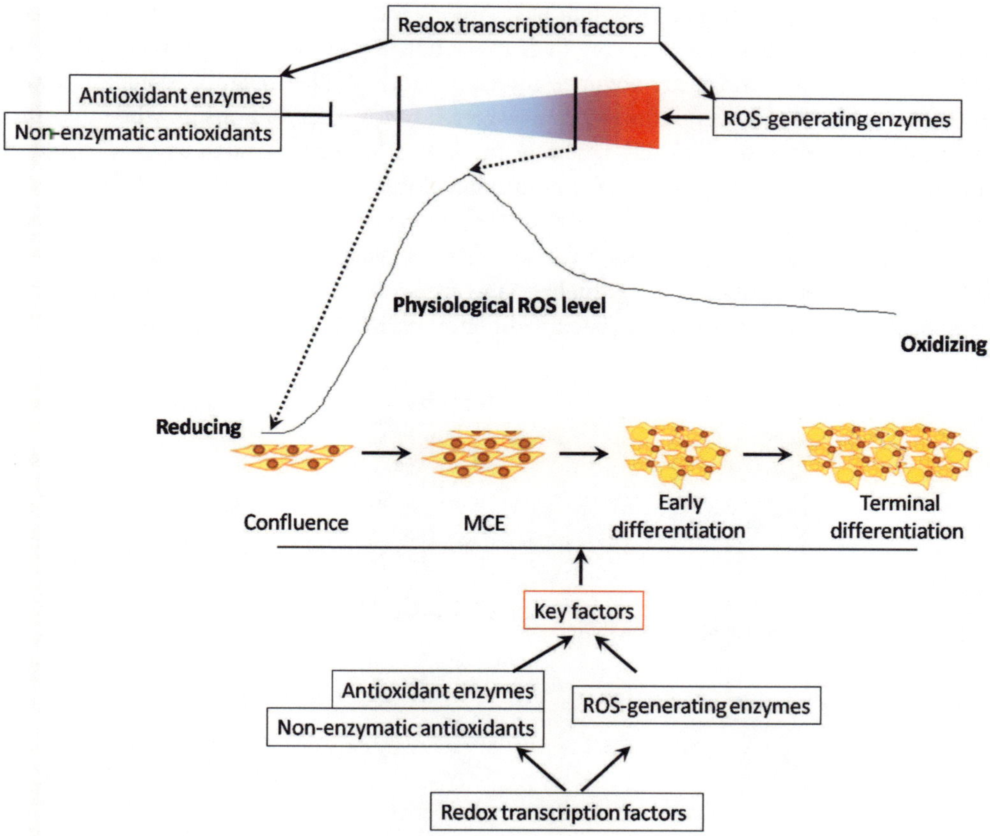

Figure 2. Role of redox system in the regulation of adipocyte differentiation.

probable "candidate" that could mediate those extensive reactions. The large amount and extensive distribution of members in redox family provide a general and microenvironment for thousands of reactions.

Adipocyte differentiation is a time-dependent process, which consists of different stages, including confluence, MCE, early differentiation, and late differentiation. In different stages of adipocyte differentiation, redox status is dramatically altered. It appears that redox system works as an "on-off switch" and orchestrates the transition from preadipocyte to adipocyte. During this process, redox regulation on the expression of key proteins for adipocyte differentiation may occur in diverse levels, including transcriptional, epigenetic, and posttranslational modulations (**Figure 3**). Indeed, the "tridimensional" redox regulatory mechanism is not limited in the process of adipocyte differentiation, but has general implications in a variety of biological processes.

A specific molecular mechanism would definitely play a role in a given background. However, one molecule or even one signaling pathway could not account for all the redox-sensitive events. Thereof, a more integrated view of redox biology should be highlighted. Considering the burst of various omics concept and technology, we suggest that redoxomics should be applied extensively and widely to help us understand the biological effects of redox alterations in a more integrated way. We need a stable and standardized "redox index" to evaluate the general redox status under a certain condition. Just like BMI, "redox index" represents a value to assess the overall situation of redox state in an organ or in a whole body. In our future work, we need to find the representative and easily obtained indicators and establish a simple equation to calculate "redox index." Moreover, adipose tissue-targeted redox interventions may be more attractive because disturbance of insulin signaling in other main organs, such as liver, could also result in insulin resistance and metabolic syndromes. Differential redox state in different organs may influence the overall metabolic end-points by using antioxidants. Our aim is to establish and maintain general and fine-tuned redox balance rather than to conduct simple prooxidant and antioxidant interventions. Individualized intervention according to their respective redox state should be emphasized to treat obesity and other redox-related metabolic disorders [13].

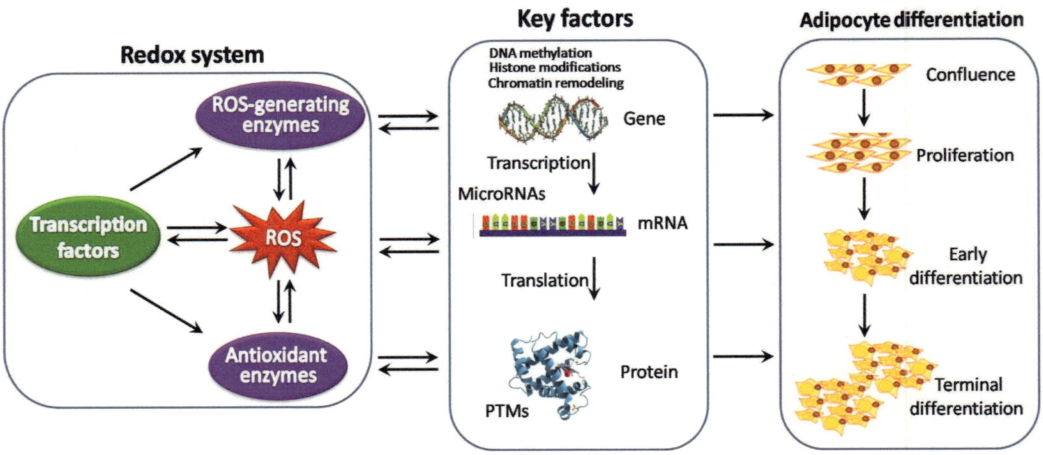

Figure 3. The "tridimensional" redox regulatory mechanism in the process of adipocyte differentiation.

Acknowledgements

This work was supported by National Natural Science Foundation of China (No. 31400724) and Natural Science Foundation of Shaanxi Province (2014JQ4135). We reused some content we published in Free Radical Biology and Medicine and got the permission.

Author details

Xin Wang

Address all correspondence to: xinwang@fmmu.edu.cn

Department of Toxicology, Shaanxi Key Lab of Free Radical Biology and Medicine, The Ministry of Education Key Lab of Hazard Assessment and Control in Special Operational Environment, School of Public Health, Fourth Military Medical University, Xi'an, China

References

[1] Tseng YH, Cypess AM, Kahn CR. Cellular bioenergetics as a target for obesity therapy. Nature Reviews. Drug Discovery. 2010;**9**(6):465-482

[2] Gimeno RE, Klaman LD. Adipose tissue as an active endocrine organ: Recent advances. Current Opinion in Pharmacology. 2005;**5**(2):122-128

[3] Rosen ED, Spiegelman BM. Molecular regulation of adipogenesis. Annual Review of Cell and Developmental Biology. 2000;**16**:145-171

[4] Rosen ED, Walkey CJ, Puigserver P, Spiegelman BM. Transcriptional regulation of adipogenesis. Genes & Development. 2000;**14**(11):1293-1307

[5] Pessler-Cohen D, Pekala PH, Kovsan J, Bloch-Damti A, Rudich A, Bashan N. GLUT4 repression in response to oxidative stress is associated with reciprocal alterations in C/EBP alpha and delta isoforms in 3T3-L1 adipocytes. Archives of Physiology and Biochemistry. 2006;**112**(1):3-12

[6] Carriere A, Carmona MC, Fernandez Y, Rigoulet M, Wenger RH, Penicaud L, Casteilla L. Mitochondrial reactive oxygen species control the transcription factor CHOP-10/GADD153 and adipocyte differentiation: A mechanism for hypoxia-dependent effect. The Journal of Biological Chemistry. 2004;**279**(39):40462-40469

[7] Wang X, Hai C. Novel insights into redox system and the mechanism of redox regulation. Molecular Biology Reports. 2016;**43**(7):607-628

[8] Wang X, Tao L, Hai CX. Redox-regulating role of insulin: The essence of insulin effect. Molecular and Cellular Endocrinology. 2012;**349**(2):111-127

[9] Wang X, Wu H, Chen H, Liu R, Liu J, Zhang T, Yu W, Hai C. Does insulin bolster antioxidant defenses via the extracellular signal-regulated kinases-protein kinase B-nuclear factor erythroid 2 p45-related factor 2 pathway? Antioxidants & Redox Signaling. 2012;**16**(10):1061-1070

[10] Hou Y, Xue P, Bai Y, Liu D, Woods CG, Yarborough K, Fu J, Zhang Q, Sun G, Collins S, Chan JY, Yamamoto M, Andersen ME, Pi J. Nuclear factor erythroid-derived factor 2-related factor 2 regulates transcription of CCAAT/enhancer-binding protein beta during adipogenesis. Free Radical Biology & Medicine. 2012;**52**(2):462-472

[11] Shin S, Wakabayashi N, Misra V, Biswal S, Lee GH, Agoston ES, Yamamoto M, Kensler TW. NRF2 modulates aryl hydrocarbon receptor signaling: Influence on adipogenesis. Molecular and Cellular Biology. 2007;**27**(20):7188-7197

[12] Wang X, Hai CX. ROS acts as a double-edged sword in the pathogenesis of type 2 diabetes mellitus: Is Nrf2 a potential target for the treatment? Mini Reviews in Medicinal Chemistry. 2011;**11**(12):1082-1092

[13] Wang X, Hai C. Redox modulation of adipocyte differentiation: Hypothesis of "Redox Chain" and novel insights into intervention of adipogenesis and obesity. Free Radical Biology & Medicine. 2015;**89**:99-125

Chapter 9

Redox-Sensitive TRP Channels: TRPA1 and TRPM2

Makiko Kashio and Makoto Tominaga

Additional information is available at the end of the chapter

http://dx.doi.org/10.5772/intechopen.69202

Abstract

The transient receptor potential (TRP) family includes cation-permeable ion channels activated by diverse stimuli ranging from chemical compounds, mechanical force and temperature. TRPA1, TRPV1, TRPC5, TRPM2, and TRPM7 are reported to be activated by reactive oxygen species (ROS). The sensitivity of TRPs to ROS is in part associated with pathogenesis caused by ROS and physiological adaptations to redox signals. The present review focuses on the well-defined ROS-sensitive TRP channels, TRPA1 and TRPM2, and summarizes recent reports regarding their activation mechanism by ROS and their relevance to pathological conditions and physiological functions in which ROS are involved.

Keywords: transient receptor potential (TRP) channel, calcium signal, pain, metabolism, body temperature

1. Introduction

Reactive oxygen species (ROS) are oxygen-derived molecules with high reactivity that are generated in diverse pathways, including by proteins of the mitochondrial respiratory chain, NADPH oxidase (NOX), dual oxidase (Duox) and cyclooxygenase (COX). Although ROS have been considered as deleterious by-products, they are generated in response to many physiological stimuli such as cytokines, hormones and mechanical force to act as signaling molecules in parallel with reactive nitrogen species (RNS), including nitrogen oxide (NO) [1]. These cellular "redox" signals are considered to play important roles in a wide range of physiological phenomena [1–3]. In the immune system in particular, H_2O_2 produced by NOX is important for removal of microorganisms and defects in redox signaling are associated with persistent infections [4]. In contrast, unregulated ROS production in pathological conditions leads to chronic inflammation and tissue damage.

The transient receptor potential (TRP) family is a superfamily of nonselective cation channels that are homologous to the founding TRP member in *Drosophila melanogaster* that is essential for phototransduction [5]. The mammalian TRP superfamily now includes 28 members that can be grouped into six subfamilies based on amino acid sequence similarity: canonical (C), ankyrin (A), vanilloid (V), melastatin (M), polycystin (P), and mucolipin (ML). TRP channels have six transmembrane domains (S1–S6), a pore region between S5 and S6 and cytoplasmic N- and C-termini [6]. The functional TRP channel is a tetramer [7]. TRP channels exhibit a remarkable diversity in their activation mechanisms and can be activated by various ligands, both cold and hot temperatures, mechanical stimuli, or in response to signal transduction pathways. Since TRP channels can be activated by a wide range of environmental stimuli, they are considered to act as cellular "sensors." Many TRP channels, including TRPA1 and TRPM2, which are the focus of the present review, show nonselective cation (Na^+, Ca^{2+}, Mg^{2+}, K^+) permeability [5], and the activation of these channels causes membrane depolarization and, most importantly, elevated intracellular Ca^{2+} concentrations ($[Ca^{2+}]_i$) that can convert environmental information to intracellular signals.

TRPA1 is the exclusive member of the TRPA subfamily and was named for its characteristic N-terminal long 18 ankyrin repeat domain (ARD), a motif that is thought to mediate protein–protein interactions [8]. TRPA1 is abundantly expressed in nociceptive primary afferent sensory neurons that detect nociceptive stimuli. TRPA1 is expressed in sensory neurons of dorsal root ganglia (DRG), trigeminal ganglia (TG), and nodose ganglia, as well as many other organs and tissues, including the brain, heart, small intestine, lung, skeletal muscle, and pancreas [9]. Because TRPA1 has broad sensitivity to reactive compounds, it can detect a wide range of hazardous compounds (see below). TRPA1 function is modulated by endogenous intracellular factors such as membrane lipids and Ca^{2+}, which has bimodal effects on TRPA1 function. EF-hand-like domains in the N-terminus between the 11th and 12th ankyrin repeat of TRPA1 are reported to form a Ca^{2+}-binding site responsible for Ca^{2+}-dependent mTRPA1 activation [10, 11]. TRPA1 desensitization can also depend on Ca^{2+} [12].

TRPM2 is a member of the TRPM subfamily that functions as a nonselective cation channel. TRPM2 is known as a "channezyme," that is, an ion channel that possesses an enzymatic region, based on its C-terminal Nudix domain that is homologous to the NUDT9 adenosine diphosphoribose (ADPR) pyrophosphatase, an enzyme that converts ADPR to adenosine monophosphate and ribose 5-phosphate [13]. However, the TRPM2 Nudix domain lacks ADPR hydrolase activity and its binding to ADPR is sufficient for channel gating [14, 15]. Adenosine diphosphoribose (ADPR), cyclic ADPR and pyridine dinucleotides, including nicotinamide-adenine-dinucleotide (NAD), nicotinic acid-adenine dinucleotide (NAAD) and NAAD-2′-phosphate (NAADP), have been reported as endogenous agonists for TRPM2. However, TRPM2 activation by pyridine dinucleotides was later shown to have been caused by ADPR contamination, and thereby, pyridine dinucleotides are likely not TRPM2 agonists [16]. TRPM2 is also activated by reactive oxygen species [17] and is therefore considered to be redox-sensitive. TRPM2 is regarded as a "metabolic sensor" because its activity is regulated by cellular and systemic metabolic states such as redox signaling, glucose levels and body temperature [18–20]. As with TRPA1, Ca^{2+} plays crucial roles in TRPM2 activation

http://dx.doi.org/10.5772/intechopen.69202

[21], and Ca^{2+} binding to the intracellular pore cavity is necessary for channel activation [22]. Meanwhile, intracellular H^+ ions reportedly inhibit TRPM2 function [23].

TRPM2 is expressed in the plasma membrane of many cell types such as neurons and microglial cells in the brain, vascular endothelial cells, pancreatic β-cells and immunocytes, as well as tissues in the spleen and liver [24]. Several reports showed that TRPM2 also functions in the lysosomal membrane [25, 26].

Despite fluctuating environmental temperatures, endothermic species such as mammals and birds can maintain a constant body temperature through the activity of peripheral thermal sensors in sensory and autonomic neurons and by thermosensitive structures within the preoptic area (POA), anterior hypothalamus, brain stem, spinal cord and possibly other places that regulate body temperatures through autonomic and behavioral mechanisms [27]. TRPM2 was recently shown to be expressed in warmth-sensitive neurons of the POA and in peripheral sensory and autonomic neurons, suggesting that TRPM2 is involved in regulating body temperature [28, 29].

2. Activation mechanism of TRPA1 by reactive oxygen/nitrogen species and diverse chemical compounds

TRPA1 is expressed in primary sensory neurons and mediates noxious pain sensations evoked by a wide range of reactive compounds. Electrophilic TRPA1 activators have environmental, dietary and endogenous origins and include acrolein (air pollutant), allyl isothiocyanate (in mustard), allicin and diallyl disulfide (in garlic), cinnamaldehyde (in cinnamon) and the proalgesic lipid peroxidation products 4-hydroxynonenal (4-HNE) and 15-deoxy-Δ12,14-prostaglandin J_2 (15d-PGJ_2) (**Figure 1A**) [30–33]. These compounds are collectively referred to as reactive carbonyl species (RCS). Many have highly reactive electrophilic carbon moieties such as an αβ-unsaturated carbonyl group. These moieties react with cysteine thiol groups in TRPA1, including those located between the N-terminal ARD and S1 (Cys621, Cys641, Cys665) of human TRPA1, which results in covalent modification by S-alkylation through a Michael addition that in turn activates TRPA1 (**Figure 1B**) [31]. In mouse, TRPA1, Cys415 and Cys422 in ARD and Cys622 between ARD and S1 are required for activation by reactive compounds [33]. In addition to cysteine residues, lysine (Lys708 in human TRPA1) is also reportedly involved in TRPA1 activation by reactive electrophilic species, although the contribution of this amino acid seems to be limited to the action of AITC [31].

TRPA1 is also activated by reactive oxygen/nitrogen species (ROS/RNS) and thus is regarded to be "redox-sensitive". TRPA1 is activated by many kinds of ROS and RNS, including hydrogen peroxide (H_2O_2), hydroxyl radical ($OH^\cdot$), hypochlorite (OCl^-), nitric oxide (NO) and peroxynitrite ($ONOO^-$) [32, 34, 35] (**Figure 1A**). Oxidative stress endogenously generates lipid peroxidation products such as 4-NHE and 15d-PGJ_2, described above as RCS. However, ROS can activate TRPA1 independently of these peroxidation products, as evidenced by the finding that H_2O_2-induced TRPA1 activation is reversed by the application of dithiothreitol

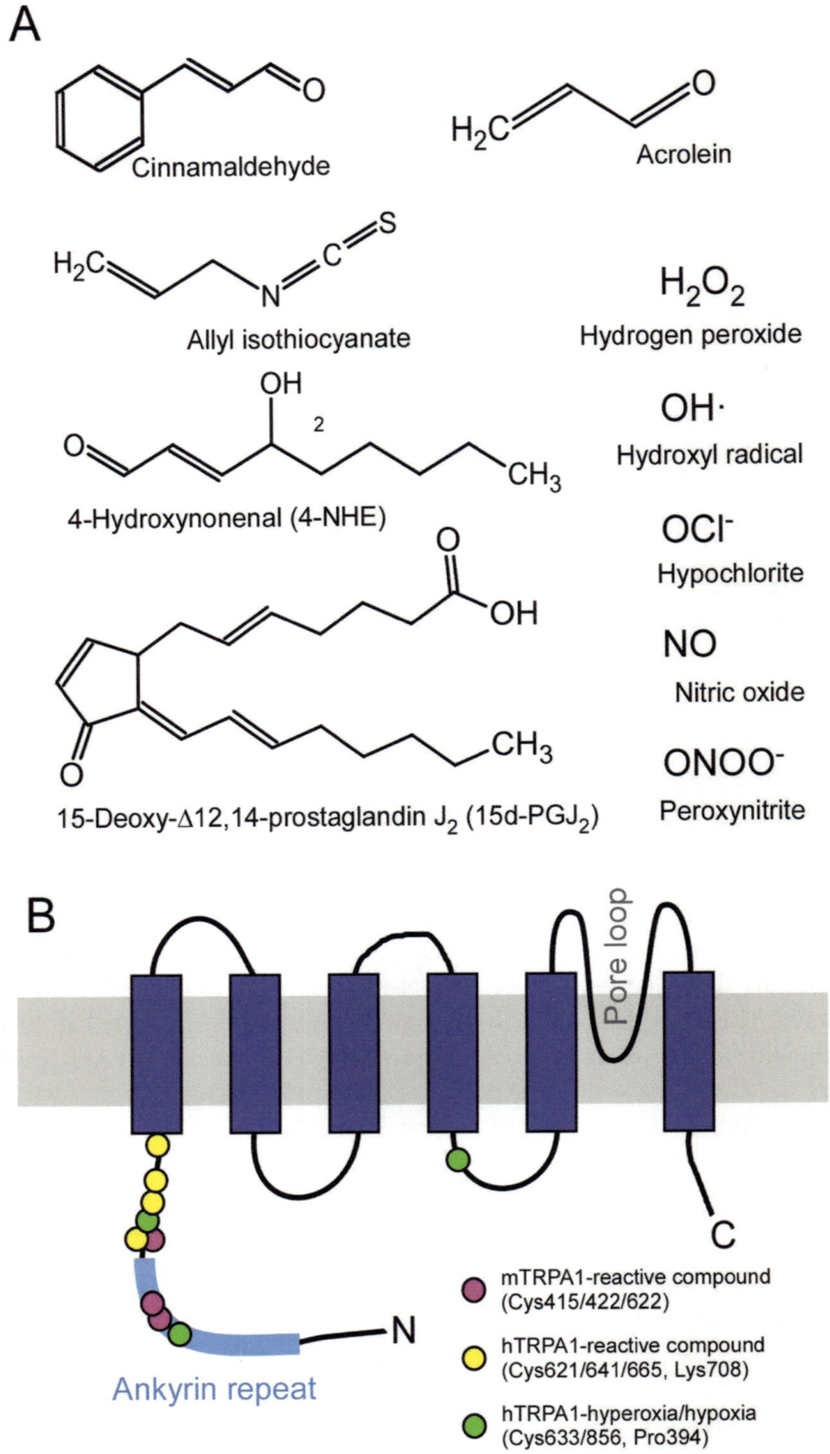

Figure 1. Chemical structure of TRPA1-activating reactive compounds (A) and reported target amino acid residues in TRPA1 (B).

(DTT), which reverses cysteine disulfide formation, nitrosylation and cysteine oxidation, but not Michael addition, without affecting the action of 4-NHE and 15d-PGJ_2 [32, 36]. This result suggests that ROS induce cysteine oxidation and disulfide formation between proximal cysteine residues in TRPA1.

TRPA1 activation by RNS is considered to be mediated by S-nitrosylation of cysteine residues, since TRPA1 activation by the NO donor SNAP was reversed by the application of DTT as described above [36]. Functional characterization of site-directed cysteine mutants of mouse TRPA1 demonstrated that modification of several cysteine residues (Cys621, Cys641, Cys665) involved in RCS activation mediates the action of ROS and RNS [34, 35]. However, mutation of cysteine and lysine residues could also affect channel activation by nonelectrophilic agonists, which would complicate the identification of amino acids responsible for covalent modification of TRPA1 [35]. TRPA1 activation by these reactive compounds leads to pain sensation that would diminish the likelihood of deleterious damages arising from adducts formed by these compounds and DNA or proteins that are associated with carcinogenesis and toxicity [37].

3. Activation mechanism of TRPM2 by redox signaling

TRPM2 channels are activated by H_2O_2 [17], and this activation could be enhanced by pretreatment with Fe^{2+} to promote formation of hydroxyl radicals generated by the Fenton reaction that are also thought to have an important role in TRPM2 activation [38]. Indirect TRPM2 activation through the production of intracellular ADPR and direct action independent of ADPR production have also been reported (**Figure 2A**). One ADPR production pathway involves nuclear poly(ADPR) polymerases (PARPs) and poly(ADPR) glycohydrolases (PARGs). PARP is activated by DNA damage upon oxidative stress and has DNA repair functions [39]. Meanwhile, PARG hydrolyzes poly-ADPR chains to generate free ADPR. The finding that treatment of cells with a PARP inhibitor decreased H_2O_2-mediated TRPM2 activation supports a role for ADPR in activation of this channel [40]. Another possible source of ADPR is mitochondria, where oxidative stress could induce the production of free ADPR [41]. Indeed, reduced mitochondrial ADPR concentrations suppressed H_2O_2-mediated TRPM2 currents [42]. ADPR can also be produced by plasma membrane/intracellular organelle CD38, which has implications for immunocyte function [43, 44]. However, the mechanism by which extracellular ADPR generated by the CD38 ecto-enzyme enters the cells is unclear [45]. In contrast, an ADPR-independent TRPM2 activation pathway may also exist based on the finding that a TRPM2 variant lacking the Nudix domain that contains the ADPR-binding domain still responded to H_2O_2 [46]. In addition, our recent results showed that H_2O_2 clearly enhanced heat-evoked responses of TRPM2 in inside-out single channel recordings in which intracellular components were completely absent (**Figure 2B**) [19]. This action was accompanied by a reduction in the temperature threshold for TRPM2 heat activation (**Figure 2C**) and was mimicked by an oxidant that reacts with methionine residues. Taken together, these results suggest that ADPR-independent actions could also be involved in TRPM2 activation; however, because H_2O_2-evoked reduction in the temperature threshold for TRPM2 activation was affected by a PARP inhibitor, there could be both direct and indirect actions in H_2O_2-mediated effects.

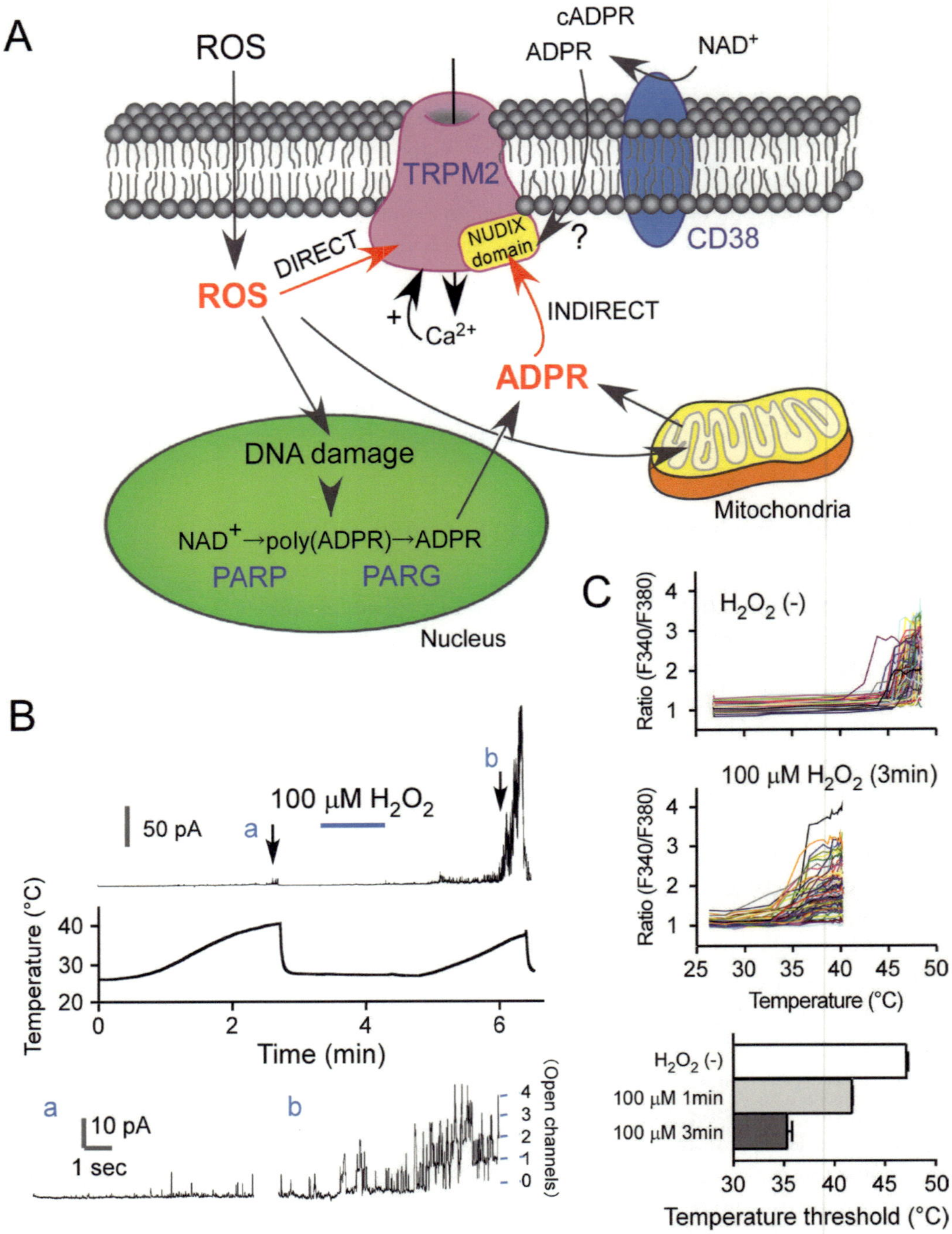

Figure 2. (A) Scheme showing the activation cascade of TRPM2 by reactive oxygen species (ROS). (B) H_2O_2-enhanced heat-evoked single channel opening observed in inside-out single channel recordings. (a,b) Magnified traces at the time points indicated in the upper column. (C) H_2O_2-treatment lowered the temperature threshold for TRPM2 activation. The temperature-fura2 ratio relationship is plotted for H_2O_2-untreated and H_2O_2-treated (100 µM, 3 min) cells. Average temperature thresholds for each group (Mean ± SEM) are shown.

4. Pathological and physiological relevance of redox signal-regulated TRPA1 activation

TRPA1 is abundantly expressed in a portion of primary sensory neurons and mediates noxious pain sensation evoked by ROS to avoid deleterious DNA damage. In an inflammatory milieu, phagocytes release large amounts of ROS that promote the production of endogenous lipid compounds, including the RCS 4-HNE and 15d-PGJ_2. These compounds activate TRPA1 to cause inflammatory pain [36, 47].

TRPA1 is readily modified by electrophilic compounds. An analysis of the redox potentials of reactive disulfides revealed that TRPA1 has the lowest redox potential threshold among TRPA1, TRPV1, TRPV2, TRPV3, TRPV4, and TRPC5, indicating its high sensitivity to electrophilic compounds [48]. Thus, TRPA1 could act as a hyperoxia sensor in vagal and sensory nerves (**Figure 3**). Heterologously expressed TRPA1 was activated through covalent modification of cysteines at high oxygen conditions in excess of atmospheric oxygen concentration (20% O_2) (hyperoxia), and the activation was diminished by point mutation of TRPA1 cysteine residues (Cys633Ser and Cys856Ser) (**Figure 1B**). In addition, hypoxic conditions also activated TRPA1, which could be suppressed by overexpression of recombinant prolyl hydroxylase (PHD). The oxygen-sensitive PHD catalyzes proline hydroxylation of TRPA1 under normoxic conditions that in turn inhibits channel function. A PHD inhibitor activated TRPA1 even under normoxic conditions, suggesting the involvement of proline hydroxylation in TRPA1 activation. Under hypoxic conditions, basal enzymatic activity of PHD is inhibited and TRPA1 is activated upon relief of proline hydroxylation-mediated inhibition. A proline residue located in the TRPA1 N-terminus (Pro394) is suggested to be involved in proline hydroxylation-mediated TRPA1 inhibition (**Figure 1B**). Indeed, vagal nerves isolated from wild type, but not TRPA1-knockout (KO) mice, showed $[Ca^{2+}]_i$ increases respond to both hyperoxic and hypoxic conditions. These data suggest that TRPA1 could function as an O_2 sensor to control respiratory, cardiac and vascular functions through cysteine oxidation by ROS and proline modification by the oxygen-sensitive enzyme PHD. Proline hydroxylation of TRPA1 is also considered to participate in cold sensitivity of this channel, although this possibility remains a subject of debate (**Figure 3**). Temperature sensitivity of mammalian TRPA1 varies among species [49] in

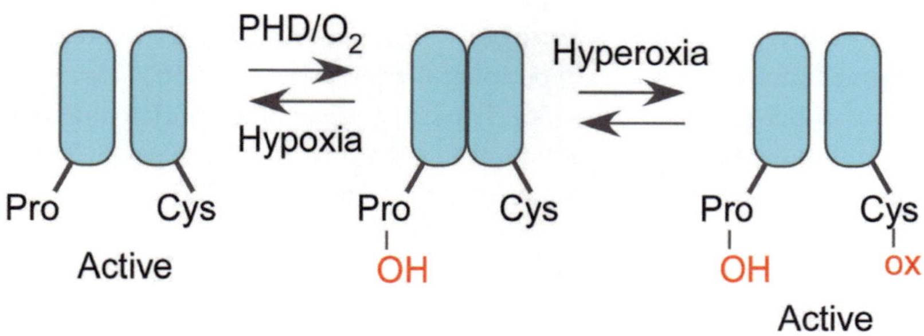

Figure 3. TRPA1 regulation by Cys-oxidation (Cys-Ox) and Pro-hydroxylation at different oxygen concentrations. PHD, proline hydroxylase.

that human TRPA1 and monkey TRPA1 are not activated by cold stimulus whereas rat TRPA1 and mouse TRPA1 are. Mutation of Gly878 in S5 of rat TRPA1 to Val that is present in human TRPA1 abolished cold sensitivity. A proline hydroxylation-deficient mutant of human TRPA1 (Pro394Ala) was activated by cold stimulation in the presence of low concentrations of H_2O_2 (0.1 μM), but wild-type TRPA1 showed no activation [50]. PHD inhibitors could result in activation of wild-type TRPA1 by cold temperature, which is attenuated by mitochondrial ROS scavenging. H_2O_2-evoked responses were significantly larger for Pro394Ala TRPA1 than for wild type, and a PHD inhibitor increased the response of wild-type TRPA1. In contrast, mouse wild-type TRPA1 was activated by cold temperature and effects of the PHD inhibitor and ROS scavengers were also observed in a mouse TRPA1 clone, confirming that the inhibitory effect of proline hydroxylation was conserved between human and mouse. Several chemotherapeutic agents can cause cold allodynia [51–53]. For instance, the drug oxaliplatin potentiates H_2O_2-induced responses in mouse DRG neurons possibly by inhibiting PHD [50].

TRPA1 is also reported to play a physiological role in artery vasodilation, although its activity varies among different types of vascular beds. In cerebral arteries, TRPA1 expressed in the endothelium mediates "endothelium-dependent" vasodilation. ROS such as superoxide anions (O_2^-) and hydrogen peroxide (H_2O_2) are known to cause vasodilation, leading to increases in cerebral microcirculation [54]. TRPA1 mRNA expression was observed in the endothelium of cerebral arteries, but not in peripheral vascular beds in renal, coronary and mesenteric arteries [55]. TRPA1 protein in the endothelium of cerebral arteries preferentially colocalizes with NOX2, a ROS-generating enzyme, in fenestration of internal elastic lamina where plasma membranes of endothelium and smooth muscle cell are in close contact. ROS generated by NADPH-induced NOX activity led to cerebral artery vasodilation following TRPA1 activation. This vasodilation could be abolished in a variety of ways, including by a NOX inhibitor, catalase-mediated degradation of H_2O_2, deferoxamine, which inhibits the Fenton reaction that generates hydroxyl radical (OH^-), the TRPA1 inhibitor HC-030031 and TRAM34, a blocker of the intermediate conductance Ca^{2+}-sensitive K channel (IK). Moreover, vasodilation could be mimicked by application of 4-hydroxy-nonenal (4-HNE), a product of lipid peroxidation. These results suggest that ROS-derived lipid peroxidation products activate TRPA1, leading to cytosolic Ca^{2+} elevation, which in turn activates intermediate conductance Ca^{2+}-sensitive K channels (IK) and membrane hyperpolarization in the endothelium. This change in membrane potential is propagated through gap junctions to smooth muscle cells to promote additional vasodilation. On the other hand, in peripheral arteries, TRPA1 expressed in primary sensory neurons is reportedly involved in "neurogenic" vasodilation. TRPA1 expressed in sensory neurons likely mediates vasodilation in peripheral arteries because TRPA1 is not expressed in the endothelium of peripheral arteries [56]. Topical application of cinnamaldehyde, a TRPA1 agonist, onto mouse ears caused vasodilation in wild-type mice, but not in TRPA1KO mice. TRPA1 agonist-induced vasodilation could be attenuated by a CGRP antagonist, a nonselective NOS or a neuronal NOS (nNOS) inhibitor, suggesting the possible involvement of CGRP and NO release from sensory neurons. In addition, TRPA1 agonist-induced vasodilation is mediated by formation of superoxide and peroxynitrite.

Taken together, TRPA1 expressed in the endothelium of central arteries is involved in endothelial-dependent vasodilation, whereas TRPA1 expressed in vagal and primary sensory neurons functions in neurogenic vasodilation in peripheral arteries.

http://dx.doi.org/10.5772/intechopen.69202

5. Pathological and physiological relevance of redox signal-regulated TRPM2 activation

TRPM2 activation induced by H_2O_2 promotes cell death due to sustained elevation of intracellular Ca^{2+} [17] and also increases inflammation and tissue injury [18]. Therefore, numerous reports have shown the implications of TRPM2 activity in oxidative stress-induced cell death of many kinds of tissues and cells [57]. Endogenous and exogenous agents that promote ROS production such as TNFα, β-amyloid, and neurotoxin MPTP also cause neurotoxicity in a TRPM2-dependent manner [17, 58–61]. In addition, ischemia/reperfusion (I/R) injury is a major pathological situation involving unregulated ROS production in stroke, brain trauma, cardiac arrest and other disorders and diseases. Ischemia is the restriction of blood flow that diminishes oxygen (hypoxia/anoxia) and glucose supplies to tissues. When blood supply is restored to the affected tissues, secondary effects associated with I/R injury can occur wherein inflammatory agents and ROS that can cause tissue damage are produced. ROS-evoked TRPM2 activity was reported to aggravate tissue damage in the presence of I/R injury [62–64], and this situation can be explained in part by increased neutrophil migration toward affected tissues [63, 65].

On the other hand, more recent reports suggest that TRPM2 could exert protective roles in I/R injury. One group reported that TRPM2 channels are expressed in the sarcolemma and transverse tubules of adult cardiac myocytes [66, 67]. They also showed that TRPM2KO heart is vulnerable to I/R injury and I/R-induced prolongation of action potential duration was enhanced in TRPM2KO myocytes compared with wild type [67]. Proteomic analysis of I/R-affected ventricles from wild-type and TRPM2KO mice revealed that, relative to wild-type heart, mitochondrial respiratory complex dysfunction in TRPM2KO heart is more severe and is associated with altered expression levels of proteins localized in the mitochondrial inner membrane [68]. Under I/R conditions, TRPM2KO myocyte mitochondria showed lower mitochondrial membrane potential, Ca^{2+} uniporter activity, ATP levels and oxygen consumption as well as higher ROS levels compared to those seen for wild type. These data suggest that Ca^{2+} supplied by TRPM2 activity can have a protective role by ameliorating mitochondrial dysfunction and diminishing mitochondrial ROS levels in I/R situations. Similar results were recapitulated in intact and TRPM2-depleted SH-SY5Y neuroblastoma cells treated with the ROS-generating chemotherapeutic agent doxorubicin [69]. Interestingly, these diminished parameters of mitochondrial function in TRPM2KO myocytes were also seen under normoxic conditions [68]. Considering these results, TRPM2 may contribute to the maintenance of basal mitochondrial bioenergetics by supplying Ca^{2+}, which has numerous functions in mitochondrial metabolism [70].

TRPM2 also has roles in inflammation and infection as evidenced by its expression in immunocytes (**Table 1**). In inflammatory situations, ROS are produced and ROS-evoked TRPM2 activity can aggravate inflammation. At sites of inflammation, phagocytes such as neutrophils and macrophages digest deleterious agents and increase oxygen consumption that enhances NOX-mediated production of ROS, phagocytized agents can then be cleared by ROS. TRPM2 is widely expressed in leukocytes, including lymphocytes, neutrophils, monocytes/macrophages, dendritic cells, microglia and mast cells [19, 25, 71–75]. TRPM2 activation by ROS is

In vivo model	The effects of TRPM2 down regulation
DSS-induced colitis [73]	CXCL2↓, neutrophil infiltration↓ inflammation↓
Listeria monocytogenes [85]	Survival rate↓, CXCL2→, neutrophil infiltration→, $iNOS^+$ monocyte↓
CLP-induced sepsis [87]	Survival rate↓, inflammation↑, IL-6↑, HMGB↑, bacterial clearance↓
PA i.t., CLP [84]	Survival rate↓, bacterial clearance↓
LPS i.p. [89]	Survival rate↓, inflammation↑, CXCL2↑, TNFα↑, IL-6↑, neutrophil infiltration↑, ROS production↑

In vitro model	Stimulus	The effect of TRPM2 down regulation
U937 monocyte [73]	H_2O_2	CXCL8 (human homolog of CXCL2)↓
Peritoneal macrophage [19]	Zymosan	CXCL2↓, GCSF↓, IL-1α↓, IL-1β↓#, TNFα→, Temp-phagocytosis↓
BMDM [76]	Liposome	Inflammasome activation↓, IL-1β↓, TNFα→, ROS production→
U937 monocyte [82]	High glucose	Inflammasome activation↓, NOX activation↓, IL-1β↓
BMDM [84]	PA, SA	Phagosomal acidification↓, Bacterial clearance↓

Table 1. Reports showing the implication of TRPM2 in immune reactivities using *in vivo* and *in vitro* models.

reported to worsen inflammation by elevating cytokine release. Indeed, TRPM2 activation in mouse monocytes elevates H_2O_2-induced release of the neutrophil-attracting chemokine macrophage inflammatory protein-2 (CXCL2) through activation of the Ca^{2+}-dependent tyrosine kinase (Pyk2) and nuclear factor κB (NFκB), and this elevation was attenuated in TRPM2KO monocytes [73]. Supporting these *in vitro* experiments, wild-type mice showed elevated CXCL2 levels and neutrophil infiltration in inflamed colon tissues with ulcerative colitis in a dextran sulfate sodium (DSS)-induced experimental colitis model. These conditions are suppressed in TRPM2KO mice, suggesting that TRPM2 activation in monocytes increased chemokine release and neutrophil migration followed by aggravation of inflammation. In our experiments, TRPM2 activity elevated release of the cytokines granulocyte colony stimulating factor (G-CSF), interleukin-1α (IL-1α), and CXCL2 from macrophages [19]. TRPM2-mediated exacerbation of inflammation also explains pathological conditions associated with chronic pain. TRPM2KO mice showed impaired pain responses in inflammatory pain induced by intraplantar injection of formalin or carrageenan in mice, and neuropathic pain models induced by partial sciatic nerve ligation or spinal nerve transaction [76]. TRPM2 activity in macrophages and microglia is suggested to aggravate chronic pain through CXCL2 release and neutrophil infiltration toward inflamed tissues.

Nod-like receptor family pyrin domain containing-3 (NLRP3) inflammasomes are activated by several conditions of cellular stress, including microbial products, particulate substances, elevated plasma glucose levels that accompany metabolic disorders, intracellular [K^+] reduction and [Ca^{2+}]$_i$ increase [77, 78]. NLRP3 inflammasome is a complex of NLRP3, apoptosis-associated speck-like protein (ASC) and caspase-1 to activate caspase-1 that in turn promotes release of pro-inflammatory cytokines, IL-1β and IL-18. TRPM2 is reported to be involved in the inflammasome activation in macrophages/monocytes. Charged liposomes evoke ROS production, caspase-1 activation and IL-1β release from bone-marrow-derived macrophages (BMDM) [79]. Caspase-1 activation and IL-1β release evoked by liposomes are attenuated by TRPM2 deficiency, although mitochondrial ROS production is comparable to that of wild type. Thioredoxin-interacting protein (TXNIP) is known to bind to NLRP3 and participate in ROS-dependent inflammasome activation [80]. TXNIP expression is up-regulated by high glucose and is involved in ROS production induced by high glucose [81]. Treatment of U937 monocytes with high levels of glucose induced TRPM2 up-regulation, ROS production, caspase-1 activation and IL-1β release, which can be attenuated by TRPM2-siRNA or TRPM2 inhibitors [82]. TXNIP up-regulation by high glucose is also inhibited by TRPM2-siRNA. Upon activation of NOX, p47phox, a cytosolic subunit of NOX, translocates to the plasma membrane. Under high glucose conditions, p47phox translocation is increased as is its co-localization with TRPM2. In addition, co-localization of TXNIP and NLRP3 is enhanced by high glucose conditions, which could also be suppressed by TRPM2-siRNA. These data suggest that Ca^{2+} influx through TRPM2 can contribute to high glucose-induced ROS production as well as inflammasome activation and amplification of TRPM2 activation to further exacerbate oxidative stress in diseases such as type 2 diabetes.

An important role of macrophages, phagocytosis, is regulated by TRPM2 activity. As mentioned above, macrophages phagocytose deleterious agents such as exogenous pathogens and digest them in phagosomes by producing ROS levels that are toxic to the pathogens. This phagocytosis function is known to be enhanced by temperatures that are associated with fever [83]. Phagocytosis induced by toll-like receptor 2 agonist was enhanced by temperature elevation to febrile range in wild-type macrophages which can detect slight temperature change through TRPM2 activity in the presence of low concentration of ROS [19]. On the other hand, TRPM2KO macrophages showed no such temperature-dependent elevation of phagocytosis, suggesting that ROS generated in phagosomes increases TRPM2 sensitivity to body temperature and that TRPM2 activation by temperature increases phagocytosis. Because TRPM2 is highly localized in phagosomal membranes [84], functional cooperation between TRPM2 and NOX is suggested to increase phagocytosis. As many reports have shown roles for TRPM2 in innate immunity, the physiological roles of TRPM2 have been studied in microbial infection models. In a *Listeria monocytogenes* (Lm) infection model, TRPM2KO mice show lower serum IL-12 and IFNγ levels and a higher mortality rate than wild-type mice; these conditions can be reversed by the application of IFNγ [85]. In contrast to the DSS-induced experimental colitis model [73], Lm-infected TRPM2KO mice have CXCL2 expression and neutrophil infiltration in the spleen that is comparable to wild-type mice. Given that the ratio of activated monocyte ($iNOS^+$) is decreased and IL-12 release is unchanged in an *in vitro* cytokine release assay using BMDMs from Lm-infected TRPM2KO, TRPM2 activation is suggested to have important roles in reciprocal activation among macrophages, natural killer cells, and $CD8^+$ T cells mediated by IL-12 and IFNγ during the early phase of Lm infection [86]. TRPM2 deficiency is also reported to increase the mortality rate in a cecal ligation and puncture (CLP)-induced polymicrobial sepsis model [87]. Moreover, TRPM2KO mice show increased bacterial burden and enhanced inflammation and injury in tissues. TRPM2-deficient BMDM show lower bacterial killing than that of wild type without noticeable effects on phagocytosis. Lipopolysaccharide (LPS) treatment of BMDM from wild-type and TRPM2KO mice revealed that heme oxygenase-1 (HO-1) induction observed in wild-type cells did not occur in TRPM2KO cells and the induction in wild-type cells is inhibited by chelating Ca^{2+}, suggesting that $[Ca^{2+}]_i$ increases mediated through TRPM2 lead to HO-1 expression. Bacterial killing in phagosomes is also reported to be regulated by TRPM2 activity [84]. TRPM2 deficiency increases the bacterial burden of *Pseudomonas aeruginosa* (PA) and *Staphylococcus aureus* (SA) in BMDM without affecting phagocytosis. TRPM2 functionally expressed in phagosomal membrane promotes acidification of SA-internalized phagosomes that is a hallmark of phagosomal maturation and is regarded to be crucial for degradation of phagocytosed particles and up-regulation of bacterial killing [88].

Contrary to reports showing that TRPM2 up-regulates inflammation/innate immunity, protective anti-inflammatory roles of TRPM2 were also reported in a LPS-induced lung inflammatory model [89]. After intraperitoneal injection with LPS, TRPM2KO mice had higher mortality with pronounced lung edema than did wild-type mice. Histopathology of tissues from TRPM2KO mice showed increased levels of tissue cytokines including CXCL2, TNFα, IL-6 and myeloperoxidase (MPO) activity, suggesting enhanced neutrophil infiltration in mice lacking TRPM2. ROS release and oxidative DNA damage in TRPM2KO cells are significantly larger than those in wild-type cells, and a TRPM2 inhibitor increased ROS release from wild-type cells that was equivalent to that of TRPM2KO, thus indicating the inhibitory effects of TRPM2 activity. As activity of the electrogenic NOX enzyme is known to be voltage-sensitive [90], membrane depolarization through TRPM2 is suggested to negatively regulate NOX activity, and the absence of this regulatory mechanism in TRPM2KO mice could enhance ROS production and aggravate inflammation. Nonetheless, the role of TRPM2 in immunity remains controversial, and its activation could depend on the stimulus type, particularly since immunogens can activate multiple cascades (**Table 1**). In addition, immune reactivity is strongly related to body temperature [83, 91]. Thus, the temperature dependence of TRPM2 must be considered [19]. Outcomes could also be affected depending on the extent of TRPM2 activity, e.g., Ca^{2+} influx through TRPM2 could be beneficial for many Ca^{2+}-dependent events, whereas large Ca^{2+} influx could cause Ca^{2+} overload.

TRPM2 is also known to contribute to the regulation of blood glucose levels (**Figure 4**) likely because of its functional expression in pancreatic β-cells [92], which secrete insulin in response to elevated blood glucose. Although the primary pathway of glucose-stimulated insulin secretion is through ATP-sensitive K^+ (K_{ATP}) channel closure and L-type voltage-gated Ca^{2+} channel activation to increase $[Ca^{2+}]_i$, many other ion channels, including TRPM2, are reported to be involved in $[Ca^{2+}]_i$-increases to evoke insulin secretion [93]. ROS are produced in response to extracellular signals such as insulin, cytokines, hormones and blood glucose elevation in pancreatic β-cells [94, 95]. In addition, expression of the antioxidant enzyme catalase and glutathione reductase in the pancreas is very low [96], indicating that the intracellular redox state of β-cells could be affected at a physiological level by the systemic metabolic state. Intracellular Ca^{2+}-oscillations evoked by high glucose concentrations are attenuated in TRPM2KO pancreatic β-cells [97]. In

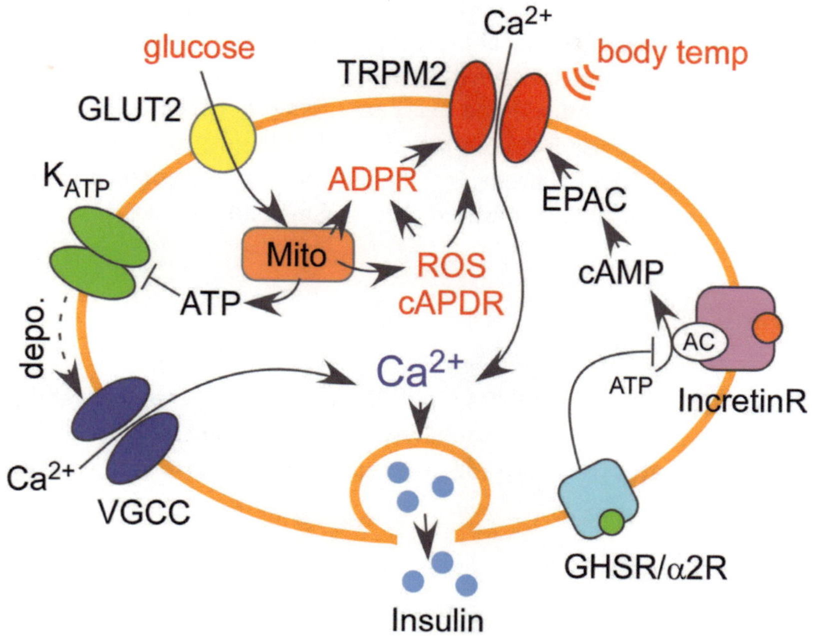

Figure 4. Functional involvement of TRPM2 in regulating insulin secretion from pancreatic β-cells. GLUT2, glucose transporter 2; K_{ATP}, ATP-sensitive K^+ channel; VGCC, voltage-gated Ca^{2+} channel; GHSR, growth hormone secretagogue receptor; α2R, α2 adrenergic receptor; IncretinR, incretin receptor; EPAC, exchange protein directly activated by cAMP; Mito, mitochondria.

addition, heat-evoked responses in Wt b-cells were enhanced by H_2O_2 in a dose-dependent manner, however, TRPM2KO cells completely lacked the response [20]. Glucose-stimulated insulin secretion from pancreatic islets was attenuated by TRPM2 deficiency [97], and the N-acetyl cysteine-sensitive fraction of glucose-stimulated insulin secretion was increased by temperature elevation from 33 °C to 40 °C in wild type, but not TRPM2KO cells, suggesting insulin secretion could be up-regulated by ROS and temperature-dependent TRPM2 activity [20]. TRPM2KO mice show blunted insulin secretion and higher blood glucose levels in oral and intraperitoneal glucose tolerance tests relative to wild-type mice, which highlights TRPM2 function in blood glucose regulation *in vivo* [97].

TRPM2 could also be involved in hormone-regulated insulin secretion. Glucagon-like peptide-1 (GLP-1) and gastric inhibitory polypeptide (GIP) are incretin hormones released from the intestine to enhance glucose-stimulated insulin secretion from pancreatic islets [98]. The up-regulation of insulin release by incretin was shown to be TRPM2-dependent [97, 99]. This increase in insulin secretion is mediated by generation of cAMP and activation of exchange protein directly activated by cAMP (EPAC). By inhibiting the same pathway, insulinostatic effects of ghrelin and adrenaline are reportedly achieved [100, 101]. Although the involvement of ROS-evoked TRPM2 activity in such hormone-regulated insulin secretion has not been defined, TRPM2 participation in glucose metabolism seems likely. In contrast, TRPM2-mediated aggravation of inflammation was reported to cause insulin resistance in peripheral tissues [102]. High-fat diet (HFD)-induced obesity, chronic inflammation and insulin resistance with elevated plasma and tissue cytokines are also reported to be related [103]. TRPM2KO mice show resistance to HFD-induced obesity, as well as lower cytokine levels and macrophage infiltration in adipose tissue compared with wild-type mice [102]. Energy expenditure and glucose uptake to muscle and heart are higher in TRPM2KO mice than in wild-type animals, and this increase is accompanied by elevated expression levels of metabolic and mitochondrial genes. These data suggest that, in addition to regulating insulin secretion, TRPM2 could control glucose metabolism in glucose-consuming tissues.

6. Conclusions

Several TRP channels, particularly TRPA1 and TRPM2, are redox-sensitive and play important roles in both physiological and pathophysiological phenomena that involve redox signals. Therefore, TRPA1 and TRPM2 would be promising targets to develop medicines to treat redox-related diseases.

Author details

Makiko Kashio[1] and Makoto Tominaga[2*]

*Address all correspondence to: tominaga@nips.ac.jp

1 Department of Physiology, Aichi Medical University, Aichi, Japan

2 Department of Cell Physiology, Okazaki Institute for Integrative Biosciences, Okazaki, Japan

References

[1] Droge W. Free radicals in the physiological control of cell function. Physiological Reviews. 2002;**82**(1):47-95

[2] Rhee SG. Cell signaling. H_2O_2, a necessary evil for cell signaling. Science. 2006;**312**(5782): 1882-1883

[3] Brown DI, Griendling KK. Nox proteins in signal transduction. Free Radical Biology and Medicine. 2009;**47**(9):1239-1253

[4] Roos D, et al. Mutations in the X-linked and autosomal recessive forms of chronic granulomatous disease. Blood. 1996;**87**(5):1663-1681

[5] Ramsey IS, Delling M, Clapham DE. An introduction to TRP channels. Annual Review of Physiology. 2006;**68**:619-647

[6] Gaudet R. TRP channels entering the structural era. The Journal of Physiology. 2008; **586**(15):3565-3575

[7] Constantine M, et al. Heterologously-expressed and liposome-reconstituted human transient receptor potential melastatin 4 channel (TRPM4) is a functional tetramer. Scientific Reports. 2016;**6**:19352

[8] Gaudet R. A primer on ankyrin repeat function in TRP channels and beyond. Molecular BioSystems. 2008;**4**(5):372-379

[9] Stokes A, et al. TRPA1 is a substrate for de-ubiquitination by the tumor suppressor CYLD. Cellular Signalling. 2006;**18**(10):1584-1594

[10] Doerner JF, et al. Transient receptor potential channel A1 is directly gated by calcium ions. The Journal of Biological Chemistry. 2007;**282**(18):13180-13189

[11] Zurborg S, et al. Direct activation of the ion channel TRPA1 by Ca^{2+}. Nature Neuroscience. 2007;**10**(3):277-299

[12] Wang YY, et al. The nociceptor ion channel TRPA1 is potentiated and inactivated by permeating calcium ions. The Journal of Biological Chemistry. 2008;**283**(47):32691-32703

[13] Perraud AL, et al. ADP-ribose gating of the calcium-permeable LTRPC2 channel revealed by Nudix motif homology. Nature. 2001;**411**(6837):595-599

[14] Iordanov I, et al. The proposed channel-enzyme transient receptor potential melastatin 2 does not possess ADP ribose hydrolase activity. eLife. 2016;**5**:e17600

[15] Toth B, Iordanov I, Csanady L. Putative chanzyme activity of TRPM2 cation channel is unrelated to pore gating. Proceedings of the National Academy of Sciences of the United States of America. 2014;**111**(47):16949-16954

[16] Toth B, Iordanov I, Csanady L. Ruling out pyridine dinucleotides as true TRPM2 channel activators reveals novel direct agonist ADP-ribose-2′-phosphate. The Journal of General Physiology. 2015;**145**(5):419-430

[17] Hara Y, et al. LTRPC2 Ca^{2+}-permeable channel activated by changes in redox status confers susceptibility to cell death. Molecular Cell. 2002;**9**(1):163-173

[18] Knowles H, Li Y, Perraud AL. The TRPM2 ion channel, an oxidative stress and metabolic sensor regulating innate immunity and inflammation. Immunologic Research. 2013;**55**(1-3):241-248

[19] Kashio M, et al. Redox signal-mediated sensitization of transient receptor potential melastatin 2 (TRPM2) to temperature affects macrophage functions. Proceedings of the National Academy of Sciences of the United States of America. 2012;**109**(17):6745-6750

[20] Kashio M, Tominaga M. Redox signal-mediated enhancement of the temperature sensitivity of Transient receptor potential melastatin 2 (TRPM2) elevates glucose-induced insulin secretion from pancreatic islets. The Journal of Biological Chemistry. 2015;**290**(19):12435-12442

[21] McHugh D, et al. Critical intracellular Ca^{2+} dependence of transient receptor potential melastatin 2 (TRPM2) cation channel activation. The Journal of Biological Chemistry. 2003;**278**(13):11002-11006

[22] Csanady L, Torocsik B. Four Ca^{2+} ions activate TRPM2 channels by binding in deep crevices near the pore but intracellularly of the gate. The Journal of General Physiology. 2009;**133**(2):189-203

[23] Du J, Xie J, Yue L. Modulation of TRPM2 by acidic pH and the underlying mechanisms for pH sensitivity. The Journal of General Physiology. 2009;**134**(6):471-488

[24] Sumoza-Toledo A, Penner R. TRPM2: A multifunctional ion channel for calcium signalling. The Journal of Physiology. 2011;**589**(Pt 7):1515-1525

[25] Sumoza-Toledo A, et al. Dendritic cell maturation and chemotaxis is regulated by TRPM2-mediated lysosomal Ca^{2+} release. FASEB Journal: Official Publication of the Federation of American Societies for Experimental Biology. 2011;**25**(10):3529-3542

[26] Manna PT, et al. TRPM2-mediated intracellular Zn^{2+} release triggers pancreatic beta-cell death. The Biochemical Journal. 2015;**466**(3):537-546

[27] Nakamura K. Central circuitries for body temperature regulation and fever. American Journal of Physiology. Regulatory, Integrative and Comparative Physiology. 2011;**301**(5): R1207–R1228

[28] Song K, et al. The TRPM2 channel is a hypothalamic heat sensor that limits fever and can drive hypothermia. Science. 2016;**353**(6306):1393-1398

[29] Tan CH, McNaughton PA. The TRPM2 ion channel is required for sensitivity to warmth. Nature. 2016;**536**(7617):460-463

[30] Bautista DM, et al. TRPA1 mediates the inflammatory actions of environmental irritants and proalgesic agents. Cell. 2006;**124**(6):1269-1282

[31] Hinman A, et al. TRP channel activation by reversible covalent modification. Proceedings of the National Academy of Sciences of the United States of America. 2006; **103**(51):19564-19568

[32] Andersson DA, et al. Transient receptor potential A1 is a sensory receptor for multiple products of oxidative stress. The Journal of Neuroscience: The Official Journal of the Society for Neuroscience. 2008;**28**(10):2485-2494

[33] Macpherson LJ, et al. Noxious compounds activate TRPA1 ion channels through covalent modification of cysteines. Nature. 2007;**445**(7127):541-545

[34] Bessac BF, et al. TRPA1 is a major oxidant sensor in murine airway sensory neurons. The Journal of Clinical Investigation. 2008;**118**(5):1899-1910

[35] Miyamoto T, et al. TRPV1 and TRPA1 mediate peripheral nitric oxide-induced nociception in mice. PLoS One. 2009;**4**(10):e7596

[36] Takahashi N, et al. Molecular characterization of TRPA1 channel activation by cysteine-reactive inflammatory mediators. Channels. 2008;**2**(4):287-298

[37] Liebler DC. Protein damage by reactive electrophiles: Targets and consequences. Chemical Research in Toxicology. 2008;**21**(1):117-128

[38] Shimizu S, et al. Sensitization of H_2O_2-induced TRPM2 activation and subsequent interleukin-8 (CXCL8) production by intracellular Fe(2+) in human monocytic U937 cells. The International Journal of Biochemistry and Cell Biology. 2015;**68**:119-127

[39] Schreiber V, et al. Poly(ADP-ribose): Novel functions for an old molecule. Nature Reviews Molecular Cell Biology. 2006;**7**(7):517-528

[40] Fonfria E, et al. TRPM2 channel opening in response to oxidative stress is dependent on activation of poly(ADP-ribose) polymerase. British Journal of Pharmacology. 2004; **143**(1):186-192

[41] Ayub K, Hallett MB. The mitochondrial ADPR link between Ca^{2+} store release and Ca^{2+} influx channel opening in immune cells. FASEB Journal: Official Publication of the Federation of American Societies for Experimental Biology. 2004;**18**(12):1335-1338

[42] Perraud AL, et al. Accumulation of free ADP-ribose from mitochondria mediates oxidative stress-induced gating of TRPM2 cation channels. The Journal of Biological Chemistry. 2005;**280**(7):6138-6148

[43] Partida-Sanchez S, et al. Chemotaxis of mouse bone marrow neutrophils and dendritic cells is controlled by ADP-ribose, the major product generated by the CD38 enzyme reaction. Journal of Immunology. 2007;**179**(11):7827-7839

[44] Rah SY, et al. ADP-ribose/TRPM2-mediated Ca^{2+} signaling is essential for cytolytic degranulation and antitumor activity of natural killer cells. Scientific Reports. 2015;**5**:9482

[45] Howard M, et al. Formation and hydrolysis of cyclic ADP-ribose catalyzed by lymphocyte antigen CD38. Science. 1993;**262**(5136):1056-1059

[46] Wehage E, et al. Activation of the cation channel long transient receptor potential channel 2 (LTRPC2) by hydrogen peroxide. A splice variant reveals a mode of activation independent of ADP-ribose. The Journal of Biological Chemistry. 2002;**277**(26):23150-23156

[47] Trevisani M, et al. 4-hydroxynonenal, an endogenous aldehyde, causes pain and neurogenic inflammation through activation of the irritant receptor TRPA1. Proceedings of the National Academy of Sciences of the United States of America. 2007;**104**(33):13519-13524

[48] Takahashi N, et al. TRPA1 underlies a sensing mechanism for O_2. Nature Chemical Biology. 2011;**7**(10):701-711

[49] Chen J, et al. Species differences and molecular determinant of TRPA1 cold sensitivity. Nature Communications. 2013;**4**:2501

[50] Miyake T, et al. Cold sensitivity of TRPA1 is unveiled by the prolyl hydroxylation blockade-induced sensitization to ROS. Nature Communications. 2016;**7**:12840

[51] Zhao M, et al. Acute cold hypersensitivity characteristically induced by oxaliplatin is caused by the enhanced responsiveness of TRPA1 in mice. Molecular Pain. 2012;**8**:55

[52] Nassini R, et al. Oxaliplatin elicits mechanical and cold allodynia in rodents via TRPA1 receptor stimulation. Pain. 2011;**152**(7):1621-1631

[53] Barriere DA, et al. Paclitaxel therapy potentiates cold hyperalgesia in streptozotocin-induced diabetic rats through enhanced mitochondrial reactive oxygen species production and TRPA1 sensitization. Pain. 2012;**153**(3):553-561

[54] Paravicini TM, Sobey CG. Cerebral vascular effects of reactive oxygen species: Recent evidence for a role of NADPH-oxidase. Clinical and Experimental Pharmacology and Physiology. 2003;**30**(11):855-859

[55] Sullivan MN, et al. Localized TRPA1 channel Ca^{2+} signals stimulated by reactive oxygen species promote cerebral artery dilation. Science Signaling. 2015;**8**(358):ra2

[56] Aubdool AA, et al. TRPA1 activation leads to neurogenic vasodilatation: Involvement of reactive oxygen nitrogen species in addition to CGRP and NO. British Journal of Pharmacology. 2016;**173**(15):2419-2433

[57] McNulty S, Fonfria E. The role of TRPM channels in cell death. Pflugers Archiv: European Journal of Physiology. 2005;**451**(1):235-242

[58] Sun Y, et al. TRPM2 promotes neurotoxin MPP+/MPTP-induced cell death. Molecular Neurobiology. 2016. Epub ahead of print

[59] Ostapchenko VG, et al. The transient receptor potential melastatin 2 (TRPM2) channel contributes to beta-amyloid oligomer-related neurotoxicity and memory impairment. The Journal of Neuroscience: The Official Journal of the Society for Neuroscience. 2015;**35**(45):5157-5169

[60] Fonfria E, et al. Amyloid beta-peptide(1-42) and hydrogen peroxide-induced toxicity are mediated by TRPM2 in rat primary striatal cultures. Journal of Neurochemistry. 2005;**95**(3):715-723

[61] Roberge S, et al. TNF-alpha-mediated caspase-8 activation induces ROS production and TRPM2 activation in adult ventricular myocytes. Cardiovascular Research. 2014;**103**(1):90-99

[62] Ye M, et al. TRPM2 channel deficiency prevents delayed cytosolic Zn^{2+} accumulation and CA1 pyramidal neuronal death after transient global ischemia. Cell Death and Disease. 2014;**5**:e1541

[63] Hiroi T, et al. Neutrophil TRPM2 channels are implicated in the exacerbation of myocardial ischaemia/reperfusion injury. Cardiovascular Research. 2013;**97**(2):271-281

[64] Gao G, et al. TRPM2 mediates ischemic kidney injury and oxidant stress through RAC1. The Journal of Clinical Investigation. 2014;**124**(11):4989-5001

[65] Gelderblom M, et al. Transient receptor potential melastatin subfamily member 2 cation channel regulates detrimental immune cell invasion in ischemic stroke. Stroke. 2014;**45**(11):3395-3402

[66] Hoffman NE, et al. Ca(2)(+) entry via TRPM2 is essential for cardiac myocyte bioenergetics maintenance. American Journal of Physiology: Heart and Circulatory Physiology. 2015;**308**(6):H637–H650

[67] Miller BA, et al. The second member of transient receptor potential-melastatin channel family protects hearts from ischemia-reperfusion injury. The American Journal of Physiology: Heart and Circulatory Physiology. 2013;**304**(7):H1010–H1022

[68] Miller BA, et al. TRPM2 channels protect against cardiac ischemia-reperfusion injury: Role of mitochondria. The Journal of Biological Chemistry. 2014;**289**(11):7615-7629

[69] Bao L, et al. Depletion of the human ion channel TRPM2 in neuroblastoma demonstrates its key role in cell survival through modulation of mitochondrial reactive oxygen species and bioenergetics. The Journal of Biological Chemistry. 2016;**291**(47):24449-24464

[70] Llorente-Folch I, et al. The regulation of neuronal mitochondrial metabolism by calcium. The Journal of Physiology. 2015;**593**(16):3447*3462

[71] Inada H, Iida T, Tominaga M. Different expression patterns of TRP genes in murine B and T lymphocytes. Biochemical and Biophysical Research Communications. 2006; **350**(3):762-767

[72] Lange I, et al. Synergistic regulation of endogenous TRPM2 channels by adenine dinucleotides in primary human neutrophils. Cell Calcium. 2008;**44**(6):604-615

[73] Yamamoto S, et al. TRPM2-mediated Ca^{2+} influx induces chemokine production in monocytes that aggravates inflammatory neutrophil infiltration. Nature Medicine. 2008;**14**(7): 738-747

[74] Kraft R, et al. Hydrogen peroxide and ADP-ribose induce TRPM2-mediated calcium influx and cation currents in microglia. American Journal of Physiology: Cell Physiology. 2004;**286**(1):C129–C137

[75] Oda S, et al. TRPM2 contributes to antigen-stimulated Ca(2)(+) influx in mucosal mast cells. Pflugers Archiv: European Journal of Physiology. 2013;**465**(7):1023-1030

[76] Haraguchi K, et al. TRPM2 contributes to inflammatory and neuropathic pain through the aggravation of pronociceptive inflammatory responses in mice. The Journal of Neuroscience: The Official Journal of the Society for Neuroscience. 2012;**32**(11):3931-3941

[77] Murakami T, et al. Critical role for calcium mobilization in activation of the NLRP3 inflammasome. Proceedings of the National Academy of Sciences of the United States of America. 2012;**109**(28):11282-11287

[78] Jin C, Flavell RA. Molecular mechanism of NLRP3 inflammasome activation. Journal of Clinical Immunology. 2010;**30**(5):628-631

[79] Zhong Z, et al. TRPM2 links oxidative stress to NLRP3 inflammasome activation. Nature Communications. 2013;**4**:1611

[80] Zhou R, et al. Thioredoxin-interacting protein links oxidative stress to inflammasome activation. Nature Immunology. 2010;**11**(2):136-140

[81] Shah A, et al. Thioredoxin-interacting protein mediates high glucose-induced reactive oxygen species generation by mitochondria and the NADPH oxidase, Nox4, in mesangial cells. The Journal of Biological Chemistry. 2013;**288**(10):6835-6848

[82] Tseng HH, et al. TRPM2 regulates TXNIP-mediated NLRP3 inflammasome activation via interaction with p47 phox under high glucose in human monocytic cells. Scientific Reports. 2016;**6**:35016

[83] Hasday JD, Fairchild KD, Shanholtz C. The role of fever in the infected host. Microbes and Infection. 2000;**2**(15):1891-1904

[84] Di A, et al. Role of phagosomal redox-sensitive TRP channel TRPM2 in regulating bactericidal activity of macrophages. Journal of Cell Science. 2017;**130**(4):735-744

[85] Knowles H, et al. Transient receptor potential melastatin 2 (TRPM2) ion channel is required for innate immunity against Listeria monocytogenes. Proceedings of the National Academy of Sciences of the United States of America. 2011;**108**(28):11578-11583

[86] Suzue K, et al. In vivo role of IFN-gamma produced by antigen-presenting cells in early host defense against intracellular pathogens. European Journal of Immunology. 2003;**33**(10):2666-2675

[87] Qian X, et al. Transient receptor potential melastatin 2 protects mice against polymicrobial sepsis by enhancing bacterial clearance. Anesthesiology. 2014;**121**(2):336-351

[88] Tjelle TE, Lovdal T, Berg T. Phagosome dynamics and function. BioEssays: News and Reviews in Molecular, Cellular and Developmental Biology. 2000;**22**(3):255-263

[89] Di A, et al. The redox-sensitive cation channel TRPM2 modulates phagocyte ROS production and inflammation. Nature Immunology. 2011;**13**(1):29-34

[90] DeCoursey TE, Morgan D, Cherny VV. The voltage dependence of NADPH oxidase reveals why phagocytes need proton channels. Nature. 2003;**422**(6931):531-534

[91] Blatteis CM. Fever: Is it beneficial? The Yale Journal of Biology and Medicine. 1986; **59**(2):107-116

[92] Togashi K, et al. TRPM2 activation by cyclic ADP-ribose at body temperature is involved in insulin secretion. The EMBO Journal. 2006;**25**(9):1804-1815

[93] Islam MS. Calcium signaling in the islets. Advances in Experimental Medicine and Biology. 2010;**654**:235-259

[94] Paulsen CE, Carroll KS. Orchestrating redox signaling networks through regulatory cysteine switches. ACS Chemical Biology. 2010;**5**(1):47-62

[95] Pi J, et al. Reactive oxygen species as a signal in glucose-stimulated insulin secretion. Diabetes. 2007;**56**(7):1783-1791

[96] Lenzen S, Drinkgern J, Tiedge M. Low antioxidant enzyme gene expression in pancreatic islets compared with various other mouse tissues. Free Radical Biology and Medicine. 1996;**20**(3):463-466

[97] Uchida K, et al. Lack of TRPM2 impaired insulin secretion and glucose metabolisms in mice. Diabetes. 2011;**60**(1):119-126

[98] Leech CA, et al. Molecular physiology of glucagon-like peptide-1 insulin secretagogue action in pancreatic beta cells. Progress in Biophysics and Molecular Biology. 2011;**107**(2):236-247

[99] Yoshida M, et al. Involvement of cAMP-EPAC-TRPM2 activation in glucose- and incretin-induced insulin secretion. Diabetes. 2014 ;**63**(10):3394-3403

[100] Kurashina T, et al. The beta-cell GHSR and downstream cAMP/TRPM2 signaling account for insulinostatic and glycemic effects of ghrelin. Scientific Reports. 2015;**5**:14041

[101] Ito K, et al. Endogenous alpha2A-adrenoceptor-operated sympathoadrenergic tones attenuate insulin secretion via cAMP/TRPM2 signaling. Diabetes. 2016;**66**(3):699-709

[102] Zhang Z, et al. TRPM2 Ca^{2+} channel regulates energy balance and glucose metabolism. American Journal of Physiology. Endocrinology and Metabolism. 2012;**302**(7):E807–E816

[103] Xu H, et al. Chronic inflammation in fat plays a crucial role in the development of obesity-related insulin resistance. The Journal of Clinical Investigation. 2003;**112**(12): 1821-1830

Chapter 10

Redox Reactions in Groundwater with Health Implications

Gunnar Jacks

Additional information is available at the end of the chapter

http://dx.doi.org/10.5772/intechopen.68751

Abstract

Mobilisation into groundwater of many trace elements are redox dependant and of health concern. A short list includes arsenic, manganese, chromium, selenium, nitrogen as nitrate and uranium. Arsenic is mobile in groundwater essentially under two conditions, under reducing conditions mobilising ferrous iron and arsenite and at high pH when the ferric oxyhydroxides lose their positive charge. Manganese is mobilised under moderately reducing conditions as Mn2+ and might affect the mental capacity at chronic exposure from drinking water. Chromium is mobile as carcinogenic chromate under oxidising conditions. Manganese oxides may oxidise Cr(III) solids. Chromium may come from natural as well as anthropogenic sources. Selenium, an essential element, rarely exceeds permissible limits but irrigation with groundwater with elevated selenium could cause toxic selenium intake via food. Selenium is mobile in groundwater under oxidising conditions. Nitrate from excess use of fertilisers may be a problem for bottle-fed children below the age of 1 year, forming methaemoglobinemia. Uranium is mobile under oxidising conditions as U(VI). Mobility and toxicity depends on numerous factors. The threat from uranium is its effect on the readsorption in the kidney of water and salts from the primary urine. Oxidation state and speciation govern uranium mobility.

Keywords: redox, groundwater, trace elements, mobility, health

1. Introduction

Groundwater is by far the largest fresh water resource on the globe. It is often a safe drinking water source from bacteriological point of view, and the development of safe wells in many countries has saved millions of children in particular. However, the development of deeper sources of groundwater has implied redox reactions that may mobilise elements that are toxic, such as arsenic and manganese. Other elements of health concern whose mobility is redox-dependant may be introduced from natural and/or anthropogenic sources such as chromium, uranium and nitrogen. This contribution will present the processes behind the mobility with examples from hydrogeology and hydrochemistry on a global scale.

To get a simple overview of the redox reactions in nature a so-called aerobic/anaerobic stair case including only major redox-sensitive species in groundwater could be used (**Figure 1**). In reality, the 'steps' are numerous if all elements, including trace elements are taken into account.

While pH is related to the activity of protons, Eh or redox potential is related to the activity of electrons (**Figure 2**). To define a specific redox level, Eh measurements with a platinum electrode and a reference electrode, commonly an Ag/AgCl electrode, are helpful but often difficult to interpret [1]. However, it is essentially the Fe(II)/Fe(III) couple that can be relied upon provided the concentrations of both species are not too small. Most redox measurements mirror a mixed potential [2]. Many species are not electrically active at the platinum electrode. ZoBell's solution is commonly used to check the electrodes [3]. The redox potential should preferably be measured in a so-called flow-through cell to avoid any air to come in contact with the groundwater. Speciation of the form of the elements is another tool that can be applied for instance for separating As(III) from As(VI). Numerous methods of speciation are published (e.g. [1, 4]).

In general, the species that are more mobile are anions as the cations in the form of most heavy metals are quite strongly adsorbed to clay minerals and organic matter at around neutral and alkaline pH while anions, adsorbed onto ferric oxyhydroxides and aluminium compounds, are less adsorbed at moderately alkaline conditions and not at all above zero points of charge (ZPC) (**Figure 3**) which is at a pH of 8.2 for ferric oxyhydroxides. Reducing conditions with the reduction of ferric iron to soluble ferrous iron is another case of the failure of adsorption of anionic species.

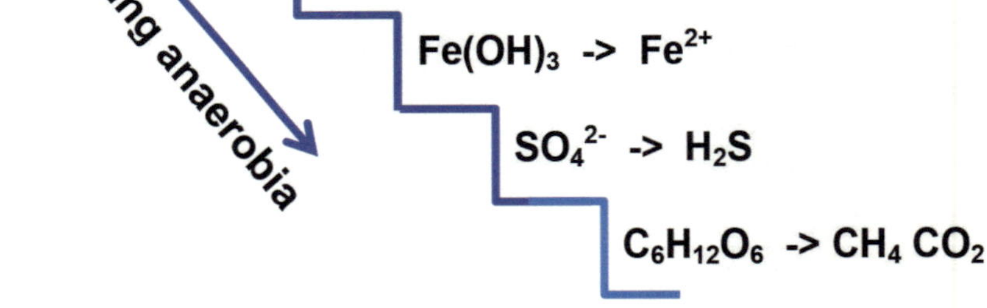

Figure 1. The aerobic/anaerobic 'staircase' for major elements in groundwater.

http://dx.doi.org/10.5772/intechopen.68751

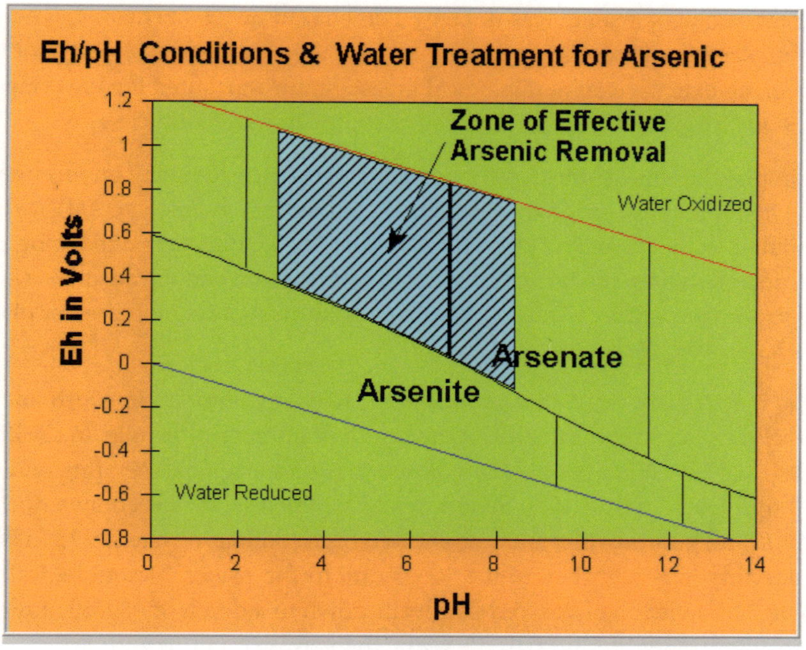

Figure 2. The Eh-pH diagram for arsenic with an overlay of iron.

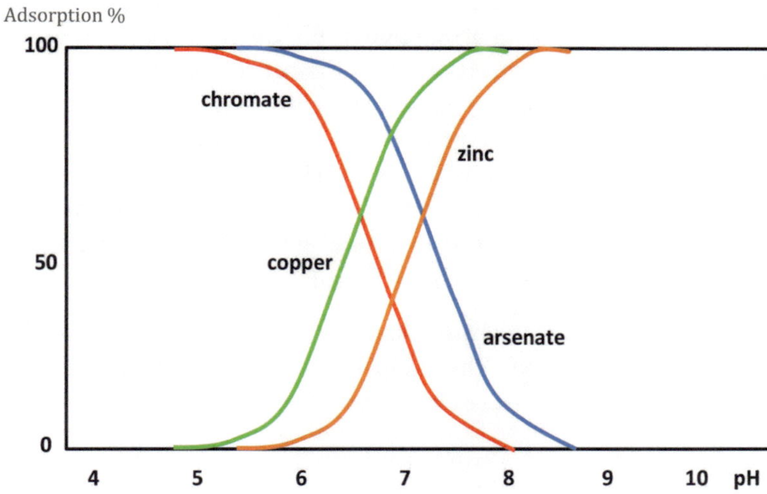

Figure 3. Adsorption of cationic and anionic species.

2. Arsenic

Arsenic is an element that has been known and used by mankind since the Bronze Age to make bronze stiffer. It has been introduced in society for different purposes such as for the removal of air bubbles in glass in medieval times. In the eighteenth century, it was used to cure different ailments. Its toxicity has been known for long time. A step forward in detection of low concentrations of arsenic was done by Berzelius, a Swedish chemist, who invented a qualitative but sensitive analysis [6]. Arsenic in groundwater has been known for quite some time; however, an extensive epidemiological investigation warranted the lowering of the safe level from 50 to 10 μg/l by WHO and that was adopted in most countries.

Arsenic is mobile in groundwater under two conditions, in a reducing environment as arsenite (As(III)) and in an oxidising environment at elevated pH as arsenate As(V). The mobility is closely related to the chemistry of iron oxyhydroxides. Under ferric-reducing conditions, the arsenite adsorbed onto the ferric compounds is released when the adsorbent is mobilised as soluble ferrous iron. Under oxidising conditions, the arsenate is mobilised at pH above 8.2, the ZPC for ferric oxyhydroxides.

In the 1990s, it was discovered that groundwater in many aquifers in south and southeast Asia had levels of arsenic that threatened the health of millions of people. In the Bengal delta in Bangladesh, 35–75 million people are exposed to excess arsenic depending on whether the 50 or the 10 μg/l level limit is used. Symptoms of arsenicosis were seen by a doctor in West Bengal in India and groundwater analysis showed high contents of arsenic [7]. In Bangladesh, the child mortality was high before the 1960s due to the use of bacteriologically polluted surface water. The switching over to cheap wells down to a depth of ~30–50 m meant a radical decrease in mortality but after 10–15 years, the slow poisoning with arsenic from those wells became evident. The discovery evoked a discussion about the mechanisms behind the elevated levels of arsenic amounting even up to mg/l. An initial hypothesis was that the introduction of wells had lowered the groundwater level and allowed oxygen to diffuse into the sub-ground level causing oxidation of arsenopyrite (**Figure 4**). However, a common feature of the polluted groundwater was high contents of dissolved iron and it turned out to be a completely internal process in the sediments where organic matter degraded under anaerobic conditions by bacteria using ferric iron as an oxidant dissolving ferric oxyhydroxides, releasing arsenic in the form of arsenite into the groundwater [8] (**Figure 4**). The arsenic content in the sediments is moderately higher than elsewhere due to sources like rocks in the Himalayas [9]. This is, however, not the cause for the mobilisation of arsenic in the groundwater in the Bengal delta and the Gangetic plain; the redox level in the sediments is the reason.

Further work has shown that deeper wells are safe with a higher redox level than the shallower. This can be traced back to glacial times. The sediments deposited during Pleistocene, before the last glacial maximum (LGM) when the sea level was lowered at a geologically rather fast rate,

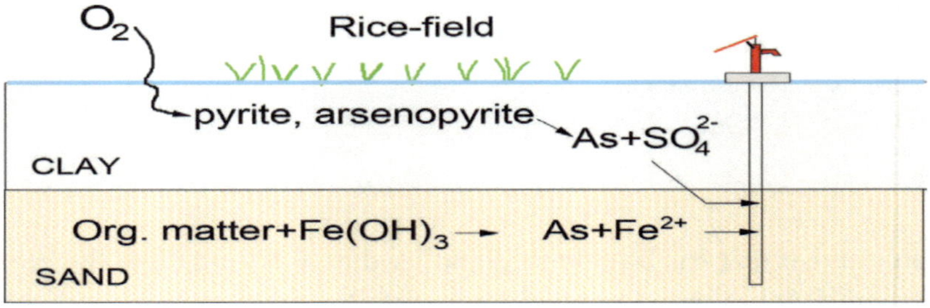

Figure 4. A hypothesis regarding the mobilisation of arsenic in the Bengal delta.

were deposited under rather oxidising conditions, and they were due to the lowering of the sea level subject erosion and resedimentation. Contrary, after the LGM when the sea level rose, there was more formation of wetlands allowing the introduction of more organic matter in the postglacial sediments.

While the removal of arsenic from the groundwater functions technically, it does not work socially as women who are expected to handle the filters are too burdened by daily tasks [10, 11]. However, it has been found that safe sediments can be identified by their colour. This is a practice that was found out by local drillers in search for low-iron groundwater. What they did not know was that low-iron groundwater is also low in arsenic (**Figure 5**) [12]. The colour code tool has been developed from a large number of sediment samples paired with many groundwater analyses. The colour scheme that has been used is the Munsell Colour code [13]. As indicated above, the Pleistocene sediments at around 100 m depth were likely to be a good target. The colour code was tested on 243 wells drilled to around that depth [14]. The prediction of safe groundwater below 10 μg/l is 91%, while low manganese, below WHOs technical guideline at 400 μg/l, can be achieved in 89% of the cases [14].

Another colour tool mirroring the redox conditions in the groundwater is the colour of the platforms at hand pumps in the Bengal delta [15]. Red precipitates of ferric hydroxide indicate an iron-reducing groundwater which has often, in the Bengal delta, elevated arsenic concentrations. A black platform with precipitates of manganese oxides mirrors a higher redox level where the arsenic concentrations are lower.

Another mechanism of mobilisation of arsenic may occur under oxidising conditions at an elevated pH above the ZPC of the major adsorbents, ferric oxyhydroxides of different types.

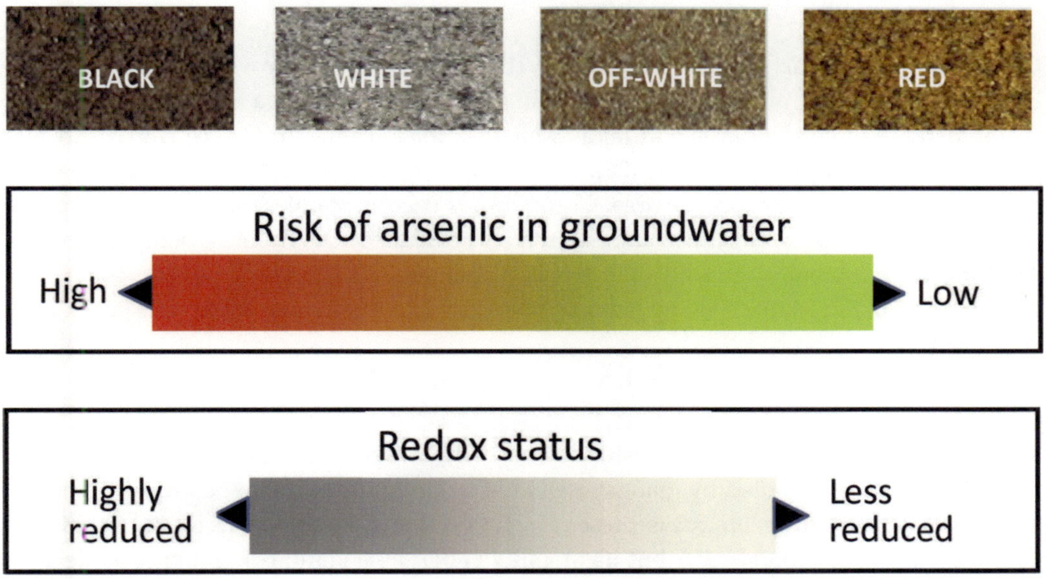

Figure 5. A simplified colour code to identify low arsenic groundwater in the Bengal delta [12].

This is common in the Andes in South America [16]. The arsenic is then present as arsenate, and the high pH of groundwater tends to be of the Na-HCO_3 type [17].

3. Manganese

Manganese in groundwater has been considered as a technical problem causing blackish flushes in tap water after oxidation of the Mn^{2+} to Mn oxides in the distribution networks. Manganese inhaled from air has neurological effects known from metallurgical industries [18]. Recently, however, reports have come to indicate that chronic exposure of children to manganese in drinking water may affect their intellectual capacity [19]. So far, there are three reports that are comparable, from Quebec in Canada [19], from Mexico [20] and from Brazil [21], and that give concern regarding the neurotoxicity of manganese in drinking water. A summary of the findings regarding toxicity is published from a conference [22]. Manganese is an essential element and the major intake is via food. Elevated intake via drinking water by women may be positive for foetal survival [23] but foetal growth may be impaired [24]. The intake of manganese via drinking water is smaller but seems unregulated. This is mirrored in hair in which the concentration increases with that in drinking water but not with the intake via food. The Mn^{2+} seems to be taken up by the divalent metal transporter 1 (DMS-1) and be affected by, for instance, iron status [25]. A hypothesis, as humans have not developed a regulation, could be that moderately reduced groundwater from larger depth has, in the history of mankind, not been a common way to get drinking water but rather from surface water and springs.

4. Chromium

Chromium is a genotoxic carcinogen and Cr(VI) is due to its similarity with sulphate taken up by the same pathways. While excess arsenic in groundwater is often a natural phenomenon, the occurrence of chromium is natural as well as anthropogenic. In Greece, a groundwater plume up to 160 μg/l was studied [26]. It was found to be essentially natural but with an anthropogenic component in one area. Chromium is used in stainless steel but major pollution by chromium is from leather tanning and electroplating. Cr(III) forms solid phases in a reducing environment in groundwater but is mobile as Cr(VI) under oxidising conditions. Cr(III) can be oxidised to Cr(VI) by manganese oxides commonly present in soils [27–30].

$$Cr\,(OH)_{2^+} + 1.5\,MnO_2 \leftrightarrow HCrO_{4^-} + 1.5\,Mn^{2+} \quad (1)$$

Kazakis et al. [31] have studied the oxidation of Cr(III) on surfaces of mafic minerals and concluded that it was mediated by manganese oxides. Chromium tends to be present at higher levels not only in ultramafic rocks present in Greece but also elsewhere in the world [32]. Cr(VI) is mobile in groundwater but its mobility is pH dependant, being higher at elevated pH levels depending on the fact that the main adsorbents aluminium and iron oxyhydroxides lose their positive charge at pH above 8 [28]. Soluble organic complexes decrease the oxidation rate due to complexation with Cr(III) [33]. These authors studied the use of tannery waste for increasing the organic matter content in soils. Cr(VI) may be oxidised to chromate even in an organic matrix like tannery sludge provided the oxygen is high enough [34]. Chromate may then be leached, and it may pollute the groundwater. A number of actions to decrease chromium leaching from leather tanning are proposed [35]. In Punjab, in northwestern India, a sand delta was detected at levels of 5 mg/l at 60 m depth, in this case from electroplating activities [36]. At a level below 2 mg/l in drinking water, chromium can be reduced to Cr(III) which is not taken up by humans.

5. Selenium

Selenium which is an essential metalloid is a part of selenoproteins in the human body. Selenium is ingested via food and drinking water. The concentration varies largely when seen in a global context. The lowest contents are seen in Sweden in the order of 0.1–0.2 μg/l while as high as 480 μg/l are seen [37, 38]. The current guideline is seldom exceeded in drinking water [39]. Nevertheless, these authors demand more studies and especially what concerns specific species of selenium in groundwater. Inorganic species are, as per their results, considerably more toxic than organic species [39]. Selenium deficiency is common in, for instance, Sweden and Finland. In Finland, action has been taken by adding selenium to commercial fertilisers [40]. This selenium is considered to reach humans mainly via the food and the serum-Se has increased while so far no obvious positive health effects are seen in humans but they are seen in animals. However, in Punjab, in northwest India, groundwater concentrations above 300 μg/l have been recorded [41]. In part, the selenium is likely to come from food as groundwater irrigation is common which has caused the accumulation of selenium in cultivated soils. Selenosis in the form of malformed nails is observed [41]. In the area, the main source of selenium is food, contributing about 90% of the total intake [42].

Selenium has a complicated redox pattern with the main species in groundwater being Se(IV) and Se(VI) [1]. Oxidising conditions increased the leaching from soils in the form of Se(VI).

6. Uranium

Uranium is present in rocks in amounts from less than 2 to a few 100 mg/kg. Lowest content is found in ultrabasic rocks while black shales of marine origin can have hundreds of mg/kg [43]. Uranium is well common in granites and is mobile under oxidising conditions. Uranium has a large number of complexes and their mobility depends on their charge, uncharged species being more mobile. Uranium is a risk from both the radiation point of view as well as a chemical risk point of view. It seems that in many cases, the chemical risk is the one that is most important [44, 45]. Its guideline values have been changed repeatedly over the last 13 years and its current provisional guideline is now, what concerns the chemical risk, 30 μg/l [46] (**Table 1**). The chemical risk is its effect on the secondary uptake of water and salts from the primary urine formed in the kidney cortex [45, 47]. The radiological guideline is different for ^{234}U and ^{238}U being, respectively,

Element	Guideline value	Note
Arsenic	10 μg/l	Provisional
Chromium	50 μg/l	Provisional
Manganese	(400 μg/l)	Technical
Nitrate	50 mg/l	
Selenium	40 μg/l	
Uranium	30 μg/l	Provisional

Table 1. Elements exhibiting redox-sensitive behaviour with guideline values. The figures are derived from WHO [5] and regard guidelines due to health reasons. For manganese only a technical guideline is established.

1 and 10 Bq/l [46]. The uptake by humans of different uranium complexes varies considerably. A speciation of uranium complexes can be calculated from an ICP-MS (Induced Coupled Plasma combined with Massspectrometer) analysis by Visual MINTEQ [48] or PHREEQC [49]. Uranium in groundwater is high especially in granitic terrains, for example, in Sweden and Finland. Another source of uranium is in the form of phosphate fertilisers as phosphate rocks have elevated uranium contents [50]. In an area in southern central Finland, the total uranium levels in groundwater were, in some samples, above 3000 μg/l while no health effects were seen [51]. This is far above the current WHO health limit of 15 μg/l. A speciation of the uranium in relatively alkaline groundwater (pH > 7.3) showed that the major portion of uranium was present as calcium-uranyl-carbonato complexes ($CaUO_2(CO_3)_3^{2-}$ and $Ca_2UO_2(CO_3) \cdot 11H_2O$). Thus, these complexes do seem to be less toxic or less bioavailable. There seems to be an interaction with the iron status such as the uranium uptake may be higher at iron deficiency [45, 52] which, by these authors, was considered as an action of divalent metal transporter 1 (DMT-1).

Mobility of uranium in groundwater is affected by a large number of factors. In general, U(IV) is less mobile than the oxidised form U(VI) [53]. This is true for several actinides and has been considered in the search for safe repositories for radioactive waste [54]. As mentioned above, uranium forms numerous complexes among then carbonate complexes. Elevated bicarbonate contents in groundwater form soluble carbonate complexes [55]. When removing the uranium ex-situ by filter, for instance, it is important to know the speciation [56].

The reduction of uranium is in part not only inorganic but also bacteriologically mediated by Moon et al. [57]. The microbial reduction results in isotope fractionation [58]. There is a considerable community of bacteria even at hundreds of meters in hard rock terrains [59, 60]. One of the substrates for uranium sequestration could be acetate [53]. Sulphate-reducing conditions are most favourable for the reduction from U(VI) to U(IV) [53, 61]. Species involved in the reduction are *Desulfobacter, Desulphoropalus* and *Desulfovibrio spp.* [53]. Oxidation of Fe(II) to Fe(III) can also reduce U(VI) to U(IV) [61].

An example of the redox behaviour of uranium is sandstone-hosted uranium deposits formed by groundwater flow with low concentrations of uranium reaching a redox barrier where the U(VI) is reduced to U(IV) and accumulates as an ore-body. These deposits are now commonly extracted by in situ recovery with injection of an oxidising solution forming U(VI) and preferably forming an uncharged complex [62]:

$$UO_2(\text{uraninite}) + \tfrac{1}{2}\,O_2 + 2H^+ \rightarrow UO_2^{\,2+}(\text{uranyl}) + H_2O \tag{2}$$

Complexation with calcium:

$$UO_2^{\,2+} + 2Ca^{2+} + 3HCO_3 \rightarrow Ca_2(UO_2)(CO_3)_3^{\,0} + 3H^+ \tag{3}$$

This technology is considerably more environment friendly compared with conventional mining which leaves tailings containing leftover uranium.

7. Nitrogen

Nitrogen is a major nutrient in soils and its cycle has been affected to a large extent by anthropogenic industrial nitrogen fixation, fertiliser production. The nitrogen fixation in nature was passed by the anthropogenic in the 1980s [63]. When organic matter like litter degrades, ammonia is formed. Ammonia is strongly adsorbed on clay minerals and to organic matter, and elevated concentrations are only found close to a point source. Nitrate, on the contrary, is mobile unless it is taken up by plants. While nitrate in surface water may be part of eutrophication, the main risk with nitrate in groundwater is as the formation of methaemoglobinemia and the decrease of the oxygen-carrying capacity of the red blood cells from the lungs to peripheral tissues [64]. This affects children below the age of about 1 year that are bottle fed with high nitrate water. Above the age of 1 year, humans develop an enzyme that recovers the normal haemoglobin. Methaemoglobinemia is not common; a few thousands of cases are reported. Nitrate converted to nitrite in the intestinal tract may be carcinogenic [65]. Nitrate is reduced to nitrite in the intestinal tract and nitrite may form nitrosamines and elevate the risk of gastric cancer [66]. Above all, nitrate in groundwater is a resource in a wrong place; it should be present in the root zone to promote crop growth.

8. Interactions between elements

In a redox reaction, there is an electron donor and an electron acceptor. Among the elements dealt with above, there are interactions. Oxygen is a common electron acceptor under oxidising conditions, for example, the oxidation of As(III) to As(V). Ferric oxyhydroxides play a crucial role for the mobility of arsenic in groundwater in two respects, under reducing conditions, they are dissolved and arsenite is released and under high pH conditions, above pH 8.2, when they are uncharged making arsenate mobile.

Manganese oxides on the surface of mafic minerals [31] and in general in soils [33] can serve as an oxidant of Cr(III), forming mobile chromate.

Selenium is mobile in groundwater under oxidising conditions, and nitrate from agriculture mobilises selenium from marine shales [67]. Another interaction between agriculture and groundwater might be found in rice cultivation. A new rice irrigation practice by using intermittent irrigation, allowing the rice field to dry up between irrigations, has several positive effects, higher yield, lower arsenic content in the rice grains and lesser methane emission [68]. It might also in the long run affect the groundwater redox conditions.

9. Conclusions

The presence of different species of elements at different redox levels makes assessments of the associated health risks complicated. In addition, the species have different levels of toxicity. A number of trace elements may pose risks to human health, arsenic, manganese, chromium as chromate, nitrate, selenium and uranium. Except for manganese, these species are either uncharged or in the form of anions and their adsorption to soils and aquifer materials is weaker than for cations such as most heavy metals like Cu, Pb and Zn. The main adsorbents for the latter are clay minerals and organic matter, and the adsorption of those increases with pH while the adsorption for anions, by aluminium and ferric oxyhydroxides, decreases with pH and approaches zero, close to ZPC at just above pH 8. Uncharged species are particularly mobile like arsenite.

Arsenic is mobilised into groundwater both under reducing and oxidising conditions. With reduction of ferric oxyhydroxides, ferrous iron and arsenite are released into groundwater. This is common especially in south and southeast Asian delta regions. While removal by filter is possible, but not always functioning due to social reasons and in addition gives a waste problem, an alternate way of supplying safe water is to identify aquifers with more oxidising environments. This can be done by examining the colour of the sediments. White, off-white and red sediments are likely to yield a low arsenic groundwater. It is also possible by the colour code to identify low manganese groundwater. In oxidising environments at neutral pH, arsenate (As(VI) is immobilised into ferric hydroxides. However, at pH above 8, these adsorbents lose their positive charge and arsenic becomes mobile.

Chromium is carcinogenic and mobile in oxidising environments. Chromium comes from both natural and anthropogenic sources. Ultramafic rocks are high in chromium. Main anthropogenic sources are electroplating and leather tanning. Manganese oxides in soils and on mafic minerals can act as oxidants of Cr(III) to form mobile chromate (Cr(VI)).

Selenium, an essential element, seldom reaches toxic levels in groundwater, but in areas where irrigation is practised with groundwater, with elevated selenium concentrations, the intake via the crops cultivated can be too high. Nitrate may act as the oxidant of Se(IV) and form mobile Se(VI).

The nitrogen cycle has been radically changed by human action mainly through the industrial production and use of nitrogenous fertilisers. The natural symbiotic nitrogen fixation was exceeded by the anthropogenic in the 1980s. Nitrate or rather nitrite formed in the human intestinal tracts could be carcinogenic by forming nitrosamines. However, the main reason for the guideline value at 50 mg/l is the risk of methaemoglobinemia in bottle-fed children below the age of 1 year. Breastfeeding efficiently counteracts this.

Uranium can be toxic from radiological point of view but more often due to its chemical action on the re-adsorption of water and salts from the primary urine formed in the kidney cortex. Uranium speciation is important both for its mobility and for its toxicity. Uncharged species are mobile. Calcium-carbonate complexes do not seem to be taken up by humans. U(IV) is largely immobilised while U(VI) is mobile. In the search for safe repositories for radioactive waste, reducing environments are preferred. These immobilise many other radioactive elements.

Author details

Gunnar Jacks

Address all correspondence to: gunnjack@kth.se

Division of Land and Water Resources, KTH Royal Institute of Technology, Stockholm, Sweden

References

[1] Ramesh Kumar A, Riyazuddin P. Speciation of selenium in groundwater: Seasonal variations and redox transformations. Journal of Hazardous Materials. 2011;**192**:263-269

[2] Bricker OP. Redox potential: Its measurement and importance in water systems. Water Analysis, Vol. 1. Academic Press Inc. pp. 55-83

[3] Nordstrom DK. Thermochemical redox equilibria of ZoBell's aolution. Geochimica et Cosmochimica Acta. 1977;**41**:1835-1841

[4] Vassileva E, Becker A, Broekert JAC. Determination of arsenic and selenium species in groundwater and soil extracts by ion chromatography coupled to inductiley coupled plasma mass spectrometry. Analytica Chimica Acta. Amsterdam, Netherlands 2001;**441**:136-146

[5] World Health Organization. Guidelines for Drinking-Water Quality. 4th ed., WHO. Geneva, Switzerland p. 564

[6] Smith R. Arsenic: A murderous history [Internet]. 2012. Available from: http://www.dartmouth.edu/~toxmetal/arsenic/history.html

[7] Guha Mazumder DN, De BK, Santra A, Dasgupta J, Ghosh N, Roy BK, Ghoshal UC, Saha J, Chatterjee A, Dutta S, Haque R, Smith AH, Chakraborty D, Angle CR, Centeno JA. Chronic arsenic toxicity: Epidemiology, natural history and treatment. In: Chappell WR, Abernathy CO, Calderon RL, editors. Arsenic Exposure and Health Effects. Elsevier Science BV. pp. 335-347

[8] Bhattacharya P, Chatterjee D, Jacks G. Occurrence of arsenic-contaminated groundwater in alluvial aquifers from delta plains, Eastern India: Options for safe drinking water supply. International Journal of Water Resources Development. 1997;**13**(1):79-92

[9] Kumar M, Ramanathan AL, Rahman MM, Naidu R. Concentration of inorganic arsenic in groundwater, agricultural soils and subsurface sediments from the middle Gangetic plain of Bihar, India. Science of the Total Environment. 2015;**573**:1103-1114

[10] Jakariya Md, von Brömssen M, Jacks G, Chowdhury AMR, Ahmed KM, Bhattacharya P. Searching for sustainable arsenic mitigation strategy in Bangladesh: Experience from two upazilas. International Journal of Environment & Pollution. 2007;**31**(3/4):415-430

[11] Singh N, Jacks G, Bhattacharya P. Women and community water supply programmes: An analysis from a socio-cultural perspective. Natural Resources Forum. 2005;**29**:213-223

[12] Hossain M, Bhattacharya P, Frape SK, Jacks G, Islam MM, Rahman MM, von Brömssen M, Hasan MA, Ahmed KM. Sediment color tool for targeting arsenic-safe aquifers for the installation of shallow drinking water tubewells. Science of the Total Environment. 2014;**493**:615-625

[13] Munsell Color Code. Munsell Color, Grand Rapids, MI 49512, USA. Available from: http://munsell.com/

[14] Hossain M. Sustainable arsenic mitigation. Ph D thesis n Land and Water Resources Engineering. Royal Institute of Technology (KTH); Stockholm, Sweden 2015

[15] Biswas A, Nath B, Bhattacharya P, Halder D, Kundu AK, Mandal U, Mukherjee A, Chatterjee D, Jacks G. Testing tubewell platform colour as a rapid screening tool for arsenic and manganese in drinking water wells. Environment Science and Technology.2011;**46**:434-440

[16] Bhattacharya P, Claesson M, Bundschuh J, Sracek O, Fagerberg J, Jacks G, Martin RA, Storniolo AR, Thir JM. Distribution and mobility in the Rio Dulce alluvial aquifers in Santiago del Estero Province, Argentina. Science of the Total Environment. 2006;**358**:97-120

[17] Nriagu JO, Bhattacharya P, Mukherjee AB, Bundschuh J, Zevenhoven R, Loeppert RH. Arsenic in soil and groundwater: An overview. In: Trace Metals and Other Contaminants in the Environment. Amsterdam, Netherlands, Vol. 9; 2007. pp. 3-60.

[18] Bouchard M, Mergler D, Baldwin M, Panisset M, Roels HA. Neuropsychiatric symptoms and past exposure in a ferro-alloy plant. Neurotoxicol. 2007;**28**:290-297

[19] Bouchard MF, Sauvé S, Barbeau B, Legrand M, Brodeur ME, Bouffard T. et al. Intellectual impairment in school age children exposed to manganese from drinking water. Environment Health Perspectives. 2011;**119**:138-143

[20] Hernández-Bonilla D, Schilmann A, Montes S, Dodroguez-Agudelo Y, Rodroguez-Dozal S, Solis-Vivanco R, Rios C, Riojas-Rodroguez H. Environmental exposure to manganese and motor function of children in Mexico. Neurotoxicology. 2011;**32**:615-621

[21] de Sousa Viana GF, de Carvalho CF, Nunes LS, Rodriguez JLG, Ribeiro NS, de Almeida DA, Dutra Ferreira JR, Abreu N, Menezes-Filho JA. Noninvasive biomarkers of manganese exposure and neuropsychological effects in environmentally exposed adults in Brazil. Toxicology Letters. 2014;**231**:169-178

[22] Roels HA, Bowler RM, Kim Y, Claus Henn B, Mergler D, Hoet P, Gocheva VV, Bellinger DC, Wright RO, Harris MG, Chang Y, Bouchard MF, Rojas-Rodriguez H, Menezes-Filho JA, Tellez-Rojo MM. Manganese exposure and cognitive deficits: A growing concern for manganese neurotoxicity. Neurotoxicology. 2012;**33**:872-880

[23] Rahman SM, Åkessson A, Kippler M, Grandér M, Hamadani JD, Streafield PK, Persson LÅ, El Arifeen S, Vahter M. Elevated manganese concentrations in drinking water may be beneficial for fetal survival. PLoS One. 2013;**8**:1-8

[24] Rahman SM, Kippler M, Ahmed S, Palm B, El Arifeen S, Vahter M. Manganese exposure through drinking water during pregnancy and size at birth: A prospective cohort study. Reproductive Toxicology. 2015;**53**:68-74

[25] Erikson KM, Aschner M. Increased manganese uptake and transport by primary astrocyte cultures with altered iron status is mediated by divalent metal transporter. Neurotoxicology. 2006;**27**:125-139

[26] Panagiotakis I, Dermatas D, Vatseris C, Chrysochoou M, Papassiopi N. Forensic investigation of chromium(VI) groundwater plume in Thiva, Greece. Journal of Hazardous Materials. 2015;**281**:27-34

[27] Bartlett R, James B. Behaviour of chromium in soils: III. Oxidation. Journal of Environmental Quality. 1978;**8**:31-35

[28] Rai D, Eary LE, Zachara JM. Environmental chemistry of chromium. Science of the Total Environment. 1989;**86**:15-23

[29] Richard FC, Bourg ACM. Aqueous geochemistry of chromium: A review. Water Research. 1991;**25**:807-816

[30] Fendorf SE. Surface reactions of chromium in soils and waters. Geoderma. 1995;**67**:55-71

[31] Kazakis N, Kantiranis N, Voudouris KS, Mitrakis M, Kaprara E, Pavlou A. Geogenic Cr oxidation on the surface of mafic minerals and the hydrogeological conditions influencing hexavalent chromium concentrations in groundwater. Science of the Total Environment. 2015;**514**:224-238

[32] McClain CN, Maher K. Chromium fluxes and speciation in ultramafic catchments and global rivers. Chemical Geology. 2016;**426**:135-157

[33] Trebien DOP, Bortolon L, Tedesco MJ, Bissani CA, Camargo FAO. Environmental factors affecting chromium-manganese oxidation-reduction reactions in soil. Pedosphere. 2011;**21**(1):84-89

[34] Apte AD, Verma S, Tare V, Bose P. Oxidation of Cr(III) in tannet sludge to Cr(VI); field observations and theoretical assessment. Journal of Hazardous Materials. 2005;**B121**: 215-222

[35] Dixit S, Yadav A, Diwedi PD, Das M. Toxic hazards of leather industry and technologies to combat them. Journal of Cleaner Production. 2015;**87**:39-49

[36] Chaudhary V, Singh KP, Jacks G, Bhattacharya P. Groundwater contamination at Ludhiana, Punjab, India. Journal of Indian Water Works Association. 2001;**23**(3):251-261

[37] Robberecht H, van Grieken R, van Srundel M, van den Berghe D, Deelstra H. Selenium in environmental and drinking water of Belgium. Science of the Total Environment. 1983;**26**:163-172

[38] Dhillon KS, Dhillon SK. Quality of underground water and its contribution towards selenium enrichment of the soil-plant system fir a seleniferous region of northwest India. Journal of Hydrology. 2003;**272**:120-130

[39] Vinceti M, Crespi CM, Bonvicini F, Malagoli C, Ferrante M, Marmiroli S, Stranges S. The need for reassessment of the safe upper limit of selenium in drinking water. Science of the Total Environment. 2013;**443**:633-643

[40] Alfthan G, Eurola M, Ekholm P, Venäläinen E-R, Root T, Korkalainen K, Hartikainen H, Salminen P, Hietaniemi V, Aspila P, Aro A. Effects of nationwide addition of selenium to fertilizers on foods, and animal and human health in Finland. Journal of Trace Elements in Medicine and Biology. 2015;**31**:142-147

[41] Bajaj M, Eiche E, Neumann T, Winter J, Gallert C. Hazardous concentrations of selenium in soil and groundwater in NW India. Journal of Hazardous Materials. 2011;**189**:640-646

[42] Dhillon KS and Dhillon SK. Selenium in groundwater and its contribution towards daily intake under different hydrogeological zones of Punjab, India. Journal of Hydrology. 2016;**533**:615-626

[43] Ielsh G, Cuney M, Buscail E, Rossi F, Leon A, Cushing ME. Estimation and mapping of uranium content of geological units in France. Journal of Environmental Radioactivity. 2017;**166**:210-219

[44] Limson Zamora ML, Zielinski JM, Moodie GB, Falconer RAF, Hunt WC, Capello K. Uranium in drinking water: Renal effects of long-term ingestion by an Aboriginal community. Archive of Environmental Occupational Health. 2009;**64**(4):228-241

[45] Konietzka R. Gastrointestinal absorption of uranium compounds. Regulatory Toxicology and Pharmacology. 2015;**71**:125-133

[46] Ansoborlo E, Lebaron-Jacobs L, Prat O. Uranium in drinking water: A unique case of guideline value increases and discrepation between chemical and radiochemical guidelines. Environment International. KTH, Stockholm Sweden 2015;**77**:1-4

[47] Berlin M, Rudell B. Uranium. In: Friberg L, Nordberg GF, Vouk VB, editors. Handbook on the Toxicology of Metals. Vol. II: Specific Metals. Amsterdam: Elsevier; 1986. pp. 623-637

[48] Gustafsson JP. 2016. Available from: https://vminteq.lwr.kth.se/download

[49] PHREEQC. US Geological Survey. 2016. Available from: http://wwwbrr.cr.usgs.gov/projects/GWC_coupled/phreeqc/

[50] Liesch T, Hinrichsen S, Goldscheider N. Uranium in groundwater – fertilizers versus geogenic sources. Science of the Total Environment. 2015;**536**:981-995

[51] Prat O, Vercouter T, Ansoborio E, Fichet P, Perret P, Kurttio P, Salonen L. Uranium speciation in drinking water from drilled wells in southern Finland and its potential links to health effects. Environment Science and Technology. 2009;**43**: 3941-3945

[52] Jiang L, Garrick MD, Garrixck LM, Zhao L, Collins JF. Divalent metal transporter 1 (Dmt1) mediates copper transport in the duodenum of iron-deficient rats and when overexpressed in iron-deprived HEK-293 cells. Journal of Nutrition. 2013;**143**:1927-2193

[53] Xu J, Veeramani H, Qafoku NPQ, Singh G, Riquelme MV, Pruden A, Kukkadapu RK, Gartman BN, Hochella MF Jr. Efficacy of acetate amended biostimulation for uranium sequestration: Combined analysis of sediment/water geochemistry and bacterial community structure. Applied Geochemistry. 2017;**78**:172-185

[54] Gimeno MJ, Auque LF, Acero P, Gômez JB. Hydrogeochemical characterisation and modelling of groundwaters in a potential geological repository for spent nuclear fuel in crystalline rocks. Applied Geochemistry. 2014;**45**:50-71

[55] Alam MS, Cheng T. Uranium release from sediment to groundwater: Influence of water chemistry and insights into release mechanisms. Journal of Contaminant Hydrology. 2014;**164**:72-87

[56] Norrström AC, Löv Å. Uranium theoretical speciation for drinkling water from private drilled wells – Implications for choice of removal method. Applied Geochemistry. 2014;**51**:148-154

[57] Moon HS, McGuinness L, Kukkadapu RK, Peacock AD, Komlos J, Kerkhof IJ, Long PE, Jaffe PE. Microbial reduction of uranium under iron- and sulphate reducing conditions: Effect of amended goethite on microbial community composition and dynamics. Water Research. 2010;**44**:4015-4028

[58] Stirling CH, Andersen MB, Warthmann R, Halliday AN. Isotope fractionation of 238U and 235U during biologically mediated uranium reduction. Geochimica et Cosmochimica Acta. 2015;**163**:200-218

[59] Pedersen K. Subterranean microorganisms and radioactive waste disposal in Sweden. Engineering Geology. 1999;**52**:163-176

[60] Pedersen K. Exploration of deep intraterestrial microbial life: Current perspectives. FEMS Microbiol Letters. 2000;**185**:9-16

[61] Zhao J, Scheibe TD, Mahadevan R. Model based analysis of mixed uranium(VI) reduction by biotic and abiotic pathways during in situ remediation. Chemical Geology. 2013;**357**:215-222

[62] Saunders JA, Pivetz BJ, Voorhies N, Wilkin RT. Potential aquifer vulnerability in regions down-gradient from uranium in situ recovery (ISR) sites. Journal of Environmental Management. 2016;**183**:67-83

[63] Vitousek PM, Aber JD, Howarth RW, Likens GE, Matson PA, Schindler DW, Schlesinger WH, Tilman DG. Human alteration of the global nitrogen cycle: Sources and consequences. Ecological Applications. 1997;**7**(3):737-750

[64] Hill MJ. Nitrates and nitrites in food and water in relation to human health. In: Hill MJ, editor. Nitrates and Nitrites in Food and Water. 2nd ed., Woodhead Publishing Series; Cambridge, UK 1996. pp. 163-193

[65] Zhai Y, Zhao X, Teng Y, Li X, Zhang X, Wu J, Zuo R. Groundwater nitrate pollution and human health risk assessment by using HHRA model in an agricultural area, NE China. Ecotoxicology and Environmental Safety. 2017;**137**:130-142

[66] Hill MJ, Hawksworth G, Tattersall G. Bacteria, nitrosamines and cancer of the stomach. British Journal of Cancer. 1973;**28**:562-567

[67] Bailey RT, Gates TK, Ahmadi M. Simulating reactive transport of selenium coupled with nitrogen in a regional-scale irrigated groundwater system. Journal of Hydrology. 2014;**515**:29-46

[68] Yang J, Zhou Q, Zhang J. Moderate wetting and drying increase rice yields and reduces water use, grain arsenic level and methane emission. The Crop Journal.2017;**5**:151-158